Interfaces, Quantum Wells, and Superlattices

NATO ASI Series

Advanced Science Institutes Series

A series presenting the results of activities sponsored by the NATO Science Committee, which aims at the dissemination of advanced scientific and technological knowledge, with a view to strengthening links between scientific communities.

The series is published by an international board of publishers in conjunction with the NATO Scientific Affairs Division

A	**Life Sciences**	Plenum Publishing Corporation
B	**Physics**	New York and London
C	**Mathematical**	Kluwer Academic Publishers
	and Physical Sciences	Dordrecht, Boston, and London
D	**Behavioral and Social Sciences**	
E	**Applied Sciences**	
F	**Computer and Systems Sciences**	Springer-Verlag
G	**Ecological Sciences**	Berlin, Heidelberg, New York, London,
H	**Cell Biology**	Paris, and Tokyo

Recent Volumes in this Series

Series B: Physics

Interfaces, Quantum Wells, and Superlattices

Edited by

C. Richard Leavens and
Roger Taylor

National Research Council of Canada
Ottawa, Ontario, Canada

Plenum Press
New York and London
Published in cooperation with NATO Scientific Affairs Division

Proceedings of a NATO Advanced Study Institute on
Interfaces, Quantum Wells, and Superlattices,
held August 16–29, 1987,
in Banff, Alberta, Canada

_____

Library of Congress Cataloging in Publication Data

NATO Advanced Study Institute on Interfaces, Quantum Wells, and Superlattices
(1987: Banff, Alta.)
   Interfaces, quantum wells, and superlattices / edited by C. Richard Leavens
and Roger Taylor.
     p.   cm.—(NATO ASI series B: Physics; v. 179)
   "Proceedings of a NATO Advanced Study Institute on Interfaces, Quantum
Wells, and Superlattices, held August 16–29, 1987"—T.p. verso.
   Bibliography: p.
   Includes index.
   ISBN-13: 978-1-4612-8307-2   e-ISBN-13: 978-1-4613-1045-7
   DOI: 10.1007/978-1-4613-1045-7
   1. Semiconductors—Surfaces—Congresses. 2. Quantum wells—Congresses.
3. Superlattices as materials—Congresses. I. Leavens, C. Richard. II. Taylor,
Roger. III. Title. IV. Series: NATO advanced science institutes series. Series B,
Physics; v. 179.
QC611.6.S9N384   1987   88-19476
537.6′22—dc19   CIP
_____

PREFACE

The NATO Advanced Study Institute on "Interfaces, Quantum Wells and Superlattices" was held from August 16th to 29th, 1987, in Banff, Alberta, Canada. This volume contains most of the lectures that were given at the Institute. A few of the lectures had already been presented at an earlier meeting and appear instead in the proceedings of the NATO Advanced Study Institute on "Physics and Applications of Quantum Wells and Superlattices" held in Erice from April 21st to May 1st earlier in the year and published by Plenum Press.

The study of semiconductor interfaces, quantum wells and super-lattices has come to represent a substantial proportion of all work in condensed matter physics. In a sense the growth of interest in this area, which began to accelerate about 10 years ago and seems to be continuing, has been driven by technological developments. While the older generation of semiconductor devices was based on adjacent semiconductors with different properties (e.g. different doping levels) separated by interfaces, modern semiconductor devices tend to be based more and more on properties of the interfaces themselves. This has led, as an example, to the field of band-structure engineering. Improved understanding of the fundamental physics of these systems has aided technological developments and, in turn, technological developments have made available systems which exhibit novel and fascinating physical properties, such as the integer and fractional quantum Hall effects.

The purpose of this ASI was to help expand the group of scientists in NATO countries with expertise in the fundamental physics of semi-conductors. Much of the expertise tends to be concentrated in a relatively small number of excellent institutions. By bringing together a very talented group of speakers we were able to provide a stimulating forum for discussion involving participants from thirteen countries and many different institutions.

The book contains 19 chapters with a mix of both experimental and theoretical topics. Chapter 1 serves as an introduction and chapter 2 then reviews the subject of Molecular Beam Epitaxy without which many of the most interesting systems, currently being studied, would not be possible. Chapter 3 discusses the calculation of electronic states in heterostructures whilst chapter 4 focuses on the experimental determination of sub-band energies. The next eight chapters discuss electronic and optical properties of systems with reduced dimensionality and quantum well or superlattice structure. These are followed by a chapter on resonant tunneling and one on polaron effects in heterostructures. Chapters 15-18 focus on high magnetic fields and the Quantum Hall Effect. Finally, chapter 19 contains a comprehensive review of the theory of Fibonacci superlattices.

We would like to thank all of the speakers for the considerable effort that they put into producing high calibre lectures. This was reflected in the fact that there was unvarying high attendance at all lectures despite the lure of fine weather and beautiful surroundings. Our chief regret concerns the fact that R. B. Laughlin's outstanding contribution to the ASI both through his lectures and his comments are not reflected in these published proceedings. Time constraints made it impossible for him to produce a manuscript before the publication deadline.

We would like to give special thanks to our co-organizers, A. H. MacDonald and E. W. Fenton, who both helped with the original NATO submission and provided much assistance and encouragement. Allan MacDonald also organized the lecture program and produced two excellent lectures. Meanwhile, Ed Fenton contributed heavily to all aspects of the organization of the ASI as well as helping out with the proofreading of several manuscripts. We would also like to thank P. J. Stiles and F. Stern for their advice and comments concerning the selection of speakers for the ASI. We are grateful to Marg Coll for her professional expertise in organizing registration, general information, sightseeing tours and the many other details necessary for a successful meeting. Finally we would particularly like to thank Kim Burke for handling all the secretarial duties, interfacing with the Banff Centre, keeping the organizers organized, handling the information desk during the second week of the ASI and retyping most of the manuscripts to the publisher's specifications.

Roger Taylor
C. Richard Leavens

CONTENTS

WHY INTERFACES, QUANTUM WELLS AND SUPERLATTICES?   SOME COMMENTS

Phillip J. Stiles

Physics Department
Brown University
Providence, RI 02912
U.S.A.

INTRODUCTION

One of the key threads that run through the study of ·the systems in this summer school is our attempt to do something different rather than find out if Schroedinger's equation describes the physics involved. We are reasonably sure of that. What we are trying to put together is the simplest description of the real situation so that we may see which aspect of the physics dominates the problem. From an historical point of view, Bloch's theorem made life simpler in that we could conceive of many problems with many fewer variables to keep track of. But alas, the real world is not a combination of the effective mass theory and a simple one dimensional potential.

Where does all the interest in two-dimensional systems come from? Is it just because the dimensionality is different or is there something due to the dimensionality that has made these more interesting systems to study? Perhaps a little of both. The latter case is certainly the most logical. Here, in these systems, one is able to obtain extremely narrow and well defined energy levels. This results in our being able to test these systems on an exceedingly fine grained energy scale. In addition, one is able in many (but not all) of these systems to vary the electron (or hole) density over a wide range in a single sample.

To obtain high densities of carriers in 3D semiconductor systems, one has a few options. One can raise the temperature and thermally excite them. By doing so one can vary the density by varying the temperature. This results in two kinds of carriers and more scattering for higher temperatures due to the presence of phonons. A similar approach would be to shine light on the semiconductor. This would again result in two kinds of carriers but fewer phonons. Problems with non-uniform absorption of light are hard to escape. The last technique would be to dope the crystal with donors or acceptors. At least this would result in a single type of carrier. However there are major disadvantages, primary among these is the fact that the carriers exist in the same space as the bare coulomb centers. The compensating charge, equal in charge and number density severely limits the mobility.

What about two dimensional systems? It is obvious that the first step is to have the carriers be in a different region of space than the compensating charge. The easiest way to envisage such a system is in terms of the parallel plates of a capacitor. If we assume that when this capacitor is charged, it is the electrons induced on the surface of one of the plates that is of interest, it is obvious that the compensating positive charge is on the other plate. The separation of this positive charge from the electrons can be as large as we like. For a given surface charge density the only limitation is the voltage applied. The scattering from this compensating charge is negligible. We will see later that in the case of certain heterostructures the separation distance cannot be so large for a given density.

Here we see in the interest from the electronic properties point of view why two dimensional systems are so attractive. First we can increase the lifetimes of a single carrier system and then we can in addition vary the density of these carriers just by varying the voltage across a capacitor. It is important not only from the point of view of this workshop but in general to consider the effects that interfaces, quantum wells and superlattices have on properties other than the electronic ones. Where else do these systems modify properties? Without trying to be complete let us consider a few, although not necessarily in the chronological order that they were first studied.

Interfaces occur naturally in nature. A simple case is that of bicrystals, where in solidification a crystal grew with planar arrangements of atoms at the interface common to both crystal segments. The capacitor interfaces of the conducting plates and the insulator are vital to the electronics industry. The phonon spectra of structures with interfaces is no longer the same in all regions of space. Specific phonons exist localized to the interface. The crystal structure is different on either side of the interface. Optical properties are not everywhere the same. It is perhaps better to think then in terms of properties characteristic of the bulk materials on either side of the interface and then those that are modified by the presence of the interface.

Quantum wells exist in nature as well, at the interface of Te bicrystals for example. Another case is a fascinating material, SiC. This material has many stacking orders of the fundamental tetrahedra. Some have long periods. For anyone interested in superlattices, one should study this material. It is easy to grow the material with stacking faults. Such a system with a reverse change in stacking is a quantum well. These can be seen with an electron microscope. There would be phonons that were characteristic of the quantum well. The optical properties as well as Raman spectra would be altered.

Although not the only case in nature, the superlattices in SiC are worth talking about. Stacking orders that run from two to hundreds have been reported. The symmetry of the lattice ranges from cubic to hexagonal to rhombohedral. The basic cubic phonon bands have gaps at the **k**-values that are appropriate for the different stackings and are Raman active. The bandgap varies from about 2.2eV to 3.3eV. Further, the first convincing demonstration of an artificial superlattice was done by looking at the modification of x-ray scattering.

We attempt here to give examples of some of the systems that under-lie the quantum wells that are the basis of the interesting electronic properties. In one sense this is a primer for an excellent review article by Ando, Fowler and Stern(AFS)[1]. In the mid sixties a conference series called Electronic Properties of Two Dimensional Systems started and meets every other year. The proceedings are published in book form and in Surface Science[2]. For a good grounding in many of the points alluded to here, consult any standard condensed matter physics textbook such as Ashcroft and Mermin[3]. The number of other conferences and sessions in larger conferences that cover these subjects abound.

In order to end up with a separation of the compensating charge and the carriers of interest, one always constructs at least one interface between the material of interest and the barrier that keeps the two kinds of charge away from each other. Historically, the metal-oxide-semiconductor (MOS) capacitive-like structure is the best example. We will go over it in detail, but first we should generalize the structure. From it all quantum wells are easy to categorize. It is then not a quantum leap to the superlattice.

Firstly, the O in MOS does not have to be an oxide, but any material that acts in the tests in question as an insulator. Hence we can have nitrides or other native insulators grown on the semiconductor or even deposited or pushed against the semiconductor. Further we can deposit non-native insulators or even push things like thin mylar films against the semiconductor. We have successfully even used air and vacuum. With this kind of approach, anything goes..... even another semiconductor with a larger bandgap. This latter approach is an exciting one because of the current abilities worldwide to grow one semiconductor on another epitaxially in a controllable fashion.

Secondly, the compensating charge does not have to be easily varied, but if it can, much more can be learned about the system. The basis of much of the modern electronics revolution is based on a structure called a MOSFET, where the FET stands for field effect transistor. It works in a manner where the application of a voltage across the structure affects the conductivity of the material at the interface, in the quantum well. Another structure in wide use is the MNOSFET. It differs from the MOSFET in that the insulator has two different kinds of materials, an oxide and a nitride layer. The interface between the two contains many states where one can rather easily vary the charge density. One can put the compensating charge at the interface. In doing so it is possible to change the conductivity in the quantum well from zero with no voltage applied across the structure to conducting with no voltage applied simply by putting compensating charge in the insulating region.

The heterostructure, which was the first major breakthrough after the MOS structure for producing quantum wells of interest, is very similar to the MNOS structure. In this case the compensating charge is located on impurity states in the bulk of the semiconductor that is acting as the insulator as far away as possible from the interface with a compromise between the density one would want and a smooth lateral potential environment. Heterostructures are also made in structures that have a metal plate (gate) on the other side of the insulator and hence can also have a variation in the density as can the MOSFET.

It is easy to see how one can make a superlattice out of hetero-structures. All we have to do is to take the heterostructure of material A (which acts as the insulator) and material B (which is the quantum well region) and make a periodic array. From the middle of material A to the middle of material B is just a heterostructure (as is the mirror image).

This look into what constitutes a superlattice is very general. One class that is very interesting is where A is doped and B is not and they are different compounds. This is the usual case for the heterostructure studied most, (Ga,Al)As:GaAs. The second kind are called nipi structures. Here both A and B are the same host material but the doping is n in one and p in the other. The 'i's in nipi are for the insulating regions where there are no carriers as in a pn junction. The latest type of superlattice is somewhat of a misnomer as it is made of amorphous materials. It can be made as a compositional variation or doping variation or both.

There is another type of classification scheme for superlattices that has to do with how the valence and conduction bands in materials A and B line up at the interface. If the narrow gap semiconductor has its conduction band (valence band) lower (higher) than that of the insulator then the quantum well for the electrons (holes) is in the narrow gap material. As you can see, if the levels line up differently the quantum well for the electrons could be in B and that for the holes in A (or vice versa). A third possibility is that both the conduction and valence bands in one of the materials lie below the valence band of the other material. This results, under some conditions, in a semi-metal interface even without doping.

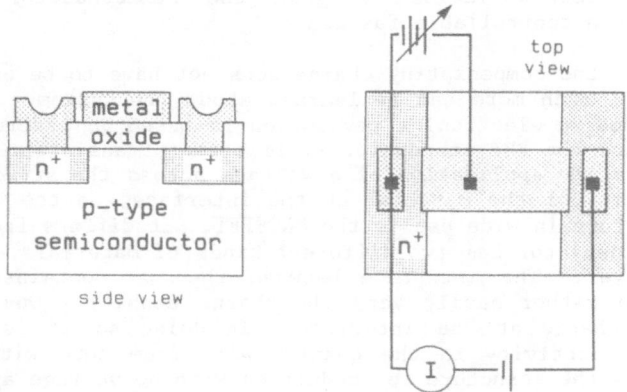

Fig. 1. Side and top views of a MOSFET.

THE PHYSICAL SYSTEMS

We discuss the physical layout of the different systems focusing on the MOSFET. It is the oldest system of interest and from its configuration we can easily extrapolate to the other systems. In Fig. 1 we show a side view and top view of a MOSFET. Because the density of carriers is directly proportional to the electric field normal to the surface and we apply a voltage not a field we should have a uniformly thick insulator. Secondly, we would like the interface to be a place where the potential is as smooth as possible. We would wish that there was no random potential variation along the surface. This requires that the surface be atomically smooth. Further there should not be compositional variation along the surface at and near the interface. All these comments hold for the other forms of quantum wells and superlattices.

We now describe a subset of structures for which we can vary the density. Our main interest in these systems is when they are at low temperatures. For this purpose let us consider them at T=0K unless noted otherwise. The n+ regions have been doped heavily enough that although the mobility is low, they are conducting at low temperatures while the bulk of the p-type semiconductor that we use as our example is not. When the upper electrode, usually called the gate, is positive relative to the n+ regions we have a net positive areal charge density on the gate and the equivalent density of electrons at the interface of the semi-conductor. These electrons are the inversion layer. They are often referred to as a two-dimensional electron gas (2DEG). A simple circuit and response are illustrated in Fig. 2.

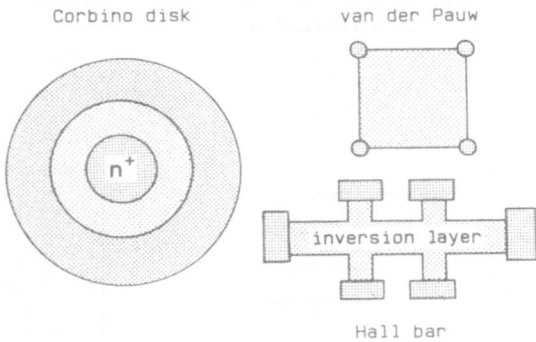

Fig. 2. Simple circuit and response diagrams.

Simple DC measurements as well as low frequency measurements are made utilizing three main structures when the dimensions along the well are defined and contacts are made to the perimeters. The two structures most often used with three dimensional samples are the van der Pauw structure and the Hall Bar. However the earliest successful work in Si MOSFETs utilized the Corbino disk structure. This structure has the advantage that the two-dimensional region 2D is entirely enclosed on the surface of the semiconductor so no possibility exists for surface leakage between the n+ regions. The disadvantage with this structure is that it is a two terminal structure where the voltage drop is measured where the current enters and leaves the sample. In addition with only two probes, one cannot measure sufficient to determine both components of the magneto-conductivity tensor. All three structures are illustrated in Fig. 3. The standard Hall bar is the ideal structure to use, except for the possible leakage along the surface. One is able to measure longitudinal and transverse voltage drops and can determine both components of the magneto-conductivity and -resistivity tensors. It is a structure which is basically a resistance structure being long and thin while the Corbino disk is a conductance structure, being short and wide. The van der Pauw structure is simple and convenient to use. Surface leakage may be a problem; certainly geometry is. The analysis uses functional relationships that may lead to more uncertainties. Yet with a structure with four contacts one can obtain the full magneto-conductivity tensor.

Simple measurement considerations for the beginning researcher may help. Often one is plagued with high resistance contacts. It need not bother the latter two structures providing the current source has sufficient output voltage and the voltage measuring circuit has an impedance higher than the total resistance of the sample and contacts.

Methods of measurement are simple enough that simple instructional laboratories routinely carry out these measurements. Basically one can use either DC or AC techniques. Two simplistic approaches are constant current and constant voltage techniques. The constant current technique uses a constant current passed through a known resistor in series with the sample. One measures the voltage drop across both and then one has the ratio of the resistances. One uses this with the Hall bar and van der Pauw structures, basically resistive structures. For the constant voltage case, one has a resistor whose resistance is low compared to that of the sample in series with the sample. A constant voltage is applied to this series and the voltage drop across the resistor is measured. This voltage is proportional to the conductance of the sample and is usually good for the Corbino structure.

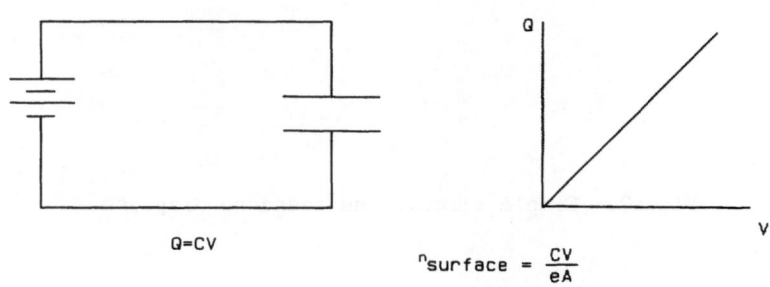

Q=CV

$n_{surface} = \dfrac{CV}{eA}$

Fig. 3. Structures commonly used for simple DC and low frequency transport measurements.

All of the structures discussed above require the technology of making reasonably good conducting contacts to the carriers at the interface. In trying out new physical systems prepared by new techniques one often wishes to get on with the measurements without having to do any more than necessary. Contacts are usually problems. A scheme that has worked very well is the capacitively coupled one. It is illustrated in Fig. 4. In this situation one has no ohmic contacts to the inversion layer and relies on having low impedance "contacts" by virtue of the large area of the pads on the end and working at high enough frequencies for this to be the case. The high resistive gate material is chosen so that under some conditions one can "charge" this capacitor but also when the measurements are made, this resistance is much larger than that of the inversion layer below. This concept can be used for Corbino, van der Pauw and Hall bar sample structures. For the cases of superlattices, it is not as easy because of the shielding of each successive quantum well by the ones above it.

THE ELECTRONIC SYSTEM

Recalling that we are idealizing the system that we want to study, we assume that it has a one-dimensional potential and treat it with simplistic approximations. We assume that the effective mass treatment applies in all three directions, that the dispersion is parabolic, that the electron has a spin but that the spin splitting may be added near the end of the discussions, that there is no spin-orbit interaction and that a semiclassical approach reveals all the essential physics. One recognizes that this is rarely the case but that adding specifics for individual problems is not difficult but perhaps tedious. Indeed the study of valence bands with their inherently more complicated band structure has been revealing of the complexity that the real world holds for us.

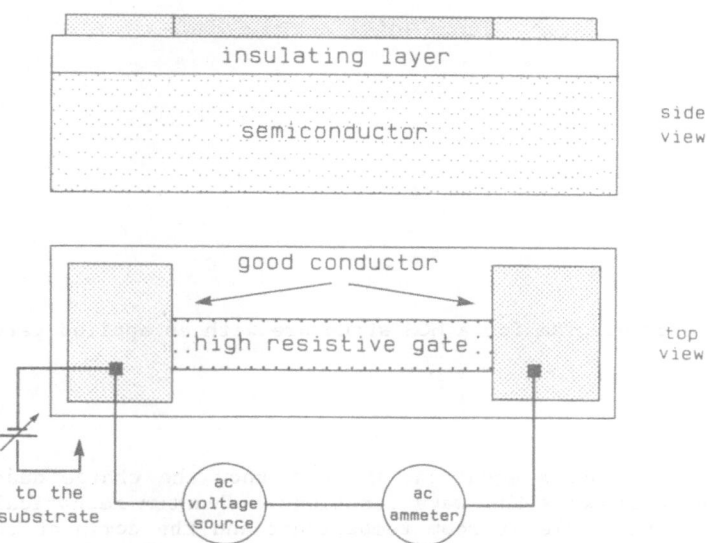

Fig. 4. Capacitively coupled scheme for making good conducting contacts to the carriers in the inversion layer.

The electric field perpendicular to the surface will be terminated on charges. These charges can be fixed, either unwanted fixed charge states or those from purposeful doping, as well as the mobile charges of interest. This is true in all systems. We must find the potential from a solution of the one dimensional Poisson equation

$$d^2\phi(z)/dz^2 = 4\pi\rho(z)/K_{sc} \tag{1}$$

where $\phi(z)$ vanishes in the bulk for the cases of a single quantum well (complete screening of the compensating charge). For the case of the MOSFET the charge is made up of depletion charge n_d (as the bulk minority carriers are being depleted) and mobile charge n_s. For the MOSFET and the heterostructure where one has a metal plate for the compensating charge one can define the mobile charge density as

$$n_s = C(V_g - V_t)/eA \tag{2}$$

where C/A is the capacitance per unit area, V_g is the gate voltage relative to the inversion layer, V_t is a threshold voltage and depends on oxide charge, thickness, workfunction differences and other things as well. Fig. 5 gives an example of a MOS structure with a gate voltage applied. We plot energy versus distance perpendicular to the surface. We have added a voltage difference between the inversion layer and the substrate of the semiconductor (the backside). It is used to indicate the cases where the potential past the two dimensional layer is not zero.

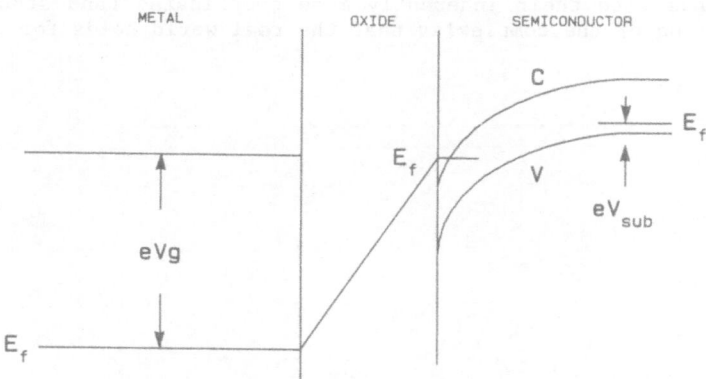

Fig. 5. Energy diagram for a MOS structure with an applied gate voltage.

To solve Poisson's equation, we must know the charge density. To know that we must know the wave functions. Quantum mechanical effects play a significant role at room temperature and the dominant one at the low temperatures of interest. It is easy to see that this is the case from the consideration of energies. The localization energies of the carriers in the quantum well are 10-100 meV. kT at room temperature is about 25 meV. The two degrees of freedom parallel to the surface have free dispersion. Appropriately we must solve an equation in the direction perpendicular to the surface, z, for the envelope of the wave function such as

$$(\hbar^2/2m_z)(d^2\Psi(z)/dz^2) + [E_i - V(z)]\Psi(z) = 0 \quad . \tag{3}$$

The energy levels are given as

$$E(k_x, k_y) = E_i + (\hbar^2/2m_z)(k_x^2 + k_y^2) \tag{4}$$

where the E_i are the energies of the localized motion.

It is obvious that we have made a significant change in the problem. If we are in a situation where the separation of the levels E_i are significantly larger than all the other energies in the problem, the carriers will act as if they are free in only two dimensions. In a sense we now have two dimensional bands. Fig. 6 illustrates the energy levels and some typical values for the Si case.

In general the energy spacing is inversely proportional to the one third power of the mass and proportional to the two thirds power of the electric field. The spatial extent of the wave function is roughly inversely proportional to the one third power of the product of field and mass.

The density of states (DOS) is given by

$$DOS(E) = (2\pi^2)^{-1} \sum_{k_x, k_y} \delta(E - E(k_x, k_y)) .$$ (5)

Assuming that the number of states is large and that we can replace the sum by an integral we have

$$n_s = g_s g_v g_1 m E_f / 2\pi\hbar^2 \text{ and } DOS(E) = g_s g_v g_1 m / 2\pi\hbar^2$$ (6)

where g_s is the spin degeneracy, g_v is the degeneracy due to band structure effects and m is the effective mass (geometric mean) of the electrons in the plane. The factor g_1 has been added to the usual description and is the number of layers contributing. This is important only for the case of many quantum wells in parallel with each other. There are certain facts worth remembering. The effective mass may be energy dependent and one has to sum up contributions from all the two dimensional bands occupied.

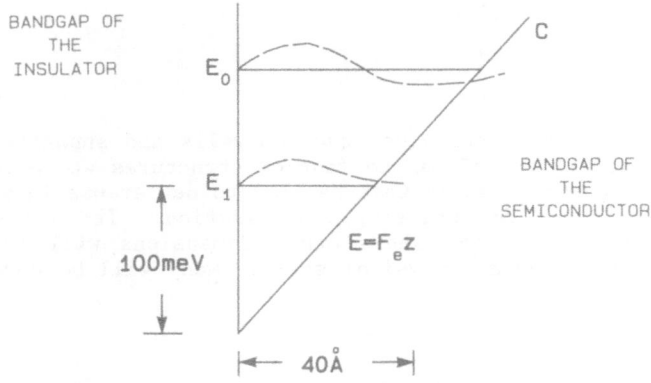

Fig. 6.   Typical   parameters   for   an   inversion   layer   in   Si.

This description leaves out the contribution of the random potential. For the case of short range potentials, both the real and imaginary parts of the energy of the state will be affected. Hence each state can have the center of its level shifted as well as broadened. In order to calculate the density of states for a particular sample we would have to know the distribution of the potential fluctuations. What is the likely distribution, white, Lorentzian, or what? We suspect that neither, that in general it is at least bimodal or multimodal, that is, there are peaks in the correlation, probably for short and long distances. We know that some doping levels vary linearly along a sample. These things would have to be taken into accounts as well as the simple energy levels.

Perhaps the major concern left out up to now is how the deviations from perfection affect the behavior of the charge carriers. If the world was as simple as assumed to now, the only limiting factor to the motion of the electrons would be scattering due to phonons. At the low temperatures that are available today, we can ignore that case. The dominate effect for the last thirty years has been impurity scattering of one kind or another. Early attempts to see the effects of quantum wells failed because the fluctuating potential was so large that the carriers were localized. The first convincing evidence that the two dimensional state was observed was for samples that had mobilities around 3000 cm^2/V-s. Since then results have been obtained on samples at least 3 times worse. However the latest rumor about results indicates a mobility in excess of 1000 times the mobility of the early samples.

The ratio of the effective masses for these differing systems only accounts for a factor of three. Si MOSFETs have improved by a factor of ten. Where is the rest of the difference? In the Si case, the insulator is amorphous. By its very nature, this insulator will have a fluctuating potential from the lack of order. On the other hand, the heterostructures have the insulating region as a single crystal oriented with the semiconductor with a smooth surface. The only difference between the regions comes from the fact that the composition is different. Attempts to grow epitaxial insulators on Si have failed in attempts to obtain a good single crystal insulator.

SUMMARY

Although natural interfaces, quantum wells and superlattices occur, it has been the ability of man to fashion structures whose perfection is high even on the atomic scale that has led to new arenas in which to test our ability to reach for the simple explanation. It is likely that we will see extensive studies where other dimensions will be made short relative to some important physical scale. Many will be covered in this summer school.

REFERENCES

1. T. Ando, A. B. Fowler and F. Stern, "Electronic properties in two-dimensional systems", Rev. Mod. Phys. 54:437 (1982).
2. The proceedings of the conference series "Electronic Properties of Two-Dimensional Systems" are published as well as special editions of "Surface Science" volumes 58, 73, 98, 113, 142, and to be published.
3. Solid State Physics, N. W. Ashcroft and N. D. Mermin, (Holt, Rinehart and Winston, New York).

MOLECULAR BEAM EPITAXY

C.T. Foxon

Philips Research Laboratories
Cross Oak Lane, Redhill, Surrey, England

ABSTRACT

Molecular beam epitaxy (MBE) is a sophisticated method of film growth capable of providing the device engineer with any desired structure. This flexibility has emerged from a thorough understanding of the fundamental factors controlling growth and dopant incorporation obtained using modulated molecular beam spectroscopy, reflection high energy electron diffraction and Monte-Carlo studies. In this article I will concentrate on the growth and interface properties of heterojunctions, multi-quantum well and superlattice structures and their use in devices such as high electron mobility transistors and lasers.

1. INTRODUCTION

The technique which has become known as molecular beam epitaxy (MBE) is, at its simplest, a refined form of vacuum evaporation. The molecular beams are produced by evaporation or sublimation from heated liquids or solids usually contained in pyrolytic boron nitride crucibles. The flux produced is thus determined by the vapour pressure of the element or compound at elevated temperatures. At the pressures used in the MBE equipment collision-free beams from the various sources interact chemically on the substrate to give an epitaxially related film. Ultra-high vacuum (UHV) techniques are used to reduce the pressure of gases from the ambient background and thus improve the purity of the layers. More recently gas sources mounted outside the equipment have been employed in what has become known variously as gas source MBE (GS-MBE), chemical beam epitaxy (CBE) or metalorganic MBE (MOMBE).

MBE began as a basic study of the chemical reactions occurring on surfaces during the growth of III-V compounds but quickly evolved into a practical method for the growth of high purity materials. The ability to start and stop the molecular beams in less than the time taken to grow a single atomic or molecular layer has led to the ability to produce complex multilayer structures. The use of UHV technology has enabled the physical and chemical properties of the films to be measured in-situ using reflection high energy electron diffraction (RHEED) and Auger electron spectroscopy (AES) studies. Modulated molecular beam spectrometry (MMBS) was developed to study the chemical processes involved and the dynamics of film growth have been investigated using RHEED. This led to the discovery of the so-called RHEED oscillation technique which can measure the growth rate in-situ, a unique feature of MBE.

In this article I will review the fundamental aspects of MBE using mostly examples from the best understood system AlGaAs. I will also discuss the techniques required to grow high purity samples such as multi-quantum-well (MQW) structures, superlattices (SLs) and high mobility two-dimensional electron gas structures (2DEGs). I will finally illustrate the practical uses of MBE for the preparation of two types of devices based on the new physical principles encountered in two-dimensional systems, namely the high electron mobility transistor (HEMT) and the short wavelength MQW or SL laser. Particular emphasis will be given in this paper to the nature of interfaces since they form the basis of the quantum well and superlattice structures.

It is appropriate to point out that the techniques developed for III-V compounds have also been applied to a whole range of other materials including elemental semiconductors[1], II-VI semiconductors, insulators and metals but these are outside the scope of the present article. Several excellent reviews have been published which deal in more depth with particular aspects of MBE for III-V compounds[2-4] and its application for the growth of superlattices[5], polar on non-polar materials[6], the concept and applications of delta doping[7], and finally devices based on GaAs and AlGaAs[8-10].

2. FUNDAMENTAL ASPECTS OF MBE

Several types of study have contributed to our present understanding of the processes controlling the growth of films and dopant incorporation in MBE: surface chemical processes were investigated using MMBS methods[11], the dynamics of film growth have been examined using the RHEED technique[12] and Monte-Carlo (M-C) simulations of growth have added to our knowledge of the factors influencing growth and interface roughness[13]. In addition thermodynamic calculations have shown fundamental limitations involved in using certain dopants and the factors governing the incorporation of unwanted impurities[14].

2.1 Growth of Binary Compounds

Modulated molecular beam studies: In conventional MBE group III elements such as Al, Ga and In are always supplied as the monomer by evaporation from the liquid. Over most of the temperature range used in MBE the group III elements have a unity sticking coefficient and therefore the growth rate and alloy composition are simply determined by relative supply rates to the surface. At high temperatures however, as Arthur first showed for Ga[15], the group III element with the higher vapour pressure will have a finite lifetime on the surface leading to a reduction in growth rate and a change in alloy composition (this point will be discussed in more detail below in relation to the growth of alloy films).

In conventional MBE As can be supplied either as the tetramer As_4 by sublimation from the solid or as the dimer As_2, by evaporating from GaAs. As_2 can also be obtained by dissociation of As_4 using a two-zone "cracker" furnace with the front end operating at about 900°C.

The first studies by Arthur[15-16] and the later work by Foxon and Joyce[17] show that As_2 is dissociatively chemisorbed on single Ga adatoms as seen in Fig. 1. The sticking coefficient of As is therefore proportional to the Ga supply reaching the surface and any excess As is lost by re-evaporation. It follows that stoichiometric GaAs is always obtained provided an excess As flux is supplied. At low temperatures, As_2 molecules can associate on the surface to form As_4 which is subsequently desorbed. This is the only direct evidence for the surface migration of As species.

For most layers grown by MBE, As_4, from elemental As has been used to avoid impurities associated with dopants in GaAs or contamination from the hot zone of the cracker. The chemical reactions involved with the tetramer are more complex, as shown in Fig 2. Pairs of As_4 molecules interact on adjacent Ga sites with the excess As atoms being desorbed[18]. The main experimental observations that led to this model are that the maximum sticking coefficient of As_4 is 0.5 even when a large excess of Ga is supplied to the surface and in addition the second order dependence of the As_4 desorption rate on the adsorption rate at low As coverages.

The behaviour with As_4 is in complete contrast to that of As_2 where a unity sticking coefficient can be achieved on a Ga rich surface. The lifetime of the tetramer is also much greater than the dimer on the GaAs surface. This in turn implies that there is a much higher population density in the mobile precursor state for equivalent fluxes of As. The different behaviour of the two As species may be expected to influence the properties of films grown under otherwise identical conditions[19].

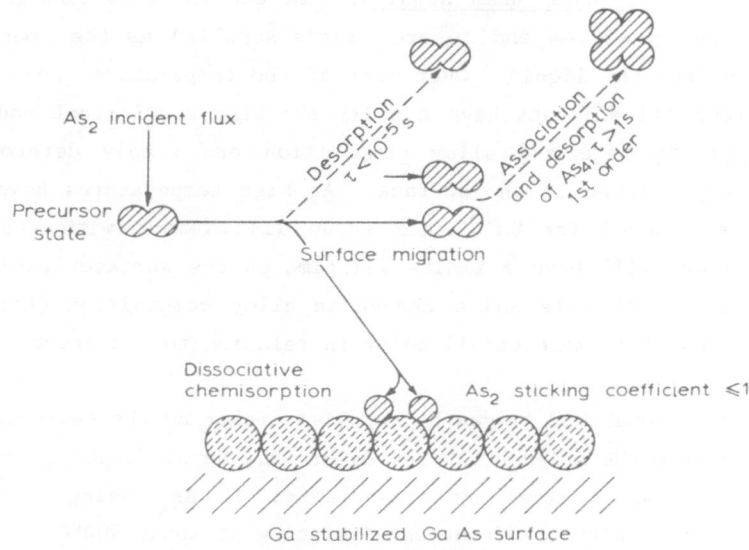

Fig. 1. The model proposed for the growth of GaAs from Ga and As_2.

This has been confirmed by a number of observations; studies of the electronic structure of the GaAs surfaces grown with As_2 and As_4[20], the lower concentration of deep levels in films grown with the dimer[21] and the improved minority carrier properties of AlGaAs-GaAs double hetero-structures obtained with As_2[22]. In this last study both the GaAs and the AlGaAs-GaAs interfaces were improved when the dimer was used. At high substrate temperatures it is possible that As_4 may be decomposed to form As_2 on the surface but there is no direct evidence for this suggestion. The fact that in a two-zone cracker furnace a temperature of 900°C is required for complete dissociation suggests that this will not occur on the GaAs surface at 700°C unless there is some catalytic process involving Ga taking place. In general, films grown with the dimer are better than those grown with the tetramer especially at low substrate temperatures.

Reflection high energy electron diffraction studies: RHEED studies during growth by MBE showed a variety of surface structures depending upon the temperature and relative fluxes of Ga and As used[23]. Subsequent work established that the crystal surface during growth is disordered in a complex manner[24-26]. The study of the dynamics of film growth resulted from the observation of oscillations in the intensity of electrons specularly reflected from the surface with a period corresponding to the time taken to deposit a single monolayer of material, a complete layer of

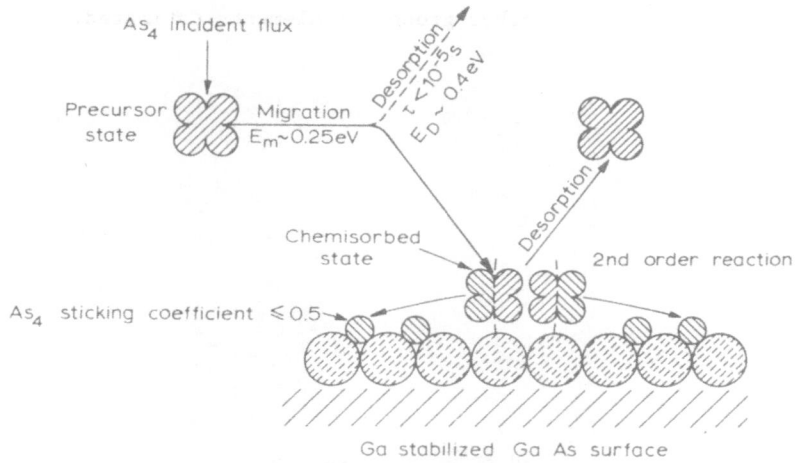

Fig. 2. The model proposed for the growth of GaAs from Ga and As$_4$.

Ga+As atoms[27]. The experimental arrangement used for such studies is shown in Fig. 3 and typical examples for GaAs and AlAs are shown in Fig. 4. The damped oscillations in the intensity of various features of the GaAs pattern can be observed over a wide range of temperatures (500-750°C) and Ga fluxes[28]. On stopping growth the intensities of the various features return gradually to their original values in a time determined mainly by the substrate temperature.

This observation has been explained using a simple model based on a single scattering optical analogue where changes in specular beam intensity are equated with corresponding variations in surface roughness on an atomic scale. This picture is qualitatively correct since the wavelength of the elecrons (0.1Å) is small compared to the step height of 2.83Å. In this interpretation the damping is thought to arise from growth of the second monolayer starting before the first monolayer is completed. The recovery after growth stops suggests that the step density on the surface can be reduced i.e. the surface is smoothing[28-30]. For this process to occur it is essential for Ga to diffuse to the existing step edges and the growth of GaAs from predeposited Ga on supplying an As_2 flux to the surface suggests that this may be the case[30]. Direct evidence for the Ga being mobile on the surface came from studies of growth on vicinal samples[31], as shown in Fig. 5. At low temperatures, where the Ga diffusion length is smaller than the distance between steps on the off-cut surface, conventional RHEED oscillations are observed. At high temperatures, however, the diffusion length is long enough for step edge growth to dominate and the oscillations disappear. From such data the activation energy for the diffusion of Ga on the surface has been deduced and the figures for other group III elements estimated.

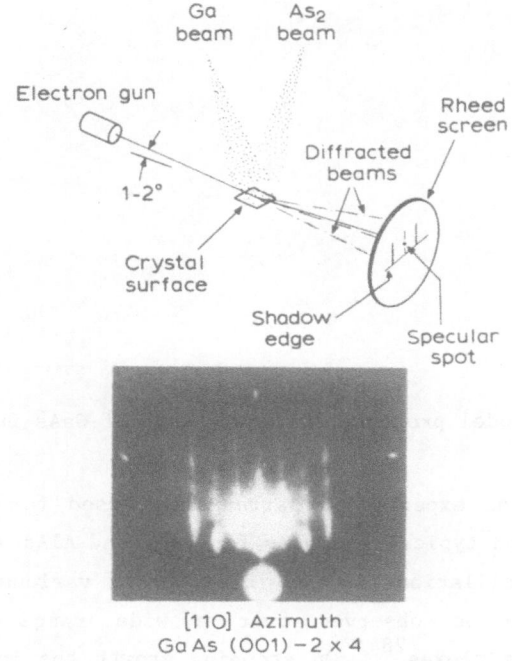

[110] Azimuth
GaAs (001) - 2 x 4

Fig. 3.   The experimental arrangement used for the study of RHEED data during MBE growth.

This simple picture of the origin and significance of RHEED oscillations and the recovery of intensity after growth suggests strongly that the growth front exists over several monolayers and that interrupting growth may lead to smoother interfaces. This last point will be discussed in more detail below.

The intensity of the oscillations for specularly reflected electrons depends strongly upon both the incidence angle and azimuth of the electron beam. Lent and Cohen[32] explained the variation with incidence angle using a simple kinematic model; electrons scattered from adjacent

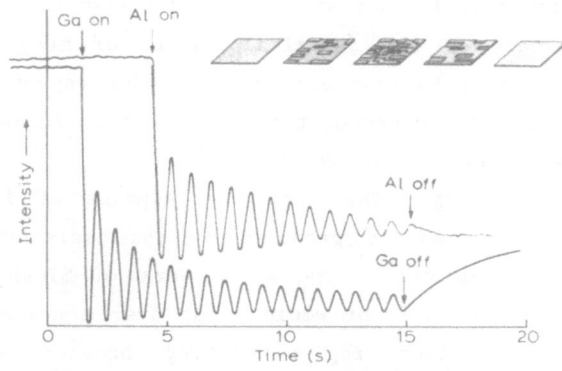

Fig. 4. Typical RHEED oscillations for GaAs and AlAs using As$_4$ at a
substrate temperature of about 600°C. The intensity of
electrons specularly reflected from the surface in a [100]
azimuth from a (001) oriented monitor slice are measured as a
function of time at the centre of rotation. This ensures
accurate determination of the true growth rate for actual
samples grown subsequently under the same conditions.

terraces in a two level system give rise to constructive or destructive
interference. When the beams from neighbouring terraces are out of phase
("off-Bragg") the intensity of the specular beam will depend more
strongly upon the step density on the surface than it will when the beams

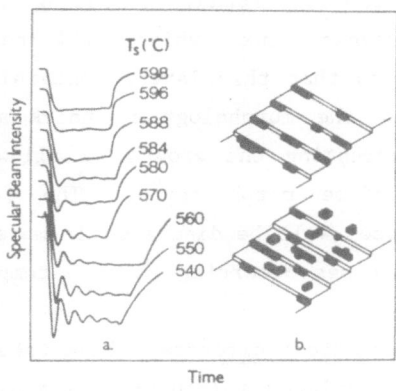

Fig. 5. RHEED intensity measurements on a vicinal plane sample showing
oscillations at low temperature corresponding to growth at
islands nucleated on the terrace. At higher temperature where
the diffusion length of the group III element is longer
oscillations disappear because growth takes place at step edges.

are in phase ("on-Bragg"). Whilst this model cannot explain the observed azimuthal dependence or the dynamical nature of many of the effects observed[33] it does nevertheless appear that under appropriate conditions and at small angles of incidence, the specular beam intensity can give a measure of the step density on the surface.

Monte Carlo studies: The first M-C studies of film growth[12,34] seemed to confirm the models proposed on the basis of a kinematical treatment of the RHEED data. The early work of Singh and Madhukar[34] suggested that the growth front would exist over a number of monolayers (ML) with the transition region getting broader with increasing temperature (3ML at 600K to 7ML at 960K).

Later, using a kinematic approximation based on the M-C predictions of the growth front morphology, the expected RHEED intensity oscillations were calculated with impressive agreement[35]. This led subsequently to a number of important suggestions. First, that the difference between the growth of AlGaAs on GaAs, the so-called "normal" interface, and the growth of GaAs on AlGaAs, the "inverted" interface, is due fundamentally to a lower diffusion length of Al on the surface compared to Ga at the same temperature[36,37]. The morphology of the growth front was shown to depend upon the mean number of sites "visited" by the cation (Ga or Al) before becoming incorporated into the lattice. Increasing the substrate temperature to improve the diffusivity, particularly of Al, should result in a smoother interface, contrary to the earlier work[34] and slowing the growth rate should improve matters. Other important results to emerge from these studies include: the observation that an optimum set of growth conditions can be found, depending primarily on substrate temperature and arsenic flux, which will result in good inverted interfaces[38]; the idea that thin layers containing only the more mobile cations will improve the morphology of thick alloy films[39]; and most crucially that interrupting the growth to allow the surface to smooth might improve interface morphology[40]. The factors controlling the roughness of interfaces will be discussed in more detail below following a discussion of processes occurring at high temperatures in both binary compounds and alloys.

All of the calculations discussed above take as a basis the kinetic data obtained using modulated molecular beam measurements and the models have attempted to include all of the processes known to occur during growth. This led Ghaisas and Madhukar to propose their configuration dependent reactive incorporation (CDRI) model in which, depending upon the model parameters they chose, the build up of As coverage is delayed with respect to Ga, the reaction limited incorporation (RLI) situation[35].

Recently a much simpler model has been used by Clarke and Vvedensky to simulate MBE growth. In their model a single species, randomly deposited on the surface, is allowed to migrate before incorporation into the growth front, to diffuse following emission from such growth islands and to re-evaporate from the surface[41]. The step density on the surface is calculated during the simulation for comparison with the time dependent RHEED intensity measurements. All of the features discussed above are reproduced by this simplified model including the disappearance of RHEED oscillations on vicinal surfaces at high temperature[42] and the improved interface smoothness resulting from slow or interrupted growth[43]. This agreement suggests that the conditions normally employed for MBE growth, namely using an excess As flux, correspond to the CDRI situation discussed by Ghaisas and Madhukar rather than the RLI situation in which the As is delayed with respect to the Ga coverage. It follows that the growth mode under such circumstances is determined by the behaviour of the group III species on the surface. The presence of the group V flux can however have a significant influence on the diffusion behaviour of the cations as shown both theoretically, by the M-C studies and experimentally, by Horikoshi et al[44], who demonstrated improved growth at low temperatures by alternately supplying Ga and As to the surface. The diffusion of Ga and Al is enhanced if a low As flux is used during cation deposition.

2.2 Growth of Alloy (mixed binary compounds) Films

There are two distinctly different situations in the growth of alloy films namely those in which different group III elements are used such as $In_xGa_{1-x}As$ or $Al_xGa_{1-x}As$ and mixed group V elements such as $GaAs_xP_{1-x}$ or $InGa_xSb_{1-x}$. For simplicity I will refer to these as InGaAs, AlGaAs, GaAsP and InGaSb respectively.

Alloys with mixed group III elements: The two examples of alloys of this type listed above have been studied in some detail using the same techniques applied to binary compounds. At low temperatures with an excess group V flux the situation is quite straightforward, both group III elements have a unity sticking coefficient and the growth rate is determined by the total cation flux reaching the surface whilst the composition depends on the relative concentrations of the group III elements.

For In containing alloys of this type the rate of loss of group V species from the surface at high temperatures has been shown both by MMBS and RHEED studies to be similar to that observed in InAs or InP[45]. To compensate for the loss of As_2 it is necessary therefore to use rather high group V fluxes. For AlGaAs, however, it has been suggested that the opposite behaviour is observed and that the alloy grows more As rich than GaAs at high temperatures. Evidence for this statement is based entirely on RHEED studies, but the surface reconstructions observed for AlAs are quite different from those seen in GaAs which may confuse the issue[46]. A much more comprehensive study combining RHEED with MMBS studies is required to resolve this question.

In both InGaAs and InGaP evidence for the segregation and subsequent desorption of the higher vapour pressure element, In, has been observed[45]. This produces both a change in film thickness and composition compared with growth at low temperatures using the same fluxes.

Similar segregation behaviour has been reported in Al containing alloys[47-49]. The evaporation of Ga during the growth of AlGaAs at high temperatures has been considered from a thermodynamic standpoint by Heckingbottom[50]. For a fixed As overpressure the calculated activation energy from an Arrhenius plot is about 4.6eV. This is significantly larger than the figure observed under equilibrium conditions for Ga over Ga[51] or Ga over GaAs[52-54] which is about 2.7eV. The difference arises because at low temperatures evaporation is occurring under As rich conditions whereas at higher temperatures there is a gradual transition to Ga stable growth. This implies that both the growth rate and composition of AlGaAs alloys will be altered by an increase in As overpressure and that the rate of loss of gallium for fixed temperature and As flux will depend upon the Al flux reaching the surface during growth.

RHEED oscillations corresponding to layer by layer sublimation of GaAs have been observed at high temperatures[55,56]. The GaAs re-evaporation rate depends upon the As flux reaching the surface as expected thermodynamically[50] and it was also established that a few monolayers of AlAs could totally suppress the loss of underlying GaAs. This last observation suggests that the observed segregation of Ga during the growth of AlGaAs[47-49] does not always occur.

Direct measurements of the rate of loss of Ga during the growth of GaAs and AlGaAs using RHEED oscillations suggest that the loss of Ga may occur from the mobile precursor state of the atoms on the surface and not by dissociation of the AlGaAs[57-58]. In this study no dependence upon the As flux or the Al fraction was observed as might be expected if Ga is lost before film growth. This is entirely consistent with the earlier work of Arthur[15] who observed a finite temperature dependent lifetime for

Ga. It is also reasonable since the binding energy of Ga in its mobile state (about 2.5-2.7eV) is much lower than the measured activation energies for the re-evaporation of GaAs 4.7eV[55] and 4.6eV[56] from the RHEED sublimation studies and calculated thermodynamically 4.6eV[50].

At present therefore there are two conflicting views of how to predict the loss of the more volatile group III element during the growth of alloys such as AlGaAs, InGaAs and InAlAs at high temperatures. The thermodynamic arguments suggest that the rate of loss will depend upon both the As flux and the alloy fraction and this will certainly occur for a situation where the loss occurs after growth of the film. If the group III element is lost from the mobile precursor state before reacting with the group V element no strong dependence upon either the anion or less volatile cation flux is expected. Further work is required to clarify this situation. The possibility of surface segregation of the more volatile group III element is also not clear at present.

Alloys with mixed group V elements: The situation for mixed group V element alloys is quite simple when growth occurs at temperatures where there is no substantial loss of material from the surface. Under such conditions the element with the lower vapour pressure is incorporated preferentially, thus for GaAsP[59] or InAsP[59] the sticking coefficient of As is much greater than P and does not depend upon whether the dimer or tetramer is used. For GaSbAs[60] or other antimony containing species Sb has the higher sticking coefficient. At low temperatures therefore the composition of the alloy can be easily controlled by limiting the amount of the preferentially incorporated element and providing an excess of the more volatile species. For example, if the As flux, J_{As}, reaching the $GaAs_yP_{1-y}$ surface is small compared to the Ga flux, J_{Ga}, and an excess P flux is supplied, the As fraction y will be given by:

$$y = 2 \; J_{As}/J_{Ga} \; .$$

This will be true for As supplied as As_2 or As_4.

At present the reason for the difference in sticking coefficients for the group V elements is not clear. It has been suggested that it may relate to the different lifetimes of the species[59-60]. This explanation must be incorrect, however, since despite the lifetime of As_2 being much shorter than that of P_4, As is nevertheless incorporated preferentially.

At higher temperatures, when rate of loss of the dimer by dissociation and re-evaporation is comparable to the supply rate composition control becomes much more difficult[60-61]. The rate of loss of the dimer does not relate in any simple way to its sticking coefficients but is determined more by the thermodynamic vapour pressure over the alloy. For GaAsP, therefore, the As fraction will decrease with increasing temperature but for InAsP the reverse will be true[61]. No simple rule can be given therefore in this situation. For Sb containing alloys similar behaviour is expected[60].

For alloys of this type there is a further complicating factor resulting from significant interdiffusion across the heterojunction[62]. Auger and SIMS (secondary ion mass spectrometry) depth profiling through layers of GaP grown on GaAs substrates suggests that the relatively abrupt interfaces observed in AlGaAs-GaAs heterojunctions may not be obtained when a change in group V element is involved. This data may have been influenced by differences in lattice constant giving rise to enhanced interdiffusion. Further work in this area is required to resolve this matter.

2.3 Interfaces in AlGaAs-GaAs Structures

The nature of the interface between GaAs and AlGaAs has been studied in great detail. Despite this, as will be clear from the discussion below, a detailed picture of structural quality of this hetero-interface is only now emerging and some of the models proposed apply, at best, to specific samples studied and should not be regarded as "typical".

If AlGaAs films are grown in the temperature range 630-690°C using As_4 with an Al fraction exceeding 33%, a surface texture is observed which will give rise to interface roughness[47,63-64]. Films grown with As_2 do not, however, exhibit this behaviour. Various causes of this roughening have been proposed including: carbon accumulated at the interface[65-66], an insufficient As surface population when using As_4[67] a reduced mobility for Al compared to Ga[64,68] and Ga surface segregation[47-49]. All of these suggestions appear reasonable and evidence for each model proposed has been presented, however, the generality of the observation lends credance to the intrinsic causes listed above. In a relatively clean growth system (see below), using As_4, we have invariably observed such surface roughness when growing thick films of AlGaAs which supports this conclusion.

More direct evidence for the quality of the interfaces in multi-quantum well (MQW) and superlattice (SL) structures has come from transmission electron microscopy, X-ray diffraction, photoluminescence (PL) and photoluminescence excitation (PLE) studies.

Petroff et al[69] studied the growth mode for GaAs-AlAs SLs as a function of temperature using a cross-section TEM technique. They found that growing at high temperatures (610°C) produced no superlattice reflections but gave rise to a diffraction pattern similar to that observed in an alloy. At lower temperatures (560°C) satellite spots corresponding to the SL period were observed with the degree of ordering decreasing with increasing temperature. The loss of ordering at higher temperatures was thought to arise from a thermodynamic roughening of the type described by Weekes and Gilmer[70] but this is unlikely since this occurs at temperatures well above those used in MBE. It could more reasonably relate to the roughening transition proposed in the M-C simulations[34] but this has not been firmly established.

The precise temperatures quoted in that study are probably incorrect since more recent work in our laboratory has shown that similar SLs can be grown at significantly higher temperatures. Fig 6 shows a typical example of a SL consisting of alternate layers of GaAs 3ML and AlAs 3ML (3+3) grown using As_2 at 650°C. One interesting feature seen in this micrograph and generally observed in such samples is the tendency for the quality of the interfaces to improve through the structure. The initial layers grown on the macroscopically rough alloy follow its texture but they become progressively smoother as growth proceeds.

High resolution lattice plane images of SLs have also been obtained using TEM[71]. There is insufficient contrast for GaAs-AlGaAs interfaces to accurately locate the position of the heterojunction but it is just

Fig. 6. A cross-section TEM photograph (J.P. Gowers) of a GaAs
(3ML) AlAs (3ML) superlattice grown (D. Hilton) by MBE using As_2
at a substrate temperature of 650°C. Note that the initially
rough interfaces become progressively smoother as growth
proceeds.

possible to resolve the boundary for GaAs-AlAs structures. In both cases it is clear that no crystallographic disorder is present at the interface since no dislocations, stacking faults or twin boundaries are observed.

Two different X-ray techniques have been used to study MQWs and SLs. The small angle scattering method relies on the difference in refractive index between GaAs and AlAs. Below a critical angle X-rays are totally reflected by the surface but beyond this point interference maxima are observed from which the periodic nature of the structure can be deduced[72-73] and in addition the individual thicknesses of GaAs and AlAs can be estimated.

More recently high angle X-ray diffractometry has been applied to similar structures[73-77]. Using the difference in the positions of Bragg reflections from the substrate and the MQW or SL an average aluminum content can be obtained. In making this calculation it is of course essential to take into account the tetragonal distortion of the AlAs or AlGaAs layers. Additional satellite peaks are observed around the main Bragg reflections of the MQW or SL layers due to the periodic modulation of the atomic scattering factor and lattice parameter. The periodicity of the MQW or SL can be deduced from the spacing of the satellite features. Finally by modelling the integrated intensities of the satellite reflections it is possible to deduce both the individual layer thicknesses and to estimate the interface roughness[77]. From such studies estimates of interface roughness from 1 to 4 ML have been proposed, somewhat better interfaces being obtained for AlAs/GaAs structures than equivalent AlGaAs/GaAs samples.

The optical properties of MQWs and SLs have been extensively studied using photoluminescence (PL) and photoluminescence excitation (PLE) spectroscopy at low temperatures. Since this subject will be discussed in detail elsewhere I will confine my remarks largely to those relevant to the study of the properties of the interface. In the case of MQWs, contrary to the situation in bulk GaAs, luminescence is observed due to the recombination of free or weakly bound excitons.

It has been shown that exciton linewidths become broader as the well width decreases. Whilst it is clear that factors other than structural imperfections, for example impurities, can cause broadening and it is also evident that the conditions used for the optical experiments are crucial, particularly the excitation density used, a comparison of line widths taken on a series of samples where only the well width has been intentionally varied can give information on the structure of the interfaces. In MQWs variations in exciton linewidth could arise from either inter-well thickness variations due to changes in growth rate through the structure or to variations in width within each individual well whilst maintaining a constant average layer thickness (intra-well fluctuations). Variations in well width will in general give rise to changes in the confinement energies of electrons and holes, together with somewhat smaller changes in the binding energy of the excitons; both factors will give rise to the line broadening observed. A simple estimate of the influence of a given degree of variation in well width can be obtained from considering the confinement energy, E, for a particle of effective mass m, in an infinite potential well of width W.

$$E = h^2/(8m \ W^2) \ .$$

Assuming a fluctuation of dW in well width the resulting change in energy dE will be given by

$$dE = -[h^2/(4m \ W^3)] \ dW \ .$$

The confinement energies of electrons and holes can be described more correctly using the envelope function approximation for confined states taking into account non-parabolicity in the conduction band and correcting for the binding energy of the excitons[78]. From this more appropriate calculation it can be shown that the expected variation at small well widths is significantly different from that given above but the general trend to broader lines with decreasing W is confirmed as shown in Fig. 7.

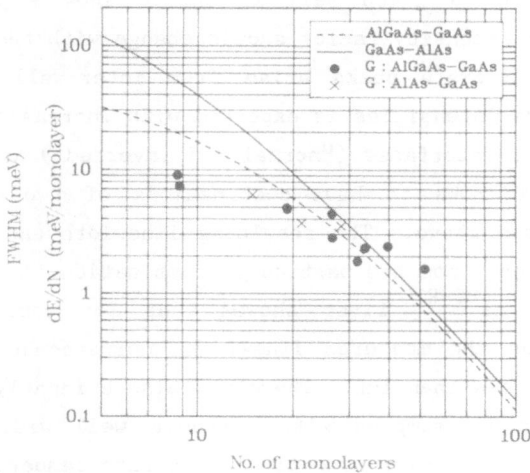

Fig. 7. The rate of change of energy per monolayer for the lowest energy electron to heavy hole transition observed in $Al_{0.33}Ga_{0.67}As$-GaAs and AlAs-GaAs MQWs (G. Duggan). Also shown are measured peak widths (P. Dawson and K.J. Moore) for 60 period MQWs grown by MBE (C.T. Foxon and D. Hilton) in a Varian GEN-II at 630°C using the minimum possible As flux.

Three physically different situations can be envisaged which determine how variations in well width will influence the measured line widths of excitons in PL and PLE spectra.

If monolayer steps occur on a lateral scale much smaller than the diameter of an exciton (~300Å in bulk GaAs), at both interfaces (type A), the energies of given peaks in PL and PLE should be identical but will not correspond to those expected for an integer number of monolayers.

If the lateral scale of monolayer steps becomes comparable to the diameter of an exciton then a so-called "Stokes shift" is expected. In this (type B) situation excitons will be created corresponding to all well widths and hence energies with equal probability, but there will be a tendency for them to thermalize to lower energy states in the wider parts of the well. This will result in a small difference in peak positions in PL and PLE for given transitions. Under such conditions the excitons may be trapped at steps (localized) which will modify the observed energy of the transitions giving rise to the observed "Stokes shift". The magnitude of this effect has been estimated to be at most a few meV and to depend upon the lateral size of the defect[79].

The third physical possibility (type C) arises when the terrace length is much longer than an exciton diameter. In this case two or three discrete peaks with energies corresponding to an integer number of monolayers should be seen in PLE. Even if PL is observed from the narrower regions of a given well it should show significantly lower intensity and the intensity ratios should change with temperature. If on the other hand discrete peaks arise from inter-well fluctuations no change in relative intensities is expected with increasing temperature.

Since the two interfaces ("normal and inverted") are not equivalent it is possible in MQWs to have combinations of two of the types of interface discussed above. The resulting linewidth and number of peaks observed will depend upon the particular combination.

Weisbuch et al[80-81] first showed that for samples grown under optimum conditions the measured linewidths corresponded to interfacial roughening of no more than 1ML. There was also evidently a trend towards better interfaces in samples with narrower well widths. They later demonstrated that the choice of optimum substrate temperature (660-690°C) was the crucial factor in determining the line width[82]. At 570°C the interface roughness was about 5-6ML (15Å) decreasing with increasing temperature, at the same time they observed a corresponding increase in photoluminescence efficiency. Above 690°C the linewidths increased but no corresponding decrease in efficiency was observed. In this study there appears to be a very small "Stokes shift" present which indicates that one of the interfaces is probably of type B and since only single peaks were observed the other interface is probably of type A.

Samples with somewhat narrower linewidths were later obtained by a number of groups, Devaud et al[83] reviewed this work and showed that for "better" samples peak splitting was observed with a spacing corresponding to an integral number of MLs with no evidence for a "Stokes shift". Their samples were grown at about 695°C using the minimum amount of As required to give As stable growth. These samples probably contain one interface of type A (GaAs grown on AlGaAs) followed by a second interface of type C (AlGaAs grown on GaAs).

In our laboratory we have also prepared samples of this type grown under rather different conditions, the substrate temperature was 630°C and again minimal As flux was used. Growth rates of $1ML \, s^{-1}$ were used for GaAs with correspondingly higher rates for the AlGaAs barriers. The low temperature linewidths are shown in Fig. 7. The interface roughness is less than 1ML and the trend observed earlier to "better" interfaces in narrower AlGaAs samples is observed[80-81]. Electrical measurements (see below) have established that the purity of our samples is among the best available to date and it is perhaps significant that good interface properties have been obtained at lower substrate temperature than reported above.

Our samples fall into several categories. For relatively wide MQWs (similar to those discussed above) double peaks are observed with a separation corresponding to 1ML difference in well width and no "Stokes shift" (<1meV) is observed. Studies of the temperature dependence of the PL from these samples indicate that the variations arise from intra-well fluctuations and not variations in growth rate. It is probable therefore that these MQWs have type A, at the "inverted", and type C, at the "normal", interfaces. The roughness produced by growing the alloy film has been sufficiently smoothed by the growth of the binary well to give a type C "normal" heterojunction.

For our narrowest AlGaAs MQWs (9ML) a quite different behaviour is seen, a single line with a half width of about 9 meV is observed in PL corresponding to a well width variation of about 1/2 ML. From a comparison of PL and PLE a "Stokes shift" of between 6 and 8 meV is observed depending upon the detection energy used for PLE. In this sample therefore the interfaces are of type A and type B and insufficient GaAs has been grown to reach the steady state step distribution needed for a type C interface.

Also included in Fig. 7 are measured linewidths for AlAs-GaAs MQWs grown under identical conditions. Because the expected energy shift per monolayer is higher for any given well width in MQWs with AlAs barriers, as shown in Fig. 7, for equivalent interface roughness one would expect broader lines. In fact the measured linewidths are almost identical in wider well samples to those observed in AlGaAs MQWs, it follows that this must mean that the interfaces obtained in the all binary structures are "better" than those observed in the AlGaAs-GaAs samples. For the narrow well samples they appear to be of comparable quality.

Much narrower line widths than those described above have been reported by Reynolds et al[84] for a series of AlGaAs-GaAs MQWs with well widths of around 100Å (36ML). The exact growth conditions were not specified in this paper but a reference to earlier work suggests that they may be the samples grown at 700°C exhibiting multiple peaks[85]. The spacing of peaks corresponded to well width fluctuations of about 1/2 ML which is physically unreasonable for an intra-well variation and the same authors later suggested[86] that they arose from inter-well variations. To see such small inter-well variations suggests that the interfaces seen by the exciton are indeed much "better" than those observed before but this result could also be explained if both interfaces are of type A, that is, with extremely small terrace widths relative to the exciton diameter[85].

In summary for conventional growth by MBE the so-called "inverted" interface produced when growing GaAs on AlGaAs or AlAs seems probably to be of type A, that is, with monolayer steps on lateral scale very short compared to an exciton diameter. The "normal" interface obtained when growing AlGaAs on GaAs appears to be of type B or C. Type B interfaces are in general obtained during growth at lower substrate temperatures or in samples where only thin layers of GaAs (<50Å or 20ML) are grown after an Al containing layer. Type C interfaces will be obtained by growth at high temperatures of a GaAs layer sufficiently thick to reach the steady state step distribution corresponding to the particular growth conditions. Finally the degradation in exciton line width observed when growth takes place above an optimum temperature may arise as a result of increasing the terrace length on the Al containing surface, converting a type A interface into type B which appears to be less perfect in PL or PLE.

The effect of modified growth procedures on interfaces: The idea that interrupting growth might lead to smoother interfaces was implicit in the model proposed by Neave et al[28] to explain observed RHEED oscillations and subsequent recovery and was explicitly proposed as a result of the M-C studies of film growth[40]. The procedure adopted is to close the shutters in front of the group III sources and allow the surface to anneal under an As flux for a period ranging from a few seconds to several minutes. During this time it is supposed that the number of monolayer steps on the surface will decrease as a result of migration from small to large islands of the group III elements.

The first direct evidence for effect of interrupts on the quality of the interfaces came from studies of the PL linewidth in samples grown using this modified procedure[87-89]. For all of these studies a low substrate temperature was used and only the GaAs surface could change during the growth interrupt. There is no evidence from RHEED recovery studies that this is expected to occur for AlGaAs or AlAs surfaces. This was later confirmed in an elegant series of experiments by Tanaka et al[90-91]. They studied the effects of interrupts at each of the two interfaces and compared the resulting PL with samples grown conventionally and with interrupts at both interfaces for the complete range of AlGaAs alloys. For Al fractions of >0.5 they propose that the "inverted" interface will be of type A with atomic steps of 40A spacing and that these will not change on annealing. For the GaAs "normal" interface, however, a 200Å step length can be increased by the interrupt resulting in a change from a type B to a type C interface. RHEED data taken in the same equipment seems to confirm this picture. For AlGaAs alloys with an Al content of <0.3 the behaviour is similar to GaAs.

A somewhat similar picture was proposed by Koteles et al[92] who used both PL and PLE to confirm the interpretation of their data. They observed multiple peaks corresponding to monolayer differences in well width but in addition observed peaks associated with bound excitons in samples where growth had been interrupted. They stressed that multiple peaks observed in PL did not prove the existence of monolayer steps.

There are several major problems in accepting this data at face value. First the linewidths obtained using the normal growth process[87-01] are far worse than those reported previously[80-83] and those illustrated in Fig. 7. Only by using the modified procedure were linewidths comparable to those obtained conventionally achieved. The second key point is that completely conflicting data for samples grown under similar conditions has been presented which seems to be equally valid[93-94]. In one of the above reports discrete peaks were again ascribed to monolayer fluctuations but a careful examination of the data suggests that the energy differences are too small and additionally a significant "Stokes shift" is observed suggesting the presence of bound rather than free excitons.

One odd fact which seems to have been overlooked is that having the As flux present during growth interrupts may be detrimental since it has been established that this can limit the diffusion length of the group

III atoms on the surface[44]. It may be that a modified procedure where both group III and group V fluxes are interrupted perhaps for different times may prove a useful alternative technique.

One other point which has received little attention so far is the role of impurities reaching the surface during the growth interrupt. This may directly influence the subsequent growth by pinning steps which might otherwise propagate in a two-dimensional mode. It may also lead to reduced optical efficiency and increased carbon incorporation[94-95].

At present we can be sure that interrupting growth is changing the nature of the interface but far from certain exactly what effect this perturbation produces. The simplified model proposed by most authors is apt to be misleading and studies may at best relate to what happens in a particular MBE machine under the conditions used and should not be taken as proof that identical results will be obtained under different conditions. Much more careful work is needed to correctly identify those conditions which will lead to genuinely smoother interfaces of type C.

3. GROWTH OF HIGH PURITY STRUCTURES

Many of the more demanding applications of MBE require very high purity layers with free carrier concentrations below 10^{14} cm^{-3}. The detailed techniques we use to obtain material of this quality have been reported elsewhere and will be briefly outlined below[96]. The MBE sources are outgassed to high temperature in the preparation chamber and loaded with the best possible sources of material. The whole MBE equipment is then baked for 2 weeks resulting in a base pressure of $<1 \times 10^{-10}$ Torr. Before each growth the sources, with the exception of As, are carefully outgassed at least 50°C above their operating point. Samples are loaded using high purity In onto Mo blocks which are then outgassed to remove water vapour before transfer to the growth chamber, impurities evolved during the removal of the native oxide are allowed to pump away before growth. Finally growth rates of 1 ML s^{-1} are used for GaAs with the As flux adjusted to a minimum required to give a 2-fold RHEED pattern on the (100) surface in the [110] azimuth. Sample rotation rates and growth times for individual layers have been adjusted to give complete numbers of monolayers.

Using this technique we have grown several thick intentionally doped GaAs films with electron mobilities at 77K of $>1 \times 10^5$ and a best value of 1.33×10^5 cm^2 V^{-1} s^{-1}, free electron concentrations were about 2×10^{14}

cm^{-3}. This suggests the background acceptor concentration in our material is below 2×10^{14} cm^{-3}. Recently we have reported an undoped GaAs film, grown using a superlattice prelayer, which at low temperature has luminescence dominated by free exciton emission[97].

For several years the best reported mobilities for 2DEG samples were approximately 2×10^6 cm^{-2} V^{-1} s^{-1} at a sheet carrier density of about 5×10^{11} cm^{-2} [99-100]. In our studies we were able to obtain similar mobility material at somewhat lower electron densities, as shown in Fig. 8[101]. We reasoned that the ionized Si atoms close to the surface could contribute significantly to the total ionized impurity scattering and by further optimizing the structure managed to obtain mobilities of 3×10^6 and 4.4×10^6 cm^2 V^{-1} s^{-1} at 4.2 and <2K respectively[102-103]. The important point first discussed by Weimann and Schlapp[100] is that for very wide spacer layers (>400Å in our case) the mobility is limited mostly by the quality of the undoped GaAs and nature of the interface. Using an MQW prelayer to improve the background scattering in the GaAs and a wide lightly doped region of AlGaAs to avoid the scattering by ionized impurities near the surface has recently enabled English et al to grow a 2DEG with a mobility at <2K of 5×10^6 cm^2 V^{-1} s^{-1} at a sheet carrier density of 1.6×10^{11} cm^{-2} [104].

The mobility of the 2DEG formed at the GaAs-AlGaAs interface depends upon the order in which the layers are grown. For the "normal" structure, as outlined above, extremely high mobilities are obtained, but for the "inverted" heterostructure the best results reported so far are more than a decade lower[105]. Three reasons have been suggested for the inferior performance: interface roughness (as discussed above), impurity build-up at the interface and Si migration from the underlying doped AlGaAs due to segregation. We have recently studied this problem in modulation doped quantum wells (QWs) thin enough for the electrons to sample both interfaces. We have observed that in as-grown and annealed samples the low electron mobility is associated with electron localization at the inferior inverted interface[106]. The localization can be removed by replacing the bulk AlGaAs layers by a short period SL of equivalent band gap. In such QWs the 2DEG mobility is, for samples with narrow undoped spacer layers, equivalent to that obtained in a conventional 2DEG formed at the "normal" interface.

Radulescu et al[107] also studied transport in QWs and found that an anisotropic mobility exists for samples grown with thick AlGaAs barriers. This was removed by replacing the first confinement layer with either a short period SL or AlGaAs grown at higher substrate temperatures (690°C). Again this points to the "inverted interface" being degraded by an intrinsic morphology problem rather than other extrinsic factors.

For our work at Philips on understanding the growth and properties of SLs we have grown four sets of samples: the first set were MQWs with well widths from 9 to 54ML with direct bandgap barriers thick enough (55ML) to provide isolated QWs, the second set had 27ML AlAs barriers with GaAs well widths from 2ML to 27ML, the third set had GaAs well widths of 9ML with AlAs barriers from 2 to 27ML and the final set are SLs with equal GaAs and AlAs thicknesses from 1 to 6ML. In each series control samples of known properties were grown to check the state of the machine.

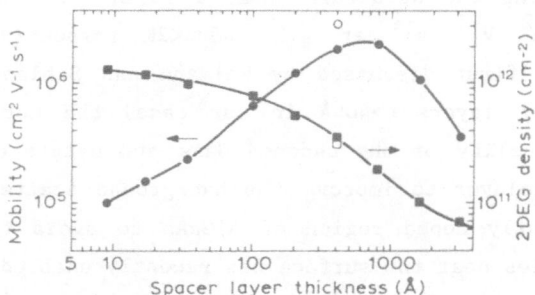

Fig. 8. Variation in carrier density and mobility of 2DEG samples at 4K as a function of the undoped spacer layer thickness. The results plotted are for samples illuminated to saturate the persistent photo-conductivity effect. Two thicknesses of doped AlGaAs were used, 400Å (solid symbols) and 500Å (open symbols).

From the first set of samples we were able to compare quantitatively the measurements of well and barrier thickness determined by RHEED, X-ray, TEM and PLE measurements with satisfactory agreement[108]. Samples grown for this study also showed for the first time clearly resolved features for the 2S state of the light and heavy hole excitons in PL and PLE[109-110] from which accurate estimates of the exciton binding energy were obtained. In addition transitions to the confined hole state of the split-off band were identified[111].

In the second set of samples we see a type I emission process from electrons and holes confined in GaAs wells >11ML thick, for narrower GaAs wells type II luminescence from electrons confined in the AlAs barriers[112] recombining with holes in the GaAs is also observed. Fig. 9 shows an example of low temperature PL from two samples of this type. Features relating to both type I and type II recombination can be identified in the spectra together with phonon replicas associated with the type II emission process.

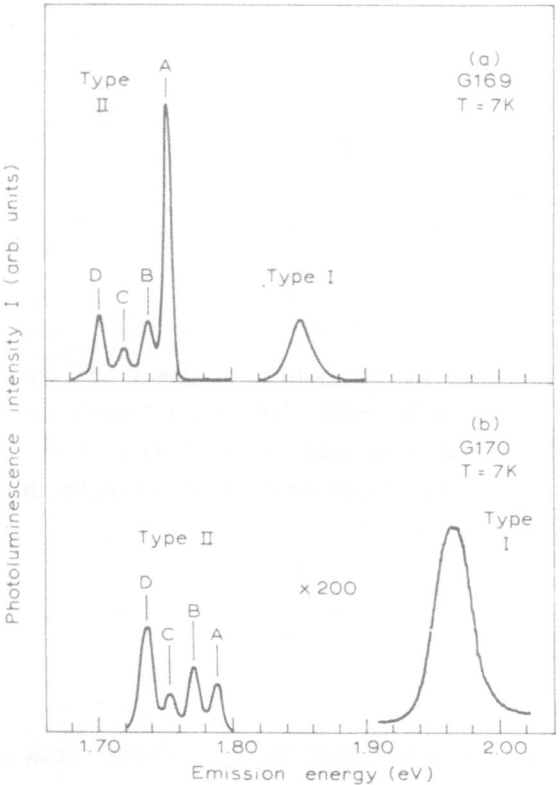

Fig. 9. Low temperature (7K) photoluminescence spectra of two GaAs-AlAs
MQWs showing emission from type I recombination of electrons and
holes localized in the GaAs and type II luminescence from
electrons at the X minimum of the AlAs recombining with holes in
the GaAs.

Standard 9ML GaAs, 27ML AlAs control samples were grown during the
course of this study under a variety of growth conditions. The spectra
obtained are extremely sensitive to the choice of growth rates and
substrate temperature. Spectra like those shown in Fig. 9 are only
obtained under optimum conditions. Much broader, less well resolved
features are observed when poorer interfaces were obtained.

From the third set of samples we were able to show a gradual
increase in confinement energy at the X point in AlAs and a decrease in
confinement energy at the Γ point in GaAs (due to coupling of electron
states in the GaAs) as the AlAs barrier width decreased. This data is
shown in Fig. 10[113]. This results in a reversion to type I emission for
samples with a GaAs well width of less than 4-5ML. PL and PLE studies from
the SL samples are not yet complete.

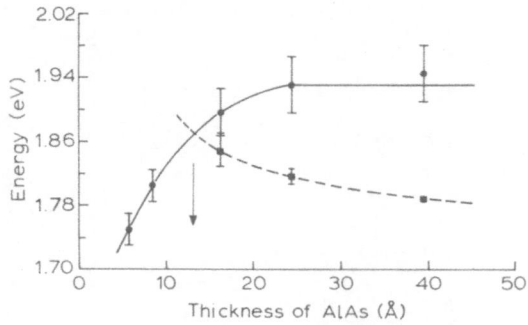

Fig. 10.   The variation in energy of the type I and type II
transitions in GaAs-AlAs MQWs as a function of AlAs thickness
for a constant GaAs well width of 9ML. A transition from type
I to type II behaviour is observed at AlAs thicknesses between
4 and 5ML.

4.   DEVICES BASED UPON LOW-DIMENSIONAL STRUCTURES GROWN BY MBE

Many potential high speed device structures have been proposed which
rely on the use of low-dimensional structures for their operation. Of
these only the high electron mobility transistor has demonstrated
improved performance over conventional devices for both microwave and
digital applications. The growth, properties and performance limitations
of this device have recently been discussed in detail by Drummond et
al[114] and only key points relating to the MBE growth process will be
outlined here.

The high electron mobility transistor (HEMT), two-dimensional
electron gas transistor (TEGFET) or modulation doped field effect
transistor (MODFET) is based on a narrow spacer 2DEG. The current
between a pair of source drain contacts is modulated by the potential
applied to a gate electrode to provide transistor action similar to that
observed in a conventional GaAs MESFET. The high frequency performance
of this new type of device, in particular the noise performance, is
better than that of the conventional FET for equivalent device
geometries. Very approximately, the performance of 0.5 micron HEMT is
equivalent to that of a 0.25 micron MESFET. The improved performance
arises in part from the higher electron velocity achieved in 2DEGs even

under high injection conditions and partly from the electron confinement to the narrow region at the heterojunction interface.

For low noise applications a 2DEG density of about $1x10^{12}$ cm^{-2} is required to reduce the source resistance to an acceptably low level. This can readily be obtained using a narrow undoped spacer (0-20Å) and a relatively highly doped AlGaAs region ($3-4x10^{18}$ cm^{-3}). Parallel conduction is avoided by using a recessed gate geometry and high frequency performance can be aided by using thick heavily doped regions of GaAs in the source and drain regions of the device. With such a structure transconductances of 300 mS/mm are readily achieved in the AlGaAs system using a 0.5 micron gate length.

For digital integrated circuit applications the main problems encountered relate to the defect density observed on the surface after growth. For example in a 4 Kb static random access memory 27000 transistors are integrated each of which must perform to specification. To achieve even a reasonable yield defect densities of <500 cm^{-2} must be obtained. On typical MBE material defects arise from poor substrate preparation or mounting, from oxidation of the group III sources and from particulates within the MBE equipment. A variety of methods has been proposed to improve these factors but this still remains a major problem. Gas source MBE does seem to offer some significant advantages in this respect over conventional MBE and may be attractive in this application.

Finally it is appropriate to point out that other materials systems such as AlInAs-GaInAs may offer some significant advantages over the AlGaAs-GaAs devices currently being produced[115].

The properties of both conventional and MQW lasers grown by MBE have been reviewed by Tsang[10] who made a major contribution in this area. He discovered that AlGaAs with high optical efficiency could be obtained by growth at high temperatures (680-700°). There are two possible reasons for the improved properties, either reduction in the number of intrinsic defects or a reduced incorporation of oxygen which is known to affect both the electrical and optical properties of AlGaAs[115]. At present no clear evidence exists to distinguish between the two possibilities.

The use of MQW structures has resulted in devices with improved threshold current, arising from higher net optical gain, and improved threshold-temperature dependence[10] compared with conventional double heterostructure lasers. It has also proved possible to achieve lasing at wavelengths as low as 710 nm using MQWs with GaAs well widths of only 4-5 MLs[116].

5. CONCLUSIONS

MBE has emerged as a powerful growth technique for the preparation of structures based mainly on AlGaAs-GaAs lattice matched alloys. Using the RHEED oscillation technique it can provide complex samples with individual layers down to a monolayer scale of known thickness and composition. Such well characterized samples are essential to study the physical properties of low-dimensional structures. A detailed picture of the AlGaAs-GaAs interface has emerged from X-ray, TEM, PL and PLE studies and novel devices such as HEMTs and MQW lasers based on low-dimensional principles have shown improved performance compared with conventional structures.

ACKNOWLEDGEMENTS

I would like to thank my colleagues at Philips Research Laboratories for their contributions to the work outlined in this article.

REFERENCES

1. J. C. Bean, J. Crystal Growth, 81:411 (1987).
2. A. Y. Cho and J. R. Arthur, Prog. Solid State Chem., 10:157 (1975).
3. C. T. Foxon and B. A. Joyce, Curr. Topics in Mater. Sci., 7:1 (1981).
4. B. A. Joyce, Rep. on Prog. in Phys., 48:1637 (1985).
5. A. C. Gossard, Treatise Mater. Sci. Technol., 24:13 (1982).
6. H. Kroemer, J. Crystal Growth, 81:194 (1987).
7. K. Ploog, J. Crystal Growth, 81:304 (1987).
8. J. J. Harris, The technology and physics of molecular beam epitaxy, Plenum Pub. Corp., 425 (1985).
9. K. Board, Rep. Prog. Phys., 48:1595 (1985).
10. W. T. Tsang, IEEE J. Quantum Electr., QE20:1119 (1984).
11. C. T. Foxon, M. R. Boudry and B. A. Joyce, Surf. Sci., 44:69 (1974).
12. P. J. Dobson, B. A. Joyce, J. H. Neave and Jing Zhang, J. Crystal Growth, 81:1 (1987).
13. A. Madhukar, Surf. Sci., 132:344 (1983).
14. R. Heckingbottom, in Molecular Beam Epitaxy and Heterostructures, edited by L. L. Chang and K. Ploog (Martinus Nijhoff, Dordrecht, Holland, 719 (1967).

15. J. R. Arthur, J. Appl. Phys., 39:4032 (1968).

16. J. R. Arthur, Surf. Sci., 43:449 (1974).

17. C. T. Foxon and B. A. Joyce, Surf. Sci., 64:293 (1977).

18. C. T. Foxon and B. A. Joyce, Surf. Sci., 50:434 (1975).

19. C. T. Foxon, J. Vac. Sci. Technol., B1:293 (1983).

20. J. H. Neave, P. K. Larsen, J. F. van der Veen, P. J. Dobson and B. A. Joyce, Surf.Sci., 133:267 (1983).

21. J. H. Neave, P. Blood and B. A. Joyce, Appl. Phys. Lett., 36:311 (1980).

22. G. Duggan, P. Dawson, C. T. Foxon and G. W. 't Hooft, J. de Phys. C5:129 (1982).

23. A. Y. Cho, J. Appl. Phys., 42:2074 (1971).

24. J. M. Van Hove and P. I. Cohen, J. Vac. Sci. Technol., 20:726 (1982).

25. P. J. Dobson, J. H. Neave and B. A. Joyce, Surf. Sci., 119:L339 (1982).

26. J. M. Van Hove, C. S. Lent, P. R. Pukite and P. I. Cohen, J. Vac. Sci. Technol., B1:741 (1983).

27. J. J. Harris, B. A. Joyce and P. J. Dobson, Surf. Sci., 103:L90 (1981).

28. J. H. Neave, B. A. Joyce, P. J. Dobson and N. Norton, Appl. Phys., A31:1 (1983).

29. J. M. Van Hove, C. S. Lent, P. R. Pukite and P. I. Cohen, J. Vac. Sci. Technol., B1:741 (1983).

30. J. H. Neave, B. A. Joyce and P. J. Dobson, Appl. Phys., A34:179 (1984).

31. J. H. Neave, P. J. Dobson, B. A. Joyce and Jing Zhang, Appl. Phys. Lett., 100:47 (1985).

32. C. S. Lent and P. I. Cohen, Surf. Sci., 139:121 (1984).

33. Jing Zhang, J. H. Neave, P. J. Dobson and B. A. Joyce, Appl. Phys., A42:317 (1987).

34. J. Singh and A. Madhukar, J. Vac. Sci. Technol., B1:305 (1983).

35. S. V. Ghaisas and A. Madhukar, J. Vac. Sci. Technol., B3:540 (1985).

36. J. Singh and K. K. Bajaj, J. Vac. Sci. Technol., B2:576 (1984).

37. J. Singh and K. K. Bajaj, J. vac. Sci. Technol., B3:520 (1985).

38. A. Madhukar and S.V. Ghaisas, Appl. Phys. Lett., 47:247 (1985).

39. J. Singh and K. K. Bajaj, Appl. Phys. Lett., 47:594 (1985).

40. A. Madhukar, T. C. Lee, M. Y. Yen, P. Chen, J. Y. Kim, S. V. Ghaisas andP. G. Newman, Appl. Phys. Lett., 46:1148 (1985).

41. S. Clarke and D. D. Vvedensky, Semi. Sci. Technol., (to be published).

42. S. Clarke and D. D. Vvedensky, Phys. Rev. Lett., (to be published).

43. S. Clarke and D. D. Vvedensky, Appl. Phys. Lett., (to be published).

44. Y. Horikoshi, M. Kawahima and H. Yamaguchi, Jap. J. Appl. Phys., 25:L868 (1986).

45. C. T. Foxon and B. A. Joyce, J. Cryst. Growth, 44:75 (1978).

46. R. Z.Bachrach, R. S. Bauer, P. Chiaradia and G. V. Hansson, J. Vac. Sci. Technol., 19:335 (1981).

47. R. A. Stall, J. Zilko, V. Swaminathan and N. Schumaker, J. Vac. Sci. Technol., B3:524 (1985).

48. J. Massies, J. F. Rochette and P. Delescluse, J. Vac. Sci. Technol., B3:613 (1985).

49. J. Massies, F. Turco and J. P. Contour, Semicond. Sci. Technol., 2:179 (1987).

50. R. Heckingbottom, J. Vac. Sci. Technol., B3:572 (1985).

51. R. E. Honig and D. A. Kramer, RCA Rev., 30:285 (1969).

52. J. R. Arthur, J. Phys. Chem. Sol., 28:2257 (1967).

53. C. T. Foxon, J. A. Harvey and B. A. Joyce, J. Phys. Chem. Sol., 34:1693 (1973).

54. C. Pupp, J. J. Murray and R. F. Pottie, J. Chem. Therm., 6:123 (1974).

55. T. Kojima, N. J. Kawai, T. Nakagawa, K. Ohta. T. Sakamoto and M. Kawashima, Appl. Phys. Lett., 47:286 (1985).

56. J. M. Van Hove and P. I. Cohen, Appl. Phys. Lett., 47:726 (1985).

57. C. T. Foxon, J. Vac. Sci. Technol., B4:867 (1986).

58. C. T. Foxon, in Heterojunctions and Semiconductor Superlattices, ed. G. Allen, G. Bastard, N. Boccara, M. Lannoo and M. Voos, Springer-Verlag 216 (1986).

59. C. T. Foxon, B. A. Joyce and M. T. Norris, J. Cryst. Growth, 49:132 (1980).

60. Chin-An Chang, R. Ludeke, L. L. Chang and L. Esaki, Appl. Phys. Lett., 31:759 (1977).

61. K. Woodbridge, J. P. Gowers and B. A. Joyce, J. Cryst. Growth, 60:21 (1982).

62. J. S. Johannessen, J. B. Clegg, C. T. Foxon and B. A. Joyce, Physica Scripta, 24:440 (1981).

63. H. Morkoc, T. J. Drummond, W. Kopp and R. Fischer, J. Electrochem. Soc., 129:824 (1982).

64. F. Alexandre, L. Goldstein, G. Leroux, M. C. Joncour, H. Thibierge and E. V. K. Rao, J. Vac. Sci. Technol., B3:950 (1985).

65. R. C. Miller, W. T. Tsang and O. Munteanu, Appl. Phys. Lett., 41:374 (1982).

66. P. M. Petroff, R. C. Miller, A. C. Gossard and W. Wiegmann, Appl. Phys. Lett., 44:217 (1984).

67. L. P. Erickson, T. J. Mattord, P. W. Palmberg, R. Fischer and H. Morkoc, Electon. Lett., 19:632 (1983).

68. M. Heiblum, E. E. Mendez and L. Osterling, J. Appl. Phys., 54:6982 (1983).

69. P. M. Petroff, A. C. Gossard, W. Wiegmann and A. Savage, J. Crystal Growth, 44:5 (1978).

70. J. D. Weekes and G. H. Gilmer, Adv. Chem. Phys., 40:157 (1979).

71. H. Okamoto, M. Seki and Y. Horikoshi, Japan. J. Appl. Phys., 22:L367 (1983).

72. L. L. Chang, A. Segmuller and L. Esaki, Appl. Phys. Lett., 28:39 (1976).

73. A. Segmuller, P. Krikshna and L. Esaki, J. Appl. Crystallogr., 10:1 (1977).

74. R. M. Fleming, D. B. McWhan, A. C. Gossard, W. Wiegmann and R. A. Logan, J. Appl. Phys., 51:357 (1980).

75. T. Ishibashi, Y. Suzuki and H. Okamoto, Jap. J. Appl. Phys., 20:L623 (1981).

76. D. A. Neumann, H. Zabel and H. Morkoc, Appl. Phys. Lett., 43:59 (1983).

77. P. F. Fewster, Philips J. Res., 41:268 (1986).

78. G. Bastard, Phys. Rev., B12:7584 (1982).

79. G. Bastard, C. Delalande, M. H. Meynadier, P. M. Frijlink and M. Voos, Phys Rev., B29:7042 (1984).

80. C. Weisbuch, R. Dingle, A. C. Gossard and W. Wiegmann, J. Vac. Sci. Technol., 17:1128 (1980).

81. C. Weisbuch, R. Dingle, A. C. Gossard and W. Wiegmann, Solid State Comm., 38:709 (1981).

82. C. Weisbuch, R. Dingle, P. M. Petroff, A. C. Gossard and W. Wiegmann, Appl. Phys. Lett., 38:840 (1981).

83. B. Deveaud, A. Regreny, J.-Y. Emery and A. Chomette, J. Appl. Phys., 59:1633 (1986).

84. D. C. Reynolds, K. K. Bajaj, C. W. Litton, P. W. Yu, J. Singh, W. T. Masselink, R. Fischer and H. Morkoc, Appl. Phys. Lett., 46:51 (1985).

85. Y. L. Sun, W. T. Masselink, R. Fischer, M. V. Klein, H. Morkoc and K. K. Bajaj, J. Appl. Phys., 55:3554 (1984).

86. K. K. Bajaj, D. C. Reynolds, C. W. Litton, J. Singh, P. W. Yu, W. T. Masselink, R. Fischer and H. Morkoc, Solid State Electr., 29:215 (1986).

87. H. Sakaki, M. Tanaka and Y. Yoshino, Jpn. J. Appl. Phys., 24:L417 (1985).

88. T. Fukunaga, K. L. I. Kobayashi and H. Nakashima, Jpn. J. Appl. Phys., 24:L510 (1985).

89. T. Hayakawa, T. Suyama, K. Takahashi, M. Kondo, S. Yamamoto, S. Yano and T. Hijikata, Appl. Phys. Lett., 47:952 (1985).

90. M. Tanaka, H. Sakaki and J. Yoshino, Jpn. J. Appl. Phys., 25:L155 (1986).

91. M. Tanaka and H. Sakaki, J. Crystal Growth, 81:153 (1987).

92. E. M. Koteles, B. S. Elman, C. Jagannath and Y. J. Chen, Appl. Phys. Lett., 49:1465 (1986).

93. F. Voillot, A. Madhukar, W. C. Tamg, M. Thomsem, J. Y. Kim and P. Chen, Appl. Phys. Lett., 50:194 (1987).

94. C. W. Tu, R. C. Miller, B. A. Wilson, P. M. Petroff, T. D. Harris, R. F. Kopf, S. K. Sputz and M. G. Lamont, J. Crystal Growth, 81:159 (1987).

95. D. Bimberg, D. Mars, J. N. Miller, R. Bauer and D. Oertel, J. Vac. Sci. Technol., B4:1014 (1986).

96. C. T. Foxon and J. J. Harris, Philips J. Res., 41:313 (1986).

97. G. W. 't Hooft, W. A. J. A. van der Poel, L. W. Molenkamp and C. T. Foxon, Phys. Rev., B35:8281 (1987).

98. S. Hiyamizu, J. Saito, K. Nambu and T. Ishikawa, Jpn. J. Appl. Phys., 22:L609 (1983).

99. J. C. M. Hwang, A. Kastalsky, H. L. Stormer and V. G. Keramidas, Appl. Phys. Lett., 44:802 (1984).

100. G. Weimann and W. Schlapp, Appl. Phys. Lett., 46:411 (1985).

101. C. T. Foxon, J. J. Harris, R. G. Wheeler and D. E. Lacklison, J. Vac. Sci. Technol., B4:511 (1986).

102. J. J. Harris, C. T. Foxon, D. E. Lacklison and K. W. J. Barnham, Superlattices and Microstructures, 2:563 (1986).

103. J. J. Harris, C. T. Foxon, K. W. J. Barnham, D. E. Lacklison, J. Hewett and C. White, J. Appl. Phys., 61:1219 (1987).

104. J. H. English, A. C. Gossard, H. L. Stormer and K. W. Baldwin, Appl. Phys. Lett., 50:1826 (1987).

105. S. Sasa, J. Saito, K. Nambu, T. Ishikawa and S. Hiyamizu, Jpn. J. Appl. Phys., 23:L573 (1984).

106. V. M. Airaksinen, J. J. Harris, D. E. Lacklison, R. B. Beall, D. Hilton, C. T. Foxon and S. J. Battersby, Semi. Sci. Technol., (to be published).

107. D. C. Radulescu, G. W. Wicks, W. J. Schaff, A. R. Calawa and L. F. Eastman, J. Crystal Growth, 81:106 (1987.

108. J. W. Orton, P. F. Fewster, J. P. Gowers, P. Dawson, K. J. Moore, C. J. Curling, C. T. Foxon, K. Woodbridge, G. Duggan and H. I. Ralph, Semi. Sci. Technol. (to be published).

109. K. J. Moore, P. Dawson and C. T. Foxon, Phys. Rev., B34:6022 (1986).

110. P. Dawson, K. J. Moore, G. Duggan, H. I. Ralph and C. T. Foxon, Phys. Rev., B34:6007 (1986).

111. G. Duggan, H. I. Ralph, P. Dawson, K. J. Moore, C. T. Foxon, R. J. Nicholas, J. Singleton and D. C. Rogers, Phys. Rev., B35:7784 (1987).

112. P. Dawson, K. J. Moore and C. T. Foxon, SPIE.

113. K. J. Moore, P. Dawson and C. T. Foxon, Proc. Montpellier.

114. T. J. Drummond, W. E. Masselink and H. Morkoc, Proc. IEEE, 74:773 (1986).

115. S. Hiyamizu, T. Fujii, S. Muto, T. Inata, Y. Nakata, Y. Sugiyama and S. Sasa, J. Crystal Growth, 81:349 (1987).

116. C. T. Foxon, J. B. Clegg, K. Woodbridge, D. Hilton, P. Dawson and P. Blood, J. Vac. Sci. Technol., B3:703 (1985).

117. K. Woodbridge, P. Blood, E. D. Fletcher and P. Hulyer, Appl. Phys. Lett., 45:16 (1984).

ENVELOPE FUNCTION APPROACH TO ELECTRONIC STATES IN HETEROSTRUCTURES

Massimo Altarelli

European Synchrotron Radiation Facility, B.P. 220
F-38043 Grenoble (France)
and
Max-Planck-Institut für Festkörperforschung
Hochfeld-Magnetlabor, B.P. 166 X
F-38042 Grenoble (France)

ABSTRACT

Envelope function calculations of electronic states in semiconductor heterojunctions, quantum wells and superlattices are reviewed. The emphasis is on states derived from coupled bulk bands. Energy levels in perpendicular and parallel magnetic fields are also discussed. Comparison with luminescence, Raman, intra- and interband magneto-optics experiments is emphasized.

INTRODUCTION

In these lectures the theoretical description of electronic states in semiconductor heterostructures through the envelope function method is reviewed at an introductory level. Although this method cannot always be applied - most notably, it has problems whenever band extrema at different k-points are mixed - it is relatively simple and versatile, can easily include external electric, magnetic, or stress fields and has provided guidance in the interpretation of many experiments on a variety of systems (GaAs-AlGaAs, InAs-GaSb, InP-InGaAs, CdTe-HgTe, etc.).

We shall proceed by discussing the motion of electrons and holes in semiconductors in a slowly varying external field via the effective-mass approximation, and then considering the appropriate boundary conditions on the envelope-functions (i.e. effective-mass wavefunctions) at the sharp interface between two semiconductors. The realistic band structure of III-V semiconductors, in particular the valence band degeneracy and the non-parabolicity of the conduction bands complicate the structure of the equations. We shall however see that numerical solutions for the band structure of superlattices, quantum wells and heterojunctions can be obtained which compare favorably with experimental information. In the cases in which substantial long-range charge transfer across the interfaces takes place, it is necessary to perform a self-consistent calculation, examples of which shall be described.

Among the various external fields which are used as a spectroscopic tool of investigation of electronic states in heterostructures, magnetic fields are especially important for their profound effects on the density of states, especially in two-dimensional systems. We shall therefore discuss the calculation of Landau levels in some detail, and compare the results with intra- and interband magneto-optical experiments.

THE ENVELOPE-FUNCTION APPROXIMATION

The Effective-Mass Equation

The envelope-function approximation is an effective-mass theory, familiar from the treatment of shallow impurities[1]. We consider, in the one-electron approximation, the Schrödinger equation for the motion of an electron in a semiconductor in presence of some additional potential $U(\vec{r})$:

$$\left[\frac{p^2}{2m_0} + V_{per}(\vec{r}) + U(\vec{r}) - E \right] \psi(\vec{r}) = 0 . \tag{1}$$

Here m_0 is the free-electron mass, V_{per} is the periodic potential of the perfect bulk semiconductor and $U(\vec{r})$ is the additional potential, which we assume to be slowly varying and weak, in the sense precisely stated below. For example, $U(z)$ could be the constant potential in one of the quantum wells of Fig. 1(a), or the band bending potential on the right-hand side of Fig. 1(c), arising from the depleted acceptors. Eq. (1) would then describe the electron in the quantum well or on the right-hand part of the heterojunction. A similar equation would hold for one of the barrier layers in Fig. 1(a), or on the left-hand side of the hetero-junction in Fig. 1(c), but with different V_{per} and U, and we shall worry later about the matching of the solutions across the interfaces. When $U(\vec{r}) = 0$, the solutions of Eq. (1) are the Bloch function, $\psi_{n\vec{k}}(\vec{r})$, with:

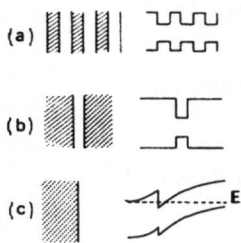

Fig. 1. Schematic representation of three types of heterostructures and, on the right, the corresponding band-edge profile. (a) Superlattice; (b) Single quantum well; (c) Single heterojunction between an n-type material (on the left side) and a p-type one (one the right side of the heterojunction).

$$\left[\frac{p^2}{2m_0} + V_{per}(\vec{r})\right] \Psi_{n\vec{k}}(\vec{r}) = E_n(\vec{k})\Psi_{n\vec{k}}(\vec{r}) \tag{2}$$

and

$$\Psi_{n\vec{k}}(\vec{r}) = e^{i\vec{k}\cdot\vec{r}} u_{n\vec{k}}(\vec{r}), \ (u_{n\vec{k}} \text{ periodic}) . \tag{2'}$$

In Eqs. (2), (2') $\vec{k}$ is restricted to the first Brillouin zone. For $U(\vec{r}) \neq 0$, we expand the solutions of Eq. (1) in terms of the complete set of Bloch functions:

$$\Psi(\vec{r}) = \sum_n \sum_{\vec{k}} \phi_n(\vec{k}) \Psi_{n\vec{k}}(\vec{r}) . \tag{3}$$

It can be shown in detail[1,2] that, if:

(a) $U(\vec{r})$ is slowly varying on the scale of the lattice parameter a of the semiconductor, i.e. its Fourier transform $U(\vec{k})$ is appreciably non-zero only in a region around k=0 of extension Δk, with $\Delta k \ll 2\pi/a$

(b) $U(\vec{r})$ is weak, i.e. if the matrix elements of U between Bloch functions are much smaller than the interband gaps at k=0, $E_n(0) - E_n'(0)$, $(n \neq n')$

(c) We are considering eigenvalues in the neighborhood of a simple parabolic band edge n, well separated from all others:

$$E_n(\vec{k}) \cong E_n(0) + \frac{\hbar^2}{2m^*} k^2 \tag{4}$$

then the Fourier transform $F(\vec{r})$ of $\phi_n(\vec{k})$,

45

$$F(\vec{r}) = (2\pi)^{-3/2} \int d^3k \, \phi_n(\vec{k}) \, e^{i\vec{k}\cdot\vec{r}} \, , \tag{5}$$

satisfies the effective-mass equation:

$$\left[-\frac{\hbar^2}{2m^*} \nabla^2 + U(\vec{r}) \right] F(\vec{r}) = (E - E_n(0))F(\vec{r}), \tag{6}$$

familiar from the theory of shallow donors. Using the first order $\vec{k}\cdot\vec{p}$ expansion[3] of the periodic factor $u_{n\vec{k}}(\vec{r})$ in Eq. (2'):

$$u_{n\vec{k}}(\vec{r}) = u_{n0}(\vec{r}) + \sum_{m \neq n} \frac{\vec{k} \cdot \langle u_{m0} | \vec{p} | u_{n0} \rangle}{m_0(E_n(0) - E_m(0))} u_{m0}(\vec{r}) \tag{7}$$

it is easy to show[2] that, in the approximation of Eq. (6), the total wavefunction $\Psi(\vec{r})$ is related to the "envelope" function $F(\vec{r})$ via the relationship:

$$\Psi(\vec{r}) = F(\vec{r}) \, u_{n0}(\vec{r}) + \sum_{m \neq n} \frac{-i\left(\vec{\nabla} F(\vec{r}) \cdot \vec{p}_{mn} \right)}{m_0(E_n(0) - E_m(0))} u_{m0}(\vec{r}) \, , \tag{8}$$

where $\vec{p}_{nm}$ is the momentum matrix element appearing in Eq. (7) as well.

Eq. (8) shows that, to lowest order, $F(\vec{r})$ is a slowly varying "envelope" modulating the rapidly varying Bloch part $\vec{u}_{n0}(r)$. The following correction term shows that there is a contribution from other $\vec{k}=0$ Bloch functions, proportional to the gradient of the envelope function $F(\vec{r})$. It is important not to forget this term if one is interested in the derivative of the wavefunction $\Psi(\vec{r})$, which enters the boundary conditions for the effective-mass equation at the sharp boundary between two semiconductors.

Boundary Conditions for the Envelope Functions

The effective-mass equation (6) has the remarkable feature that all reference to the microscopic structure of the host semiconductor is condensed in the effective-mass m^* and the band energy $E_n(0)$. This is possible when the potential $U(\vec{r})$ is weak and slowly varying. The two parameters m^* and $E_n(0)$ assume different values in the two semiconductors, say A and B, making up an interface system. Given the high quality of state-of-the-art heterostructures the transition region includes only a few atomic layers. One could think of writing a more general form of Eq. (6), in which the effective-mass and the band edge vary as a function of z. The z-dependent $E_n(0)$ can be attached to $U(\vec{r})$ to form a new effective potential $U(\vec{r}) + E_n(0,z)$. The modified equation

46

(6) would then read:

$$\left[-\frac{\hbar^2}{2}\vec{\nabla} \cdot \left(\frac{1}{m^*(z)}\vec{\nabla} \right) + U(\vec{r}) + E_n(0,z) \right] F(\vec{r}) = E\, F(\vec{r}) \; . \qquad (9)$$

The kinetic energy term has been rewritten, for a z-dependent mass, in a way which restores the hermitian character of the Hamiltonian, following Harrison[4] and Ben Daniel and Duke[5]. For z well to the left of the interface, $m^* = m^*_A$, $E_n(0) = E_n^A(0)$, and for z well on the right side $m^* = m^*_B$ and $E_n(0) = E_n^B(0)$.

We cannot take Eq. (9) seriously, however, because the variation in $E_n(0,z)$ between $E_n^A(0)$ and $E_n^B(0)$, which is typically as large as 0.1-1eV, takes place over a few lattice distances. The potential term in Eq. (9), therefore, varies much too rapidly for the effective-mass formalism to be valid. Nevertheless, we can learn something about the boundary conditions from this differential equation. Assuming the continuity of the envelope-function (see Eq. (11) below), taking the limit in which the effective-mass and band edge energy variations occur over an infinitesimal thickness 2ε and integrating Eq. (9) between z = -ε and z = +ε we obtain

$$F^A(-\varepsilon) = F^B(+\varepsilon)$$

$$\frac{1}{m^*_A}\frac{\partial F^A(z)}{\partial z}\bigg)_{-\varepsilon} = \frac{1}{m^*_B}\frac{\partial F^B(z)}{\partial z}\bigg)_{+\varepsilon} \; . \qquad (10)$$

In order for the boundary conditions (10) on the envelope functions to make physical sense, we must see their implications on the total wave-function as described by Eq. (8). In order for Ψ to be continuous when F is, one must assume that:

$$u^A_{n0} \cong u^B_{n0} \; , \qquad (11)$$

and that the second term of the wavefunction (8) be small, i.e. that the $\vec{k}$-dependence of the Bloch function $u_{n\vec{k}}$ about $\vec{k}=0$ be weak, as emphasized by Ben Daniel and Duke[5]. Eq. (11) is plausible, in the III-V semiconductor family, as long as we are considering the same band edge on both sides (e.g. the conduction band direct minimum). Then by looking at pseudopotential wavefunctions one sees that Eq. (11) is reasonably verified.

Consider now the probability current operator. The existence of stationary states (probability density constant in time) implies that the z-component of the current be the same on all planes parallel to the interface, therefore, also that its average over a microscopic volume Ω, including one or few unit cells, be the same on both sides of the interface. Let us calculate this average on the A side. To do this, we use the wavefunction as given by Eq. (8) and recall that

$$\int_{cell} u_{n0}^* \frac{\partial}{\partial z} u_{m0} = \frac{i}{\hbar} p_{nm} \tag{12}$$

and that this matrix element vanishes for m=n, in the conduction and valence band edges of cubic semiconductors, for symmetry reasons[6]. We find, for the average current J

$$J_A = \frac{\hbar}{m_0} \text{Im} \int_\Omega d^3r \phi_A^* \frac{\partial}{\partial z} \phi_A = \frac{\hbar}{m_A^*} \text{Im} \left(F_A^*(0) \frac{\partial}{\partial z} F_A(0) \right) , \tag{13}$$

with the use of the well-known $\vec{k}.\vec{p}$ expression for the effective-mass of band n:

$$\frac{1}{m^*} = \frac{1}{m_0} + \frac{2}{m_0^2} \sum_{m \neq n} \frac{p_{nm}^{(z)} p_{mn}^{(z)}}{E_n(0) - E_m(0)} , \tag{14}$$

and of Eq. (8). Therefore, the continuity of J implies:

$$\frac{\hbar}{m_A^*} \text{Im} \left(F_A^*(0) \frac{\partial}{\partial z} F_A(0) \right) = \frac{\hbar}{m_B^*} \text{Im} \left(F_B^*(0) \frac{\partial}{\partial z} F_B(0) \right). \tag{15}$$

It is now apparent that the boundary conditions (10) on the envelope-functions imply that the average of the probability current is constant (see Eq. (15)). They are, therefore, meaningful on physical grounds and compatible with the limit behavior of the effective-mass equation for z-dependent mass and band edge, and we shall adopt them in our treatment of heterostructures. A further discussion of the boundary conditions is given at the end of this section.

Coupled Bands

The conditions for the validity of the simple-band approach are certainly violated in many situations of interest. This may happen because of various reasons:

a) Band degeneracy near an extremum, as in the case of the valence band maxima at Γ in all cubic semiconductors.

b) Coupling between bands producing deviations from parabolicity, as in the conduction band of direct gap semiconductors. For narrow-gap materials, like InAs or InSb the non-parabolicity of the conduction band, due to coupling with the valence bands, is quite large for energies very near the band minimum[3], but even in GaAs it has a sizeable effect on levels with energy ≥0.1eV above the band minimum.

c) There are situations specific to heterostructures in which the single band approach fails; if the two materials have a "staggered" energy gap configuration, then, in a large and interesting energy range, the wavefunction has conduction band character on one side of the junction and valence band character on the other. InAs-GaSb super-lattices provide an example of this situation.

This simple-band case was treated in a way modelled on the theory of donor impurities; the case in which many bands contribute with comparable weight to the formation of the eigenfunctions[2] is modelled on the theory of acceptors[1]. We start by identifying the bulk bands which are to be treated on the same footing, and describe their behavior near k=0 by the generalization of Eq. (4), i.e. we write[3]

$$H_{1m}(\vec{k}) = E_1(0)\delta_{1m} + \sum_{\alpha=1}^{3} P_{1m}^{\alpha} k_{\alpha} + \sum_{\alpha,\beta=1}^{3} D_{1m}^{\alpha,\beta} k_{\alpha} k_{\beta} \, , \qquad (16)$$

where 1,m=1,2,..., n, and α,β run over the x, y and z directions. Given a $\vec{k}$-vector, the n band energies $E_1(\vec{k})$ are given by the eigenvalues of the nxn matrix $H_{1m}(\vec{k})$. The direct $\vec{k}.\vec{p}$ coupling between the n bands is thus retained in the terms $P_{1m}^{\alpha} k_{\alpha}$, where the matrix P^{α} is given by:

$$P_{1m}^{\alpha} = \frac{\hbar}{m_0} < u_1 |p^{\alpha}| u_m > \, . \qquad (17)$$

The k-quadratic terms proportional to the matrix $D^{\alpha,\beta}$, on the other hand, represent the indirect $\vec{k}.\vec{p}$ coupling between two of the n bands via the other bands (n=1 to ∞) not included in the set. They have indeed an expression very similar to the r.h.s. of Eq. (14). A specific example is the 6x6 matrix which represents the conduction and the upper spin-orbit-split component of the valence band. This is a very good description of these bands for materials with large spin-orbit splittings. We have, therefore, a conduction band with s-like character at Γ and two spin states, s↑ and s↓. The effective-mass m* appearing in the corresponding diagonal terms originates from coupling to the bands not included in the set, and primarily from the split-off valence band[7]. The valence band has p-like character and the upper spin-orbit manifold corresponds to J=3/2 states, classified by the four possible J_z values from -3/2 to 3/2.

The P matrices are expressed entirely in terms of a single parameter P, defined as the interband momentum matrix element, $iP=<s|p^x|p_x>$, where p_x indicates the p_x-like valence wavefunction and s the conduction wavefunction at Γ. Then it turns out that $m^{*-1}=1+2P^2/3(E_c-E_v+\Delta)$, where Δ is the valence band spin-orbit splitting. If we wish to consider the energy region close to the valence band top, we can ignore the conduction band altogether. The resulting 4x4 matrix is the Luttinger Hamiltonian[8], which describes the valence band top and the splitting in light- and heavy-hole bands at $k\neq0$. The D matrices are expressed in terms of the three Luttinger parameters, γ_1, γ_2 and γ_3, specific to each material.

We are now ready to generalize Eq. (6), (8) and (10) for the many-band case, in analogy with the many-band effective-mass equation written by Luttinger and Kohn[9] for acceptor impurities. We obtain a system of n differential equations:

$$\sum_{m=1}^{n} \left[H_{lm}(-i\vec{\nabla}) + U(\vec{r})\delta_{lm} \right] F_m(\vec{r}) = EF_1(\vec{r}) \tag{18}$$

for the n-component envelope-function $F_1(\vec{r})$, $l=1,2,\ldots,$ n. The total wavefunction, $\Psi(\vec{r})$, is expressed, in analogy to Eq. (8), as:

$$\Psi(\vec{r}) = \sum_{l=1}^{n} \left[F_1(\vec{r})u_{l0}(\vec{r}) + \sum_{m>n} \frac{-i(\vec{\nabla}F_1(\vec{r}))\cdot\vec{p}_{1m}}{m_0(E_1(0) - E_m(0))} u_{m0}(\vec{r}) \right]. \tag{19}$$

Note that the "kinetic energy" part is just obtained by replacing $\vec{k}$ with $-i\vec{\nabla}$ in the $\vec{k}\cdot\vec{p}$ matrix (16) and the potential energy term is diagonal in the band index. This is a consequence of its slow spatial variation: as it can be taken as a constant in each unit cell, its off-diagonal matrix elements vanish by Bloch function orthogonality. It is also straightforward to generalize the boundary conditions. One finds

$$F^A_1(-\varepsilon) = F^B_1(+\varepsilon) \qquad 1 = 1,2,\ldots, n. \tag{20}$$

$$\sum_{m=1}^{n} \left[P^z_{1m} -i \sum_{\alpha=1}^{3} \left(D^{z\alpha}_{1m} + D^{z\alpha}_{1m} \right) \nabla_\alpha \right] F_m \quad \text{continuous}$$

at z=0 for $l=1,2,\ldots,$ n. One must, of course, assume Eq. (11) for the n bands of interest. Let us write down these boundary conditions for an ideal planar interface between lattice matched semiconductors A and B and for the nxn coupled-band case of Eq. (16). In this situation the potential $U(\vec{r})$ depends only on z, and k_x and k_y are good quantum numbers. That means that we can write

$$F^{A,B}(\vec{r}) = e^{ik_x x} e^{ik_y y} F^{A,B}(z) , \tag{21}$$

so that the second of Eqs (20) reads

$$\sum_{m=1}^{6}\left[P_{1m}^{z} + \sum_{\alpha=x,y}\left(D_{1m}^{z\alpha} + D_{1m}^{\alpha z}\right)k_{\alpha} - 2iD_{1m}^{zz}\frac{\partial}{\partial z}\right] F_{m} \quad \text{continuous} .\qquad(22)$$

A further simplification of the boundary conditions occurs by noticing that, if Eq. (11) holds for $1=1,\ldots,$ n, then also the P_{1m} matrix elements between these Bloch functions must be equal, as they are momentum matrix elements connecting the same conduction and valence band functions. It is empirically known[10] that such matrix elements depend essentially only on the lattice constant and, therefore, are indeed about equal in two lattice-matched materials. Thus the continuity of the $P_{1m}^{z}F_{m}$ terms follows from the continuity of the F_{m} alone, and Eq. (22) further simplifies to:

$$\sum_{m=1}^{6}\left[\sum_{\alpha=x,y}\left(D_{1m}^{z\alpha} + D_{1m}^{\alpha z}\right)k_{\alpha} - 2iD_{1m}^{zz}\frac{\partial}{\partial z}\right] F_{m} \quad \text{continuous} .\qquad(22')$$

This is the form of boundary conditions[2,11,12], used e.g. in the specific examples to be discussed soon.

Discussion of the Envelope-Function Approximation

In deriving the equations for the envelope-function approximation, the many and sometimes very restrictive assumptions which were needed were stated in detail. It is quite clear that there are situations in which they are untenable. Apart from the cases of strong or rapidly varying potentials $U(\vec{r})$, which cause the breakdown of effective-mass theories also for impurity states, the condition of Eq. (11) cannot be satisfied for many interesting heterostructures. This occurs for all cases in which the eigenstates we want to compute are derived from the mixing of Bloch waves from different points of the Brillouin zone (e.g. the conduction-related states in $Si-Si_{x}Ge_{1-x}$ heterostructures, or in $GaAs-Al_{x}Ga_{1-x}As$ with $x \gtrsim 0.4$, where the alloy is indirect bandgap, etc.). In these situations, one must resort to other methods, such as the empirical tight-binding scheme[13], or the pseudopotential method[14,15]. In either approach, some of the attractive features of the envelope-function method are lost. Alternatively, one can devise an extension of the envelope-function method suitable for such heterostructures, but at the price of introducing some additional parameters in the theory. Ando[16] has proposed such an approach for GaAs to indirect- AlGaAs heterostructures.

Also in the cases where the two semiconductors have relevant band edges which, because of their similarity in symmetry and chemical origin, can be paired to satisfy Eq. (11) approximately, a few words of caution on the accuracy of the envelope-function calculations are necessary. Our discussion of boundary conditions is based on the idea of perfect periodicity of both media up to a geometrical plane defining the interface. This is certainly an idealization and, in the very important case in which one of the components is an alloy (e.g. GaAs-Al$_x$Ga$_{1-x}$As) it is impossible even to define such a plane. It would be more realistic to talk about an interface layer, comprising several atomic planes, separating the two semiconductors, and characterized by given reflection and transmission coefficients. These would be again parameters of unknown value to be used as input.

Finally, a word of caution concerning the use of $\vec{k}.\vec{p}$ matrices to describe the bulk band edges was recently put forward by Schuurmans and 't Hooft[12]. They pointed out that a Hamiltonian like Eq. (16) can have unphysical solutions; in the case of the 8x8 description of GaAs including conduction, upper- and split-off valence bands, for example, they find bulk solutions with E in the gap and very large real k-values. Care must be exerted to prevent such unphysical branches from being included in the construction of the superlattice solution.

In spite of these caveats the envelope-function results are in overall good agreement with experiments, sometimes even more than one would expect. When comparisons with e.g. tight-binding calculations - which are free from the problem of boundary conditions - are performed in detail, the agreement is very good[12,13,16,17]. These facts can be regarded as valid a posteriori justifications for the approximations of the method.

EXAMPLES OF RESULTS FOR SOME SYSTEMS

GaAs-Al$_x$Ga$_{1-x}$As

The envelope-function method is applicable to this system for all states derived from Γ-point bulk Bloch functions. This excludes the conduction subbands high enough in energy to approach the secondary minima at X and L; this is the case at all energies for x>0.4, where the alloy has an indirect gap. The valence subbands, on the other hand, are accessible to the method at all alloy compositions.

An important input parameter of the calculations is the band offset. Even for an extensively investigated system like GaAs-AlGaAs, this quantity was very controversial until recently. There seems to be an emerging consensus for a value of the conduction band discontinuity between 60% and 70% of the difference in band gap between AlGaAs and GaAs[18-22]. Theory can only be of limited help, given the uncertainty (>0.1eV) of state-of-the art band theory in bulk materials.

We start by some general considerations. The growth axis of MBE heterostructures is in a <001> direction[23]. In a superlattice, one has, therefore, a Brillouin zone which is very thin in the z direction ($2\pi/d$, where d is the period) and has the usual $\sim 2\pi/a$ size, where a is the lattice constant, in the x and y direction. One has, therefore, a very asymmetric band structure with a bulk-like bandwidth (~10eV) in the k_x, k_y plane and $\sim (a/d)^2$ x 10eV, typically 10meV, bandwidth in the k_z-direction.

The dispersion in the k_z-direction is relatively simple to handle even in the 6x6 coupled-band model, because most of the off-diagonal terms vanish for $k_x, k_y = 0$. In detail, we see that $J_z = \pm 3/2$ states, corresponding to the heavy holes, completely decouple and behave as simple particles with effective mass $m_{hh} = 1/(\gamma_1 - 2\gamma_2)$. We have a Kronig-Penney type of eigenvalue problem, with the boundary condition (10), and, as shown e.g. by Bastard[24], one finds the implicit dispersion relation:

$$\cos(k_z d) = \cos(k_{zA} d_A) \cos(k_{zB} d_B)$$

$$-\frac{1}{2}\left(\frac{m_{hh}^A k_{zB}}{m_{hh}^B k_{zA}} + \frac{m_{hh}^B k_{zA}}{m_{hh}^A k_{zB}}\right) \sin(k_{zA} d_A) \sin(k_{zB} d_B) \qquad (23)$$

where

$$k_{zA,B}^2 = 2m_{hh}^{A,B}\left(E - E_v^{A,B}\right), \qquad (23')$$

and the superlattice period d is made up of thickness d_A (d_B) of material A (B). As for light-holes and conduction electrons, one has the 2x2 matrix between s↑ and $J_z = +1/2$, or an equivalent one for s↓ and $J_z = -1/2$:

$$\begin{pmatrix} E_c + \dfrac{1}{2m^*} k_z^2 & -i\sqrt{\dfrac{2}{3}}\, P\, k_z \\[4mm] i\sqrt{\dfrac{2}{3}}\, P\, k_z & E_v - \dfrac{1}{2}(\gamma_1 + 2\gamma_2)\, k_z^2 \end{pmatrix}. \qquad (24)$$

One could solve directly the superlattice problem for these coupled bands as discussed previously or, for energies very near the band edges, find the eigenvalues of (24), $\xi_c(k_z)$, $\xi_v(k_z)$ and describe their non-parabolic dispersion via energy dependent effective-masses. For a complete discussion of k_z-dispersion in GaAs-AlGaAs superlattices, we refer the reader to the treatment of Schuurmans and 't Hooft[12]. In general, the larger the energy distance from the band extremum, the more accurate the description of the non-parabolicity should be.

Consider now the dispersion for k_x, $k_y \neq 0$. This is much more complicated, but also very interesting. The main features can be investigated in a model in which the conduction band is uncoupled from light- and heavy-hole bands, but the latter are coupled to each other via the Luttinger Hamiltonian[8]. For the conduction subbands one has again a simple Kronig-Penney like problem with possible non-parabolic effects embodied in an energy-dependent effective-mass.

For the valence bands we have to deal with the 4x4 Luttinger Hamiltonian, which can be written in the form:

$$
\begin{array}{ccccc}
 & 3/2 & 1/2 & -1/2 & -3/2 \\
3/2 & a_+ & b & c & 0 \\
1/2 & b^* & a_- & 0 & c \\
-1/2 & c^* & 0 & a_- & -b \\
-3/2 & 0 & c & -b^* & a_+
\end{array}
\tag{25}
$$

with

$$
a_\pm = E_v - \frac{1}{2}(\gamma_1 \pm \gamma_2)(k_x^2 + k_y^2) - \frac{1}{2}(\gamma_1 \mp 2\gamma_2) k_z^2
$$

$$
b = \sqrt{3}\ \gamma_3 (k_x - ik_y) k_z \tag{25'}
$$

$$
c = \frac{\sqrt{3}}{2} \gamma_2(k_x^2 - k_y^2) - 2i\gamma_3 k_x k_y\ .
$$

The bulk solutions of (25) give the spin-degenerate light- and heavy-hole bands. This degeneracy is actually lifted in non-inversion symmetric zincblende materials, like GaAs, by terms linear in the $\vec{k}$-vector[25]. The terms are, however, extremely small and we neglect them here. The valence bands are anisotropic and, therefore, one has to expect that also for superlattices the band dispersion will be different for different directions in the k_x, k_y plane. However, this warping of the bands is small and a good approximation is obtained by the "axial model" in which[26,27] γ_2 and γ_3 in the c matrix elements (see the last of Eq. (25')) are replaced by $\gamma = (1/2)(\gamma_2 + \gamma_3)$. With this replacement, Eq. (25) acquires cylindrical symmetry about the z-axis, and the bands are isotropic in the k_x, k_y plane.

The solution of the four-component equation with the boundary conditions must be pursued numerically[27]. At $k_x = k_y = 0$ there is no mixing of the light and heavy components. It is therefore possible to identify each subband as heavy or light (this is a consequence of the envelope function approximation). For a more complete description, see Ref. (15). As we move out of the $k_x = k_y = 0$ axis, the mixing grows rapidly and produces the non-parabolic behavior, first pointed out by Nedorezov[28], who solved the problem exactly in the limit of infinitely high barriers. A particularly striking feature is the positive in-plane effective mass of the first light-hole subband, which changes sign and becomes hole-like only for $k \sim \pi/L$, where L is the well thickness.

In the example of GaAs-AlGaAs, one has to deal with a system in which no significant charge rearrangement across or near the interfaces takes place, if we consider superlattices involving intrinsic semiconductors. In contrast, in modulation doped superlattices, filling of the quantum wells takes place, with carriers released from impurities located in different layers.

Whenever such transfers take place, they contribute to the definition of the one-electron potential $U(\vec{r})$ appearing in the envelope-function equations (6) or (18). The potential depends on the charge density of the states filled by the transferred carriers; and, because the corresponding wavefunctions depend in turn on the potential $V(\vec{r})$, the problem must be solved self-consistently, within some scheme of approximation, e.g. the Hartree approximation, the local density approximation, etc. These procedures are familiar from the treatment of Si MOS systems[29]. The relative ease of performing such self-consistent calculations is one of the advantages of the envelope-function methods.

From the experimental point of view, the complexity of the valence bands is best revealed by intra- and interband optical experiments. Analysis[30-32] of Raman scattering[33] results on p-type modulation doped wells and of polarized luminescence[34] in n-type structure confirms the strong light - heavy hole mixing and the upward curvature of the light-hole first subband. Further confirmation is offered by magneto-optical experiments, to be discussed later.

InAs-GaSb Superlattices and Quantum Wells

The InAs-GaSb system is particularly interesting[35] because of the peculiar band line-up of these materials. The valence band top of GaSb is estimated to lie ~0.15 eV higher than the bottom of the InAs conduction band[35,36]. Electrons tend to transfer to the InAs layers, unless the layers are so thin that the particle-in-a-box quantization energy reverses the level order, locating empty InAs quantum well conduction states above the full GaSb valence states. Therefore, charge transfer occurs only for InAs layers thicker than ~90 Å, and in this case self-consistency plays a crucial role in determining the level structure. Such a calculation[37], was performed in the Hartree approximation and in the 6x6 band model which neglects the split-off valence band. The results show strong valence-conduction subband mixing, with small (few meV) hybridization gaps opening up at subband crossing points. For a 120 Å InAs - 80 Å GaSb superlattice, the density of transferred carriers is ~ 0.9×10^{12} cm^{-2}, thus providing good justification for the Hartree approximation. Magneto-optical experiments, to be discussed later, indicate good agreement with the envelope function results. Single quantum wells of InAs between GaSb barriers, however, show a concentration of electrons in InAs much larger than that of holes in the barriers[38]. The origin of these extrinsic charges is not clear, although a plausible explanation is that the well is in a strong band-bending region, deriving from the proximity of the GaSb free surface of the outer barrier[39].

CdTe-HgTe Superlattices

Superlattices of II-VI compounds have recently attracted much attention, and among them the CdTe-HgTe system is especially interesting. This is a consequence of the zero-gap character of HgTe and of the band line-up, which puts the HgTe Γ_8 edge above the corresponding edge of CdTe. The value Δ of this band offset is controversial, the estimate $\Delta = 0.04$ eV from magneto-optics[40,41] being confirmed by Raman measurements[42], but contradicted by photoemission results[43,44] which suggest a much larger value $\Delta \sim 0.35$ eV.

These systems provide an excellent testing ground for the boundary conditions, Eq. (10) and (22'). This is because the Γ_8 "light holes" are in fact Γ_8 "electrons" in HgTe, i.e. they form an unoccupied band with positive mass in this material, due to the interchange in the energy position of the Γ_6 and Γ_8 edges. There are therefore three types of Γ_8 states in these superlattices: Γ_8 "electrons", Γ_8 heavy holes, both confined in the HgTe wells, and in addition unusual interface states[17,45] arising in the gap between Γ_8 light holes in CdTe and Γ_8 electrons in

HgTe and decaying exponentially on both sides of an interface. If one ignores other bands for simplicity, then it is easy to see that the boundary conditions, Eq. (10), allow for such states if m_A, m_B have opposite signs.

The good agreement[17] between envelope-function calculations, in which the boundary conditions are so crucial in predicting the interface states, and tight-binding results, provides strong a posteriori confirmation of the envelope-function method. On the other hand, the fact that the band offset value carefully measured by photoemission has to be drastically reduced to account for infrared and Raman results leads to the conclusion that some aspects of the physics of this system are not understood. More theoretical and experimental work will be needed to sort out this point.

EFFECT OF EXTERNAL MAGNETIC FIELDS

Landau Levels in Heterostructures: Perpendicular Fields

One of the advantages of the envelope-function method is that it is easily extended to include external fields. There are various kinds of such fields which are of great experimental interest. One is a strain field, which is almost invariably present because lattice match is never perfect. Here the effect of an external magnetic field perpendicular to the layers is discussed. Many of the most revealing experiments concerning 2-dimensional electronic systems are performed with an external magnetic field, which has a dramatic influence on the energy spectrum. In a 2-dimensional system, a perpendicular magnetic field quantizes both available degrees of freedom, producing an entirely discrete spectrum. This leads to an enrichment and sharpening of the optical structures, and to striking transport phenomena, like the quantum Hall effect. Our main motivation in describing the Landau level spectrum is the stringent test of electronic calculations provided by the quantitative interpretation of magneto-optical experiments.

To include a magnetic-field in the many-band envelope-function formalism[45], we follow the lines of the classical work of Luttinger[8] on the cyclotron resonance of holes in semiconductors. The field $\vec{B}=(0,0,B)$ is described by the vector potential $\vec{A}$. (It is convenient to choose a gauge with $A_z=0$.) In the $\vec{k} \cdot \vec{p}$ bulk Hamiltonian, Eq. (16), $\vec{k}$ is to be replaced by $\vec{k}'= \vec{k} + (e/c) \vec{A}$. Then it is easy to see that the x and y components of this new operator do not commute, but instead:

$$\left[k_x', k_y'\right] = -i(e/c)B \ . \tag{26}$$

Also, new diagonal terms arise, representing the direct coupling of the electron and hole spins to the field. The conduction band diagonal elements in the 6x6 Hamiltonian now become

$$H_{cc} = E_c + \frac{1}{2m^*} (\vec{k} + (e/c)\vec{A})^2 + \frac{e}{2c} g^* s_z B \tag{27}$$

where s is the electron spin, m^* is the effective-mass defined previously in the discussion of Eq. (16-17) and the effective g-factor[7] g^* is simply $2/m^*$. For the valence band diagonal elements, one must add the term

$$\frac{e}{c} \kappa J_z B + \frac{e}{c} q J_z^3 B \tag{28}$$

where J_z is the spin 3/2 operator and κ and q are two material parameters[8]. Actually q is very small for the semiconductors of interest here and the second part of (28) can be neglected. It is easy to see that as a consequence of (26) we can define operators a, a^+

$$a = \sqrt{\frac{c}{2eB}} \left(k_x' - k_y'\right)$$
$$a^+ = \sqrt{\frac{c}{2eB}} \left(k_x' + k_y'\right) \tag{29}$$

with commutator

$$[a, a^+] = 1 \tag{29'}$$

so that all the terms in k_x or k_y in the Hamiltonian can be expressed in terms of these harmonic oscillator raising and lowering operators.

It is not possible to write a general solution of this Hamiltonian in closed form. Fortunately, however, it is possible to do it if we take the axial approximation in the valence-band portion of the 6x6 Hamiltonian[46] (see the discussion following Eq. (25)). This procedure was applied to the GaAs-AlGaAs, to the InAs-GaSb and to the CdTe-HgTe systems. The computed Landau levels are directly comparable to magneto-optical experiments, after establishing the selection rules for optical transitions and, for a more stringent analysis, after computing the matrix elements from the envelope-functions.

Magneto-optical experiments in the interband region for GaAs-AlGaAs multiquantum well systems were shown[47-49] to provide a large amount of clearly resolved structure. The calculated[47] Landau levels show a striking contrast between conduction-band related and valence-band related levels. Conduction band Landau levels are nearly linear in the B field intensity, small deviations being related to the slight non-parabolicity. Valence band Landau levels, on the other hand, show striking deviation from linearity, corresponding to the large deviations from parabolicity in the subband structure for B=0.

A detailed comparison of theoretical calculations with experiment was recently performed by Ancilotto et al.[50]. The calculation of Landau levels and optical matrix elements was supplemented by a simplified inclusion of excitonic effects. The valence band intricacies are shown to be essential for the interpretation of experiments. The comparison reveals also that the electron in-plane effective mass is more non-parabolic than expected from the six-band model (including corrections for the contribution of the split-off band), pointing out the need of a more complete treatment. Recently, Ekenberg[51] showed that indeed a better description of the non-parabolicity can produce a large effect on the slope of Landau levels.

The structure of the valence band Landau levels is more directly probed with intraband experiments. Cyclotron resonance experiments[52] on p-type GaAs-AlGaAs single heterojunctions were interpreted by several groups[53-56] in terms of Landau level calculations in the B=0 self-consistent potential of the heterojunction. The agreement can be good, although this must involve some accident, because of the use of the Hartree approximation (inadequate for hole densities ~5 x 10^{11} cm^{-2}), of the axial approximation, etc.. Comparisons of the cyclotron resonance in p-type quantum wells are less favorable[57], whereas satisfactory semi-quantitative agreement is found for intraband Raman scattering experiments[58] on p-type quantum wells. In general, while theory seems to be correct in predicting the main features of the valence band Landau levels, more work is needed to establish quantitative agreement with interband magneto-optics.

In the InAs-GaSb superlattices, magneto-optical experiments[36] were crucial in establishing the basic properties of the electronic structure. The calculated[46] transition energies are in good agreement with experiment. More recently[59], the theory was shown to provide a reasonable account of magneto-optical experiments performed under hydrostatic pressure, which modifies the band offset; due to the peculiar nature of this system, the reduction of the overlap between the GaSb valence band and the InAs conduction band with hydrostatic pressure affects the electronic states very much, and produces a change in the character of prominent transitions from interband-like to intraband-like.

Landau Levels in Heterostructures: Parallel Fields

Interesting information on the electronic properties of quantum wells and superlattices can be obtained by performing magneto-optical experiments in a field parallel to the layers of the heterostructure. In an intuitive picture, the field forces the particles to perform cyclotron orbits in a plane intersecting both the well and the barrier materials. If the radius of such orbits (i.e. the magnetic length $a_M = (c/eB)^{1/2}$) is large compared to the thickness of the layers, then the completion of the cyclotron orbit requires traversal of the barriers. This condition is easily met experimentally as, e.g., a 10 T field corresponds to $a_M \sim 8$ nm. Parallel field experiments probe therefore intriguing properties such as tunneling, perpendicular transport, etc., and are particularly fruitful in the investigation of superlattice band structure in the growth direction[60], which is more elusive than the in-plane dispersion discussed in the previous sections.

To keep the notation of the previous section, we now denote the growth direction by x, and choose the gauge $\vec{A}=(0,Bx,0)$ for the field $\vec{B}$ parallel to z. Consider first an electron in a bulk semiconductor in a simple parabolic band with mass m^*. The Schrödinger equation can as usual be cast in the form

$$\left[\frac{p_z^2}{2m^*} + \frac{p_x^2}{2m^*} + \frac{1}{2} m^* \omega_c^2 (x - x_0)^2 \right] \Psi = E\Psi \quad , \tag{30}$$

where $x_0 = -a_M^2 k_y$, $\omega_c = eB/m^*c$. If we now consider a heterostructure, e.g. a superlattice, we must add in Eq. (30) a periodic potential $V_s(x)$. It is then apparent that the degeneracy of the free-electron Landau levels with respect to x_0 is now lifted, because it makes a difference if x_0 is, e.g., in a barrier or in a well.

The general problem of a particle in a field plus a periodic potential cannot be solved exactly. However, if the magnetic length a_M is large compared to the superlattice period d, one can use the semiclassical quantization scheme[61]. The approach is discussed in great detail by Zilberman[62]. One must first consider the subband structure problem for the superlattice in the absence of the field, and construct the sections of the constant energy-surfaces in $\vec{k}$-space with the $k_z =$ constant planes. For the $k_z=0$ case, these are schematically shown in Fig. 2. This figure reflects the fact that the bandwidth in the super-lattice direction, x, is much smaller than in the in-plane directions. The closed curves correspond to the energies between the subband bottom and the bandwidth in the x direction. For energies exceeding this value,

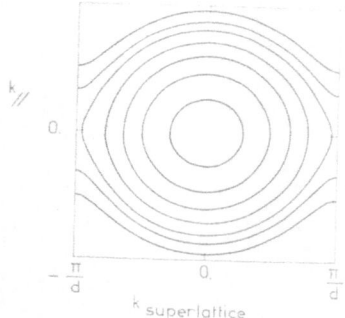

Fig. 2. Constant energy surfaces for a subband in a superlattice of period d. The Brillouin zone in the $k_{||}$ direction extends to much larger values (of order π/a, where a is the lattice constant of the constituent semiconductors).

the constant energy surface touches the edge of the superlattice Brillouin zone, corresponding to the "open orbit" case (one must think of a periodic repetition of the zone in the superlattice direction x).

When the field in the z direction is turned on, both open and closed curves correspond to $\vec{k}$-space trajectories of the representative point in the usual semiclassical Bloch dynamics[61]. To get a deeper picture of the quantized spectrum, one can regard these $\vec{k}$-space trajectories as phase-space trajectories of some kind. Indeed, Eq. (26) shows that k' and k' are like canonical variables (i.e. like "y" and "k_y"). We then construct the potential in the "y" direction that would produce the classical phase space trajectories of Fig. 2. Clearly, it is a periodic potential, with potential wells in which closed phase space orbits occur for low energies and unbound orbits for energies higher than the bandwidth. The corresponding quantized spectrum is obtained by solving the motion in this potential, in a WKB-like approach.

Then, the energies within the x-direction bandwidth (closed orbit region) correspond to essentially discrete levels, or to very narrow bands, as the probability of $\vec{k}$-space tunneling to the next periodically equivalent well is small[62]. The gap between these discrete levels is $\approx \omega_c$. Their energies are given by the Onsager prescription:

$$A_k(E) = \frac{2\pi eB}{c}\left(n + \frac{1}{2}\right) , \qquad (31)$$

where $A_k(E)$ is the $\vec{k}$-space area of the closed orbit. As the energy gets close to the bandwidth, the orbits get larger and the tunneling

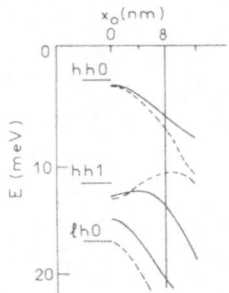

Fig. 3. Energy levels of holes in a 16 nm quantum well of GaAs in a 10 T
         parallel field, as a function of x_0, the orbit center position
         (see Eq. (30)). Infinite barrier height and axial symmetry of
         the valence bands in the plane of the figure were assumed. The
         position of the first few heavy (hh) and light (lh) subbands at
         B=0 are shown for comparison.

probability increases. The discrete levels begin to broaden appreciably.
At larger energies, the nearly flat open orbits correspond to nearly-free
phase space motion, therefore to a weak periodic potential, in which wide
bands separated by narrow gaps occur. Zilberman[62] shows that a "band"
extends approximately over the energy region:

$$\frac{2\pi e B}{c} n < A_k(E) < \frac{2\pi e B}{c} (n + 1) \tag{32}$$

where $A_k(E)$ is the area in the first Brillouin zone between two open
orbits with energy E, and is separated by gaps of order $\omega_c(d/a_M)^2$ from
the neighboring bands.

   The results of this analysis are in excellent agreement with the
magneto-optical interband experiments of Belle et al.[63] in GaAs-AlGaAs
superlattices with 5 nm period (as well as with their numerical results
for the Kronig-Penney model). A series of sharp absorption peaks is
observed in the region corresponding to transitions between the quasi-
discrete energy levels of the valence and conduction subbands. Above
this energy range, no sharp structure is observed, corresponding to the
broad "bands" discussed above. The intraband results of Duffield et
al.[64] in n-type samples are also in agreement with this general picture.

   It is however important to point out that the semiclassical analysis
cannot be invoked for the valence subbands, because of their origin from
a degenerate bulk band. Although one would suspect that the identifi-
cation of the k_x-bandwidth with the discrete portion of the spectrum
should hold (it just says that a hole in that energy range does not "see"

the barriers), a complete quantum-mechanical calculation for a Luttinger hole in a superlattice in a parallel magnetic field would be needed. We are not aware of any such calculation for a superlattice; as a first step, a calculation for a single quantum well was recently reported[65]. It shows a complicated x_0-dependence (see Fig. 3) of the energy levels, with anti-crossings and non-parabolicities strongly reminiscent of the $k_{||}$-dependence of the corresponding subbands.

ACKNOWLEDGEMENT

It is a pleasure to thank J.C. Maan for many contributions to this article.

REFERENCES

1. W. Kohn, in: "Solid State Physics: Advances in Research and Applications", F. Seitz and D. Turnbull, eds., (Academic, New York, 1957) vol. 5, p. 257.

2. M. Altarelli, in: "Heterojunctions and Semiconductor Superlattices", G. Allan, G.Bastard, N. Boccara, M. Lannoo, M. Voos (Springer, Berlin, 1986) p. 12.

3. E.O. Kane, in: "Semiconductors and Semimetals", R. K. Willardson and A. C. Beer, eds., (Academic, New York, 1966) vol. 1, p. 75.

4. W. A. Harrison, Phys. Rev. 123:85 (1961).

5. D. J. Ben Daniel and C.B. Duke, Phys. Rev. 152:683 (1966).

6. See e.g. E. O. Kane, ref. 3.

7. L. M. Roth, B. Lax and S. Zwerdling, Phys. Rev. 114:90 (1959); L. M. Roth, in: "Handbook of Semiconductors", W. Paul, ed., (North-Holland, Amsterdam, 1982) vol. 1, p. 451.

8. J. M. Luttinger, Phys. Rev. 102:1030 (1956).

9. J. M. Luttinger and W. Kohn, Phys. Rev. 97:869 (1955).

10. M. Cardona, in: "Atomic Structure and Properties of Solids", E. Burstein, ed., (Academic, New York, 1972) p. 514.

11. M. Altarelli, in: "Applications of High Magnetic Fields in Semiconductor Physics", G. Landwehr, ed., (Springer, Berlin, 1983) p. 174.

12. M. F. H. Schuurmans and G. W. 't Hooft, Phys. Rev. B31:8041 (1985).

13. J. N. Schulman and Y. C. Chang, Phys. Rev. B24:4445 (1981); ibid. B27:2346 (1983).

14. M. Jaros, K. B. Wong and M. A. Gell, Phys. Rev. B31:1205 (1985).

15. C. Mailhiot and D. L. Smith, Phys. Rev. B35:1242 (1987).

16. T. Ando, in: "Proceedings of the 3rd Brazilian School of Semiconductor Physics", (World Scientific, Singapore, 1987, in press).

17. Y. C. Chang, J. N. Schulman, G. Bastard, Y. Guldner and M. Voos, Phys. Rev. B31:2557 (1985).

18. H. Kroemer, Surf. Sci. 174:299 (1986).

19. G. Duggan, J. Vac. Sci. Tech. B3:1224 (1985).

20. J. Menéndez, A. Pinczuk, D. J. Werder, A. C. Gossard, J. H. English, T. H. Chiu and W. T. Tsang, Superlattices and Microstructures, 3:163 (1987).

21. W. I. Wang, E. E. Mendez and F. Stern, Appl. Phys. Letters 45:639 (1984).

22. G. Duggan, H. I. Ralph and K. J. Moore, Phys. Rev. B32:8395 (1985).

23. See e.g. K. Ploog, Ann. Rev. Mat. Science 12:123 (1982).

24. G. Bastard, Phys. Rev. B24:5693 (1981).

25. E. O. Kane, in: "Handbook of Semiconductors", W. Paul, ed., (Academic, New York, 1982) vol. 1, p. 193.

26. D. A. Broido and L. J. Sham, in: "Proc. of the 17th International Conference on the Physics of Semiconductors, San Francisco, 1984", J. D. Chadi and W. A. Harrison, eds., (Springer, New York, 1985) p. 337.

27. M. Altarelli, U. Ekenberg and A. Fasolino, Phys. Rev. B32:5138 (1985).

28. S. S. Nedorezov, Soviet Phys. Sol. State 12:1814 (1971).

29. See e.g. T. Ando, A. B. Fowler and F. Stern, Rev. Mod. Phys. 54:437 (1982).

30. M. Altarelli, in: "Festkörperprobleme", P. Grosse, ed., (Vieweg, Braunschweig, 1985) vol. XXV, p. 381.

31. T. Ando, J. Phys. Soc. Japan, 54:1528 (1985).

32. Y. C. Chang and G. D. Sanders, Phys. Rev. B32:5521 (1985).

33. A. Pinczuk, D. Heiman, R. Sooryakumar, A. C. Gossard and W. Wiegmann, Surf. Sci. 170:573 (1986).

34. R. Sooryakumar, D. S. Chemla, A. Pinczuk, A. Gossard, W. Wiegmann and L. J. Sham, J. Vac. Sci. Tech. B2:349 (1984).

35. See e.g. L. L. Chang, in: "Heterojunctions and Semiconductor Superlattices", G. Allan, G. Bastard, N. Boccara, M. Lannoo and M. Voos, eds., (Springer, Berlin, 1986) p. 152.

36. J. C. Maan, Y. Guldner, J. P. Vieren, P. Voisin, M. Voos, L. L. Chang and L. Esaki, Solid State Commun. 39:683 (1981).

37. M. Altarelli, Phys. Rev. B28:842 (1983).

38. E. E. Mendez, L. Esaki and L. L. Chang, Phys. Rev. Lett. 55:2216 (1985).

39. M. Altarelli, J. C. Maan, L. L. Chang and L. Esaki, Phys. Rev. B36, in press (1987).

40. Y. Guldner, G. Bastard, J. P. Vieren, M. Voos, J. P. Faurie and A. Million, Phys. Rev. Lett. 51:907 (1983).

41. G. S. Boebinger, J.P. Berroir, Y. Guldner, J. P. Vieren, M. Voos and J. P. Faurie, in: "Proceedings of the MSS III Conference, Montpellier, 1987", (Les Editions de Physique, Les Ulis, 1987, in press).

42. D. J. Olego, J. P. Faurie and P. M. Raccah, Phys. Rev. Lett. 55:328 (1985).

43. S. P. Kowalczyk, J. T. Cheung, E. A. Kraut and R. W. Grant, Phys. Rev. Lett. 56:2755 (1986).

44. Tran Minh Duc, C. Hsu and J. P. Faurie, Phys. Rev. Lett. 58:1127 (1987).

45. Y. R. Lin-Liu and L. J. Sham, Phys. Rev. B32:5561 (1985).

46. A. Fasolino and M. Altarelli, Surf. Sci. 142:322 (1984); and in: "Two-Dimensional Systems, Heterostructures and Superlattices", G. Bauer, F. Kuchar and H. Heinrich, eds., (Springer, Berlin, 1984) p. 176.

47. J. C. Maan, G. Belle, A. Fasolino and M. Altarelli, Phys. Rev. B30:2253 (1984).

48. N. Miura, Y. Iwasa, S. Tarucha and H. Okamoto, in: "Proc. of the 17th International Conference on the Physics of Semiconductors, San Francisco, 1984", J. D. Chadi and W. A. Harrison, eds., (Springer, New York, 1985) p. 359.

49. D. C. Rogers, J. Singleton, R. J. Nicholas, C. T. Foxon, K. Woodbridge, Phys. Rev. B34:4002 (1987).

50. F. Ancilotto, A. Fasolino and J. C. Maan, Superlattices and Microstructures 3:187 (1987).

51. U. Ekenberg, in: "Proceedings of the MSS III Conference, Montpellier, 1987", (Les Editions de Physique, Les Ulis, 1987, in press).

52. H. L. Störmer, Z. Schlesinger, A. Chang, D. C. Tsui, A. C. Gossard and W. Wiegmann, Phys. Rev. Lett. 51:126 (1983).

53. A. Broido and L. J. Sham, Phys. Rev. B31:888 (1985).

54. U. Ekenberg and M. Altarelli, Phys. Rev. B32:3712 (1985).

55. E. Bangert and G. Landwehr, Superlattices and Microstructures 1:363 (1985); Surf. Sci. 170:593 (1986).

56. U. Ekenberg, Surf. Sci. 170:601 (1986).

57. Y. Iwasa, N. Miura, S. Tarucha, H. Okamoto and T. Ando, Surf. Sci. 170:587 (1986).

58. D. Heiman, A. Pinczuk, A. C. Gossard, A. Fasolino and M. Altarelli, in: "Proceedings of the 18th International Conference on the Physics of Semiconductors, Stockholm, 1986", E. Engström, ed., (World Scientific, Singapore, 1987).

59. M. L. Claessen, J. C. Maan, M. Altarelli, P. Wyder, L. L. Chang and L. Esaki, Phys. Rev. Lett. 57:2556 (1986).

60. J. C. Maan, in: "Festkörperprobleme, vol. 27", P. Grosse, ed., (Vieweg, Braunschweig, 1987).

61. See e.g. J. C. Slater in: "Insulators, Semiconductors and Metals", (McGraw-Hill, New York, 1967).

62. G. E. Zilberman, Soviet Physics JETP 5:208 (1957); ibid. 6:299 (1958).

63. G. Belle, J. C. Maan and G. Weimann, Solid State Commun. 56:65 (1985).

64. T. Duffield, R. Bhat, M. Koza, F. De Rosa, D. M. Hwang, P. Grabble and S. J. Allen, Jr., Phys. Rev. Letters 56:2724 (1986).

65. M. Altarelli and G. Platero, Surface Sci., to be published.

THE DETERMINATION OF SUBBAND ENERGIES

F. Koch

Physik-Department E 16
Technische Universität München
D-8046 Garching
Federal Republic of Germany

ABSTRACT

The electronic energy states of mobile carriers in 2-dimensional systems are referred to as subbands. We review in this paper the various experimental schemes that have been used to determine these 2D bands in a variety of physical semiconductor-structures. The cases considered are exemplary and include those studied in past years as well as current challenges in the field of subband spectroscopy.

INTRODUCTION

Two-dimensional (2D) systems, be they in the form of single hetero-structure interfaces, of quantum wells or the sheet of donor atoms refer-red to as the δ-doping layer, are of great current interest in semi-conductor physics and applications. They are the theme of this workshop. Just as the electronic properties of bulk semiconductors are linked with their band structures, so the distinct electronic features of 2D systems relate to their specific electronic level distribution in the subband structure. Subbands are the energy eigenstates of carriers confined to move in a plane. They are derived from the volume energy bands by taking proper account of the quantum-mechanical confinement in one of the dimensions. The band structure $E(\vec{k})$ translates into a set of distinct subbands $E_n(\vec{k}_{||})$, which for the simplest case of a parabolic, isotropic energy-momentum relation are a set of nested parabolas of the form $E_n(\vec{k}_{||}) = E_n + (\hbar^2/2m^*)k_{||}^2$. Their separations are $E_{nm} = E_n - E_m$, with each of the E_n an eigen-energy of the confining potential. We consider here various experimental schemes that have been employed to measure the E_{nm} together with representative results.

The present lecture is intended both to review the classical examples of subband spectroscopy and to give some indication of current work and outstanding challenges. The purpose is to acquaint the reader, who perhaps nowadays may be engaged with studies of a complex multilayer heterostructure of compound semiconductors, with the basic notions and approaches developed earlier for work on the classic 2D system of electrons on (100) Si. Many aspects of subband spectroscopy are today being rediscovered and are assigned fancy new names. Thus the infrared excitation of electrons in a quantum well has been newly christened as "giant electric dipole resonance". The electric field tuning of levels in a well via a Schottky-gate contact has been hailed as a new effect in a band-to-band optical context. Band-gap renormalization in populated quantum wells is the new name for the many-body energies that are a familiar fact of subband physics for electrons on the Si surface. All of this has prompted my effort here to put things into perspective, to explore the useful analogues with previous work on Si.

This is not a complete and unbiased history of subband spectroscopy. Nor does it properly weigh the many individual contributions to the subject. When I choose examples that have come out of the Munich laboratories, I do so because I am most familiar with these and have available ready-made figures and numbers. I absolve my feelings of guilt regarding this biased view by pointing the reader to the very competently researched history written by Ando, Fowler, and Stern [1] on the subject.

2D SYSTEMS AND THEIR SUBBAND STRUCTURE - IN THEORY

Of the various 2D systems currently in vogue in semiconductor struc-tures, the electron inversion layer (100) Si interfacing to a SiO_2 barrier deserves a special place as a prototype forerunner, the one that has captured the attention of pioneer workers in the field and has nourished the journals for something like two decades, since that famous discovery of 2D magnetotransport [2] mentioned by P. Stiles in his lecture in this volume. It is sketched in Fig. 1a. A simple variant of this is the remotely-doped AlGaAs-GaAs heterostructure in 1b where the positive charge instead of residing on the gate electrode is embedded in the AlGaAs "dielectric". Despite all the claims to fame of this high-mobility, modulation-doped heterostructure, the principle of separating the negatively charged channel carriers from their positive donor ions stems from the Si metal-oxide-semiconductor (MOS) structure. Yet another heterostructure with a claim to fame, particularly in optoelectronics and modern tunneling devices, is the quantum well in Fig. 1c. When the latter is undoped and consequently unoccupied, the subband energies E_n are the square-well states familiar from elementary quantum mechanics. Using the techniques and methods for modern semiconductor layer growth, it has recently been shown that dopant impurities can be incorporated in atomically sharp defined sheets. For such a δ-doping layer [3] the electronic states are subbands in the V-shaped potential in Fig. 1d.

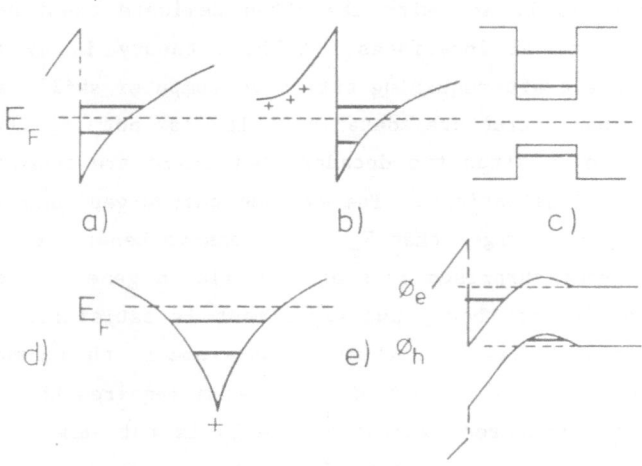

Fig. 1. Various 2D systems and their typical subband levels E_n as discussed in the text.

Yet another very novel 2D system, that arises dynamically under conditions when surplus electrons and holes are generated under illumination, is shown in Fig. 1e. As discovered originally by Altukhov et al. [4], a new luminescence line appears and can be interpreted as a recombination between a surface electron bound in a gate-voltage controlled channel and a nearby quantized hole. The microscopic origin of the hole binding potential remains to be explored. According to Ref.

[4] it is suggested to arise from a quantum effect named "overscreening" in which the constraints of wave mechanics and the Poisson equation are thought to give rise to an overshoot effect of the band edge energy in trying to match surface and volume Fermi levels.

The systems listed in Fig. 1 are exemplary of currently considered 2D structures, but they are by no means exhaustive. They are all reasonably well described by existing theory. In fact, the calculation of subband structures has long since progressed from crude estimates based on the triangular potential well and particle-in-a-box quantum mechanics to the fine art of multiband, self-consistent potential, envelope function computations as they are represented in this volume in lectures by M. Altarelli, F. Stern and G. Bastard. One has learned to cope with many-body effect self-energies, with the band-mixing induced by the surface potential, and with the often delicate question of how to match wavefunctions at interfaces. Subband theory is by and large a mature science, a craft requiring extensive computer skills and diligent attention to band-structure details. It is not really uncharted territory any longer after two decades, but there are some concerns for details in special situations. The extreme narrow-gap case when surface band energies E_n are larger than E_g and extensive Zener-tunneling occurs, may still hold some surprises in store. While in general the comparison of subband energies in theory and experiment is satisfactory, there are subtle second order effects that are troublesome in this comparison. In general, the spectroscopic probe of the system requires the system to be perturbed and the measured excitation energy is not immediately related to the levels E_n or their spacings, E_{nm}. Thus it is necessary to account for such things as excitonic final state interactions, depolarization shifting as well as selection rules and matrix element effects. For the more complex spectroscopies, such as photoconductive response and luminescence, lateral inhomogeneities of the layered system play a role. It is this variety of detailed information contained in the measurement that still represents a challenge to proper interpretation.

WAYS AND MEANS TO MEASURE SUBBAND ENERGIES - TRANSPORT EXPERIMENTS AND SURFACE LAYER CAPACITANCE

With this section we begin to examine the various techniques and methods that have been employed to measure subband energies. We choose to start the discussion with the dc transport experiments, then to progress to infrared excitation-spectroscopy, and finally to present some optical interband schemes. This approach reflects the historical evolution of methods, apparatus, and available sample materials.

It is natural to think of electrical transport by the 2D electrons as a means of learning something about their energy level structure. Not only are such experiments easy to realize in practice, but the original 2D electron gas system, as we know it in the Si MOS field-effect-transistor, is actually designed with transport in mind. Nothing is more natural than to look for subband evidence in the electronic conduction. As we shall see later in this section, subband energies can be obtained from simple dc transconductance experiments under favorable circumstances. It is only that the Si case does not belong to the favored few and the history of subband measurement took a devious route.

The transport experiment that did make subband spectroscopy history is one that makes use of a tiny leakage current through the insulator, a current flowing normal to the plane of the 2D system. D. Tsui in 1970 [5] made use of a tunneling arrangement in order to show that for an n-type accumulation layer on degenerately doped InAs the electronic levels were quantized. The experiment elegantly demonstrated the existence of 2D bands and measured their energies. This publication represented the birth of subband spectroscopy and it was achieved by simple and straightforward means, by an apparatus available in any laboratory. The secret to success lies in the skill and know-how of the experimentalist in making the delicate, thin insulating layer necessary for the tunneling work.

Tsui's tunneling experiment worked because he employed a degenerate semiconductor, thus being able to use simple substrate and gate contacts in the tunneling diode circuit. His experiment gave the quantized energies of occupied levels in the accumulation layer configuration. Noteworthy of more recent approaches to tunneling spectroscopy is the work of U. Kunze [6] who devised a special triode Si-MOS structure. Kunze's arrangement has a separate contact to the surface inversion layer on a p-type substrate. The electron tunneling current can be biased to flow from the surface channel to the gate, thus measuring occupied subbands, or from the gate electrode into the semiconductor, thus probing unoccupied bands above E_F in the inversion layer potential. The additional substrate electrode allows him to study the influence of the depletion layer on the level structure.

An experiment that is conceptually related to the Kunze-configuration for Si, but that was derived to probe the subband levels of the δ-doping layer (Fig. 1d), is shown in Fig. 2. The tunneling structure used by M. Zachau et al. [7] allows tunneling through a Schottky-barrier layer at the surface. Electrical contact is made to the doping layer independently of the substrate. Both occupied and empty subbands are probed when there is a depletion region in the substrate. In the presence of band-gap light quasi accumulation conditions apply in the substrate, and features related to the tunneling to unoccupied levels disappear in the spectrum of Fig. 2. The Zachau experiment has all the features that can be expected for an up-to-date tunneling measurement on a single barrier. Tunneling and devices based on this effect are currently much in vogue. The reader will find additional material on tunneling and how it relates to subband levels in the lecture by E. Mendez in this volume.

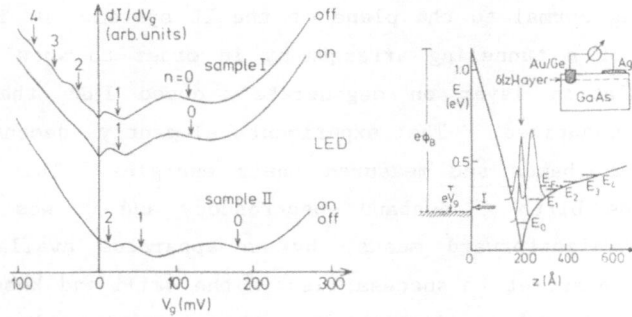

Fig. 2. Tunneling current characteristic for electrons through a Schottky-barrier on GaAs. The arrows mark energies of various subband levels in the δ-doping layer potential. Samples I and II have p- and n-type substrates and different layer densities N_s (after Zachau et al., Ref. [7]).

There is one feature of the tunneling experiment that makes it less than the ideal subband spectroscopy. It is a major disadvantage for the analysis that the subband-related features are each measured under the conditions of band-bending and surface charge density N_s for the voltage at which they are recorded. The sweep of the voltage as in Fig. 2 continually changes the system to be measured. It requires considerable calculation to relate the voltage separation for a pair of levels such as E_0 and E_1 to the energy separation E_{01} at a fixed N_s. For the complex multi-barrier structures studied in resonant tunneling experiments the change of tunnel current has dramatic consequences. When the resonant level has been reached by the bias voltage, additional charge accumulates in that state and its energy is modified. The result is a shift and hysteresis in the current-voltage characteristic [8].

Subband-related features can also be seen in the transconductance curves $\sigma(V_g)$ that probe lateral interface transport. With increasing V_g, as more and more electrons occupy the surface layer a higher-lying subband may become occupied. The sudden increase in the density of states at the Fermi level, when such a higher band E_n coincides with E_F, will usually be registered as a distinct structure in the conductance.

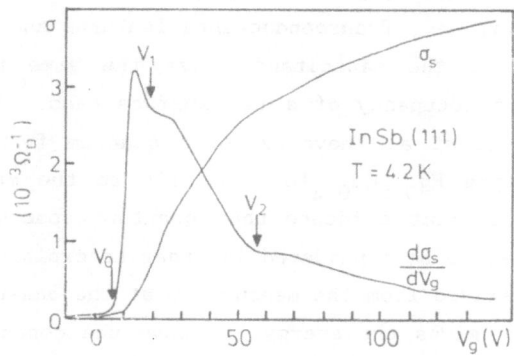

Fig. 3. Conductance and its derivative (arb. units) for electrons on InSb(111). The arrows mark voltages for the onset of occupation of the n = 0, 1, and 2 subbands and defining the condition E_F = E_n (1 V = 6.4 x 10^9 e/cm^2) (after W.Q. Zhao et al., Ref. [10]).

Such a signal was first reported by I. Eisele in pressure experiments on Si(100) FETs [9]. It was interpreted in terms of a coincidence of the ground state of the electrons in a higher valley with E_F. In narrow-gap semiconductors like InSb, InAs, and $Hg_{0.8}Cd_{0.2}Te$ where additional subbands become filled in rapid succession with rising N_s, the occupancy-onset anomalies are a regular feature of transconductance data. We show an example from work on InSb in Fig. 3 [10]. The subband energy E_n equals the Fermi energy E_F at the value of surface charge density N_s where it is observed. If E_F is known independently from N_s and the density of states in the subbands, the experiment gives a value for E_n. For parabolic subband structures the density of states is well known and simply related to the effective mass m^*, but unfortunately just in those systems where the anomalies are readily seen the band structures are not at all parabolic. It is necessary to invoke some model of the energy-dependent density-of-states to translate a measured N_s-value into an energy E_n. The conductance anomalies therefore have limited values in a precision measurement of subband energy. They can, however, serve as a critical test of an existing subband calculation.

Another quantity that relates in an easy and straightforward way to quantum features of the subband electrons is the capacitance. It has become fashionable to employ the surface layer capacitance to measure the density of states, in particular its variation with surface layer occupation. As such it is clear that the capacitance measurement, provided it is not seriously perturbed by localized interface states, can provide subband energies. Transconductance features and density-of-states structures in the capacitance have the same physical origin, namely the onset of occupancy of a new surface band. In a very recent paper [11] V. Mosser et al. have explored quantum features in the MIS capacitance of p-type $Hg_{0.8}Cd_{0.2}Te$. The fit to the measured $C(V_g)$ in Fig. 4 provides clear-cut evidence that quantum aspects of the surface inversion layer need to be considered in order to explain the data. Thus the band-bending derived from the measured C at the onset of surface-band occupation directly gives the energy E_0 above the conduction-band edge. The variation of $C(V_g)$ above the onset is related to the rising density-of-states in the E_0 subband. The dependence is linked in a complicated way with changes in the binding length of the groundstate $<Z_0>$, changes in E_0 with N_s, and changes in the depletion layer charge. The transconductance curve in the lower part of the figure shows an anomaly at voltage V_1 where the n=1 subband begins to be occupied. The expected structure in $C(V_g)$ is not really resolved. It is masked by a small ($\sim$5%) random variation of oxide charge density. Thus the really useful subband information in $C(V_g)$ is E_0. Nevertheless, since in the other resonance spectroscopies, to be discussed in the following two sections, only energy differences E_{nm} are determined, such an independent groundstate energy measurement proves a valuable complement.

There is mentioned in the capacitance literature yet another variant of the subband tunneling spectroscopy. J.D. Beck et al. in Ref. [12] have observed structures in the imaginary part of the $C(V_g)$-measurement. They interpret the signal as band-to-band tunneling into surface subbands. This Zener-type of tunneling can serve to charge the surface inversion layer through the bulk material.

The final example in this category of dc subband spectroscopies is the Shubnikov-de-Haas experiment. In a magnetic field B applied perpendicular to the surface, the conductance is found to oscillate periodically in 1/B. The period measures directly the occupation N^i of each of the subbands i. With this information, the energy separation between each pair of partially occupied levels can be derived provided the density-of-states function is known independently. In the simplest case of a constant D(E) the energy $E_{nm} = (N^m - N^n)/D(E)$. When the nonparabolicity is well known such as for GaAs, it is straightforward to make the necessary corrections to this simple formula.

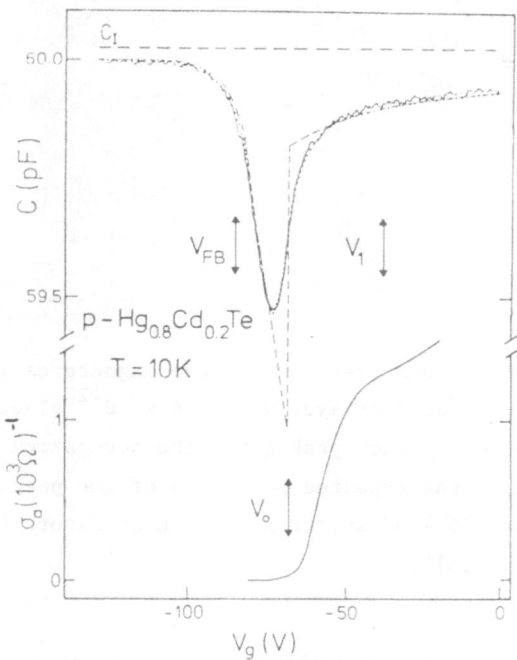

Fig. 4. Capacitance and conductivity of a surface layer on $Hg_{0.8}Cd_{0.2}Te$. The capacitance has been fit by a model calculation describing the quantum features of the surface carriers (after V. Mosser et al., Ref. [11]).

The magnetotransport approach to subband energy separations has been used in many cases in the literature. Energy band information has been gained in InSb, InAs, PbTe, $Hg_{0.8}Cd_{0.2}Te$ and Ge in this way. Even the first measurement of the famous 0-0' separation for Si(100) electrons came from magnetoconductance oscillations [13]. The separation of hole

subbands in grain boundaries on Ge has been determined from Shubnikov-de-
Haas data [14]. As a very recent example of subband energy measurements
using the oscillatory transport we cite the work of Zrenner et al. [15]
on the GaAs δ-doping layer. Zrenner not only measured energies in the
layer, but was able to reconstruct a model potential and show that the Si
donor ions appear to spread out over a finite distance in the direction
of MBE growth. We show in Fig. 5 an example of this work.

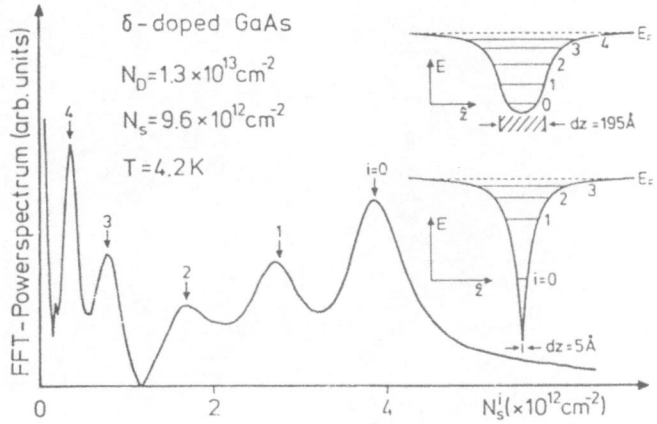

Fig. 5. Fourier transform spectrum for the magnetoresistivity
oscillations of a δ-layer with 9.6 x 10^{12} electronically active
Si atoms cm^{-2}. Each peak gives the occupation of a subband.
Arrows mark the expected positions of the peaks for a potential
well with 195 Å of spreading of the Si donors (A. Zrenner et
al., Ref. [15]).

The dc experiments, be it in the form of tunneling, of transconductance, capacitance or magnetotransport, have had their merits in subband
energy determination, but they are not universally applicable and in many
cases do require very special structures to be prepared. Compared to the
resonance spectroscopies that we shall discuss in the next two sections,
they are relatively simple to do and their data relate usually straight-
forwardly to subband energies. The system is only little perturbed from
equilibrium. There are no final state interactions such as the excitonic
effects in infrared and optical excitation experiments. Also the strong
depolarization shifting that accompanies the infrared dipole active mode
or the collective excitation in Raman spectroscopy is missing in transport experiments. In the case of the capacitance measurement an absolute
energy for the groundstate above the band edge can be determined.

When it comes to energy level determination, the name of the game is spectroscopy. What is more natural than to try to excite the one-dimensionally bound surface carriers by a properly polarized electromagnetic wave? The energies required are known from calculations to be in the range 10-100 meV, so that far-infrared sources and techniques are called for. The surface layer is a system that is easily tuned or modulated by a properly designed electrode configuration. For the Si MOS structure this is the insulated gate that controls N_s, or the substrate contact that will change the depletion field. The remotely-doped heterojunctions and the δ-layer can be adjusted to suit an external

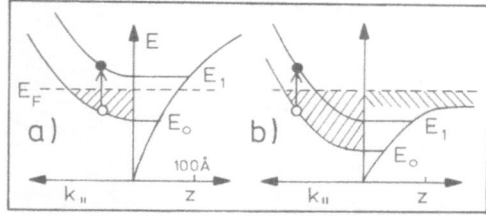

Fig. 6. Resonant intersubband transitions (0→1) for inversion- and accumulation-layer configurations. When several bands are occupied, $k_{||} = 0$ transitions are excluded.

source frequency by a Schottky-barrier electrode. The basic requirement for an intersubband excitation experiment is only that there is a filled lower state. As few as 10^{10} electrons in a single occupied E_0 state will do under favorable circumstances (Fig. 6a). Even if all the levels are occupied as in the accumulation configuration in case b) of figure 6, the resonance can be excited. The reader will note in the figure that provided a pair of subbands is equidistantly spaced, then in a) all carriers will participate. The electric dipole moment for a transition 0→1 can be as large as e x 100 Å for all the occupied states participating in the resonance. This is truly a giant electric dipole resonance! The only case not accessible to infrared excitation is that of empty bands such as for the quantum well in Fig. 1c.

With the good, fixed frequency molecular gas laser sources available in the early 1970's, it was natural to try to observe the gate-voltage

tuned resonance of electrons on (100) Si MOS samples. The first attempt
by R. Wheeler and co-workers [16] made use of FET devices. They
attempted to find a resonant response in the channel conductivity when
the transistor was illuminated by a far-infrared source. This very
clever approach eventually succeeded to give reliable data. It proved
necessary to measure at low temperatures, at low carrier densities, and
in such samples that have a significant temperature dependence of the
transconductance. The effect was demonstrably linked to resonant carrier
heating [17]. We give in Fig. 7 a sample of photoconductive response
data that shows the excitations from an occupied n = 0 groundstate to the
next higher lying 7 levels.

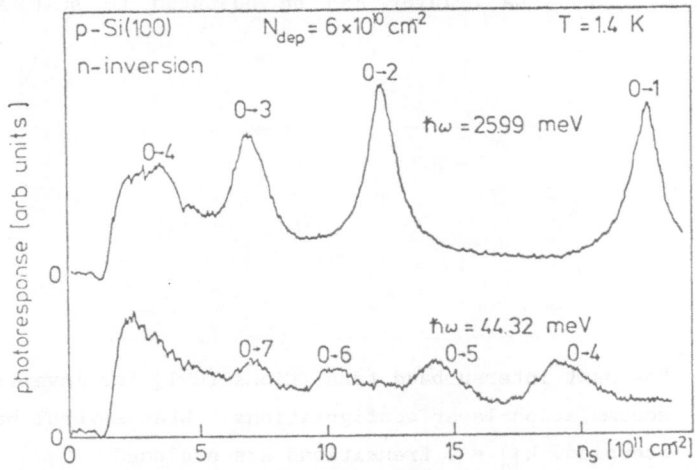

Fig. 7. Subband transitions for a Si(100) MOS device observed as a
photoconductive signal at two different excitation energies.
The signal is actually a decrease in conductivity at resonance
because of a heating effect. Peaks mark various transitions
from the occupied n = 0 groundstate (after F. Neppl et al., Ref.
[17]).

The more general approach to infrared subband spectroscopy was that
pursued by A. Kamgar et al. [18] working with large area MOS capacitor
structures. To achieve the correct sense of polarization of the electromagnetic field, as well as to observe sensitively an absorption signal
from the relatively few surface electrons, a transmission line arrangement was used. In this configuration the infrared laser radiation was
guided along the $Si-SiO_2$ interface for a length of several mm. The
signal was observed as a decrease of the transmission at resonance. Such
an absorption spectroscopy scheme has proved useful to measure the
resonances for electrons and holes in all major symmetry planes of Si.

Numbers have been obtained for the excitation energy of the subbands and have been carefully examined in the light of existing calculations. Various physical properties and dependences of the absorption resonance have been noted in the literature. One of the most striking ones to my mind has been the temperature dependence as studied by F. Schäffler et al. [19] and shown in the adjoining Fig. 8. It proved possible to follow the resonance signal over the entire range of temperatures from liquid He to room temperature There were found significant shifts of the peaks, as new levels such as the 0'-state of (100) Si became thermally occupied. Only a relatively small additional broadening of the resonances was observed. Most surprising was the fact that the strength of the resonances did not much decrease with rising temperature T. This fact is easily understood from the diagram sketched in Fig. 6. With increased T the carriers do spread out over various $k_{||}$-states of a given subband. For parallel bands with their constant energy separation the contribution to the resonance remains the same. Fig. 8 also shows vividly that the quantized energy states are very much a feature of the Si(100) FET device at ~300K where it is usually operated.

The transmission line is only one approach to the problem of achieving the required perpendicular polarization of the electromagnetic field. This polarization is an essential selection rule for exciting the carriers in a parabolic subband when the parallel and perpendicular components of the motion are uncoupled. Various and sundry other schemes have been employed and are being tried in current work on subband spectroscopy. Much of the current work is being done with frequency tunable sources such as grating and Fourier-transform IR spectrometers which have their own particular requirements for guiding the radiation to and from the sample. In Fig. 9 we show a selection of coupling schemes that have since evolved. Thus B. McCombe and co-workers introduced the prism coupling scheme in 9b), where the sample is clamped to a prism. 9c) is a variant of this where sample and prism coupler are "monolith-ically" fabricated from the same piece of Ge material. The non-normal incidence 9d) has been employed successfully to excite resonance in InAs. J. Kotthaus and D. Heitmann have exploited the near-field of a grating coupler lithographically engraved on the SiO_2 insulator to provide the proper polarization (9e)). With currently available MBE and CVD techniques the sequence of layers can be grown to the dimensions required to guide a TEM mode surface wave in the semiconductor structure. Figure 9f) shows how the latter is excited with an external prism in the attenuated total reflection (ATR) configuration.

As an electric-dipole active mode associated with the collective vibration of a layer of charge normal to the interface, the subband resonance is linked with a strong self-screening effect. At resonance, the "giant-dipole" moment of the vibrating charge very efficiently screens out the exciting electric field. Resonance is not really observed at the condition $\hbar\omega = E_{01}$. When the resonance experiments were first being presented as the means of precise and clear-cut measurement of subband structure, E. Burstein and co-workers in Ref. [20] pointed out

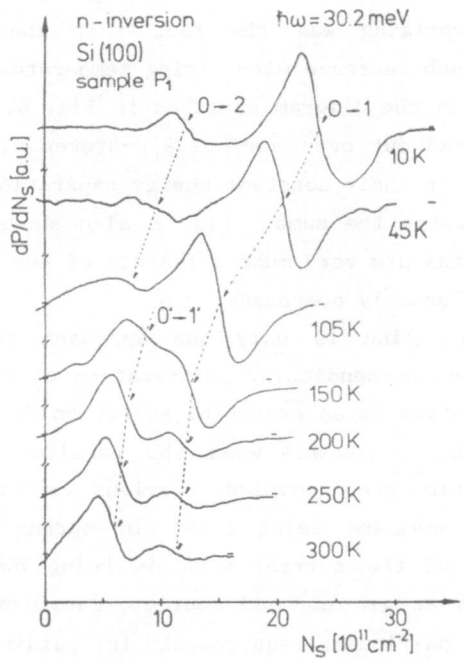

Fig. 8. Subband resonances recorded in terms of the V_g - derivative of the power transmitted through the strip transmission line: With rising temperature subband 0' becomes thermally populated and a new line marked 0' → 1' appears (after F. Schäffler et al., Ref. [19]).

this uncomfortable fact. Their warnings went unheeded and the first publication on the subject was largely ignored, because the observed resonances nicely matched the sophisticated many-body calculations that T. Ando had presented [21]. The depolarization shift, as it is now known after its rediscovery in Ref. [22], is quite a sizeable effect. The resonance absorption occurs when the dielectric function of the surface layer has a zero. Depending on the density N_s and the size of the dipole moment of each vibrating electron, the expected shift can be as much as 50% of the energy E_{01}. Depolarization shifts the resonance to an energy above the subband separation and is linked to the intensity of the resonance. Because of the generally ill-defined geometries of excitation this quantity is not easily obtained in the experiments with great precision. The spectroscopically measured, shifted energy E_{01} can only be compared with calculations when the correction is included.

For the work on the Si subbands with their large many-body energies, there was yet another correction that needed to be considered. The excitation of an electron to a higher band involves a rearrangement of charge that leaves behind a hole in the groundstate. Thus the excitation

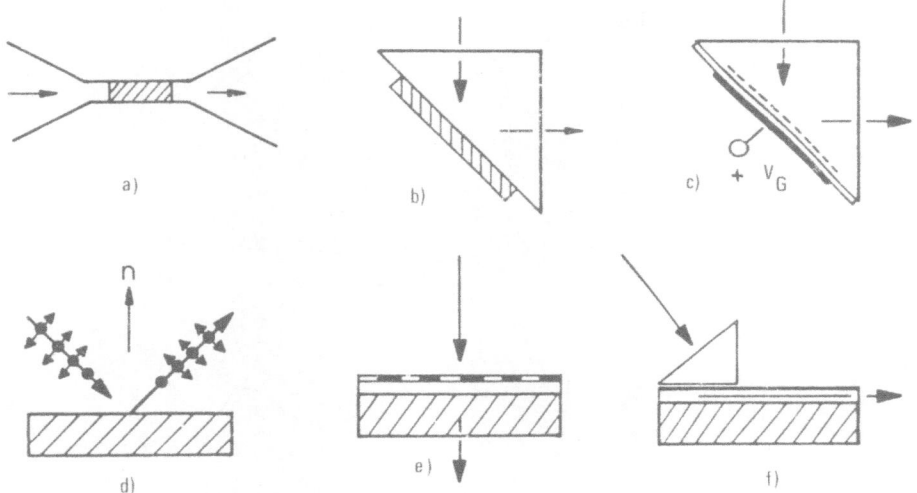

Fig. 9. Experimental configurations used in infrared subband
         spectroscopy in order to achieve a perpendicular component of
         the electric field.

as sketched in Fig. 6 has an associated final state interaction energy which also needs to be accounted for. T. Ando [23] has named this the exciton-like shift. It is inherently negative and reduces the resonance energy. For Si its value is comparable to the depolarization energy and partially compensates for the latter. We have come to express these facts after Ando in the simple approximate formula $\tilde{E}_{nm}(1 + \alpha_{nm} - \beta_{nm})^{1/2}$, where α_{nm} is the depolarization-shift factor, β_{nm} the exciton-like correction. Above all, those of us who have worked on the experimental side of infrared subband resonance, have come to respect the absorption peaks as dynamical excitation modes of a complex many-body system that need careful interpretation in order to relate them to a subband energy.

Looking back nowadays at the subband excitation experiments of years gone by, there were a great many unexpected results and puzzles that in more recent times have found quantitative physical explanation. One of these was the excitation of subbands using normally incident radiation. It was found for InSb that an IR beam passing through a thin slab of this material could excite transitions between the subbands. This type of excitation has become known as the nonparabolicity mechanism and arises

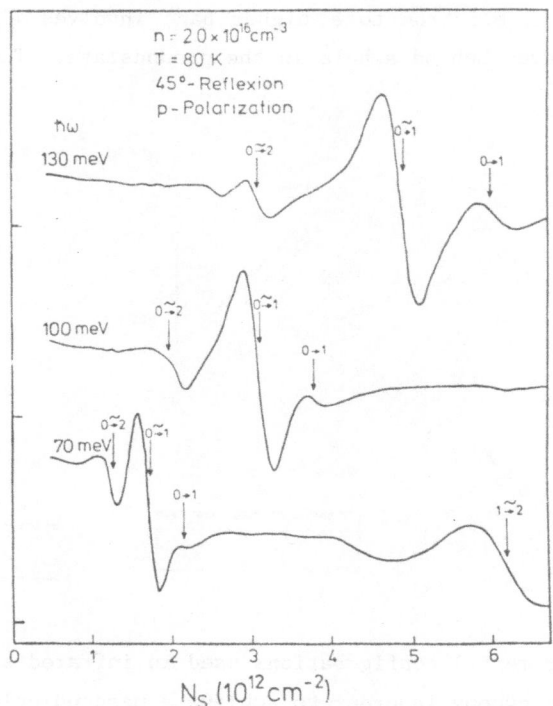

Fig. 10. Reflection derivative spectra calculated for radiation incident at 45° on an InAs MIS structure. The spectra compare favorably with the experiments (after H. Reisinger, Ref. [25]).

basically because this deviation from free-electron-like behavior couples the parallel and perpendicular degrees of freedom. The parallel excitation mode only involves states with finite $k_{||}$ in Fig. 6. States with $\pm k_{||}$ contribute out of phase to the $\hat{z}$-dependent motion and therefore have no net electric dipole moment associated with the excitation. Because of the very light mass and high dielectric constant also the many-body effect contained in Ando's β_{nm} parameter is negligibly small in a material like InSb. The resonant mode excited at normal incidence there-

82

fore most nearly represents a subband separation energy such as E_{01}. The very puzzling appearance of doublet peaks for InSb, separated by as much as 30% in the excitation energy, has since been proven to result from simultaneous excitation of the ordinary electric-dipole mode with its expected polarization shift and the mode excited by the parallel radiation component [24]. The two modes E_{01} and $\tilde{E}_{01}$ can appear simultaneously when the excitation geometry is poorly defined.

H. Reisinger in his as yet unpublished thesis research [25] explored the nonparabolicity mechanism in quantitative detail. His experiments deliberately used the tilted-angle incidence (Fig. 9d)) and careful control of the polarization in order to identify for InAs the two modes. The so-to-say "clean" configuration of the infrared electromagnetic field and carefully computed subband structure using W. Zawadski's multiband formalism [26] led him to nearly perfect quantitative agreement of calculated and experimental resonance spectra. In the accompanying Fig. 10 is shown an example from this work, where the reflection derivative spectrum from degenerately doped InAs ($N_D \sim 10^{17}$ cm^{-3}) is calculated at three different $\hbar\omega$ values. Tracing back all the steps that have gone into this result one discovers that the only electronic parameter in the calculation, aside from the dielectric constants and the well-known band gap of InAs, is a phenomenological damping time for intersubband relaxation. The splitting between the $\tilde{E}_{01}$ and E_{01} excitation modes is directly linked with the observed signal amplitude. Reisinger's work, in particular his careful comparison of the calculated and measured spectra, for the first time made subband resonance a quantitative spectroscopy.

There are numerous results that have followed more recently and make use of the ideas developed in the work on the classic "nonparabolic" materials InSb and InAs. Thus the electron surface bands of $Hg_{0.8}Cd_{0.2}Te$ have been measured, as well as holes on all the major symmetry planes of Si. Parallel excitation spectroscopy has also proved possible for electrons on (111) and (110) Si where the mechanism is the ellipsoid tilting and consequent coupling of the parallel and perpendicular motions. In this case it is easy to see how the E_{01} mode is excited without the depolarization effect. Electrons move on symmetry-related pairs of ellipsoids in such a way as to cancel the z-related displacement [27]. When confinement in the plane of the sample occurs together with the z-directed confined motion, the electron system becomes a 1-dimensional linear conductor. For this case the parallel excitation mode is described by J. Kotthaus in his lecture in this volume.

As a final note in this chapter on the infrared spectroscopies, we mention the emission experiments of E. Gornik et al. [28]. In this very elegant arrangement a dc current pulse is used to heat the carrier system. Relaxation between the subbands has been found to generate infrared radiation at the $\bar{E}_{01}$ energy in Si(100).

INTERBAND SPECTROSCOPIES. THE OPTICAL APPROACH TO SUBBAND ENERGY DETERMINATION.

The finely honed and sensitive optical spectroscopy techniques have also been applied successfully to subband energy determination. In this chapter we come to examine this optical approach to the subband structure of 2D systems. In doing so, the intent is to be exemplary and to point out in principle how a measurement can be done and how the information to be derived relates to the infrared experiment. Needless to say, the limits of space and time do not permit a review of all the far-ranging optical work that has been done, nor to mention the many individual contributions to the field.

The optical approach is essentially an interband experiment in that in one way or another both the conduction and valence band states are involved. The optical experiments involve a transition matrix element in which the perturbation potential $(\vec{p}.\vec{A})/2m_o$ acts on the Bloch part of the subband wavefunction. Whereas for the usual infrared excitation the dipole operator appears with the prefactor $1/(2m^*)$ and resonance involves matrix elements of the type $\langle \chi_n | \vec{p}.\vec{A} | \chi_m \rangle$, the optical transition is based on matrix elements of the type $\langle \chi_n | \chi_m \rangle \langle c | \vec{p}.\vec{A} | v \rangle$. Here the χ_n represent the envelope functions in the effective mass formulation, the c and v are the Bloch functions for the conduction and valence bands respectively. It follows that the selection rules for the transitions are quite different, that optical experiments and infrared spectroscopy can prove to give complementary information. In addition, optical excitation of additional electrons and holes can greatly modify the interface potential and, as in the Rogachev experiment [4], create subband levels that are absent in equilibrium.

Any discussion of subband optics must necessarily begin with the classical optical absorption experiment of Dingle et al. [29] on the square-well potential heterostructure as in Fig. 1c). This was the prototype of experiments capable of detecting the quantized energy band structure in the unoccupied square well. Dingle and his co-workers worked with a GaAs well embedded in $Al_xGa_{1-x}As$. Distinct peaks were observed for transitions from the filled valence quantum states to the empty electronic levels in the conduction band. It was recognized quite early that matrix element effects determine precisely which subbands contribute to the signal and that the optical excitation involves an excitonic final state interaction.

There are several things to note immediately from these pioneering experiments as they relate to subband energy determination. It is clear that interband optics is one way to get information on the conduction band levels when these are not occupied and for which the infrared spectroscopy experiments cannot be done. Interband optics necessarily involves also the hole subbands with all their complex nonparabolicity effects and strong mixing of the spin-orbit-split heavy and light hole valence band states. The reader will find a discussion of these in the lectures of G. Bastard and M. Altarelli in this volume. Thirdly, the interband excitation involves an exciton final-state energy that is different than the exciton-like effect in intraconduction-band inter-subband spectroscopy but that has the same physical origin. Also the interband excitation is not depolarization-shifted as was the intersubband resonance. Finally, the interband experiment is most easily done with an optically active material such as GaAs.

In modern optics work on quantum well systems experiments have recently been done under conditions of strong excitation, when a sizeable number of electrons occupy conduction-band subbands and a correspondingly large number of holes exist in valence band subbands. It is interesting to note that under these circumstances a many-body energy similar to that calculated by T. Ando [21] for the electrons on (100)Si applies. The major contribution for a material like InGaAs will come from the hole bands. The shifts observed for high excitation conditions in quantum wells has received the noble name "band-gap renormalization". It is a fact of subband life that is thoroughly familiar in infrared spectroscopy where it amounts to a near 50% effect for the subband energies on (100)Si. It is also clear that optical excitation of a large density of electrons and holes will modify the subband structure because of space charge effects.

There are a number of variants of the optical absorption experiment that give similar information on the subband structure when the electron states are unoccupied. Thus in Fig. 11 we show an example of how the absorption can also be registered in terms of photoconductive response [30] involving conduction parallel to the layered structure. The latter is most sensitively observed when the structure is intrinsic. The figure serves to demonstrate a specific feature of the photoconductivity experiment, namely that the excitation of a carrier into the subband only results in a signal if the carrier is mobile. At the lowest T in the figure optically excited carriers become trapped by interface potential fluctuations. The subband excitation resonances decrease in amplitude. At the same time the absorption edge signal that results from excitation to the top of the potential well, where the electron is free to spread out into the neighboring InP, becomes stronger and more sharply defined. The experiments serves to measure precisely the band offset between the two materials.

Yet another way to obtain information on the absorption between subbands is the excitation photoluminescence variant. In this arrangement one monitors the radiative recombination between the electron and hole groundstate level, while sweeping the excitation energy through the spectrum of the various higher-lying subbands.

That optical techniques can also serve to determine the separation

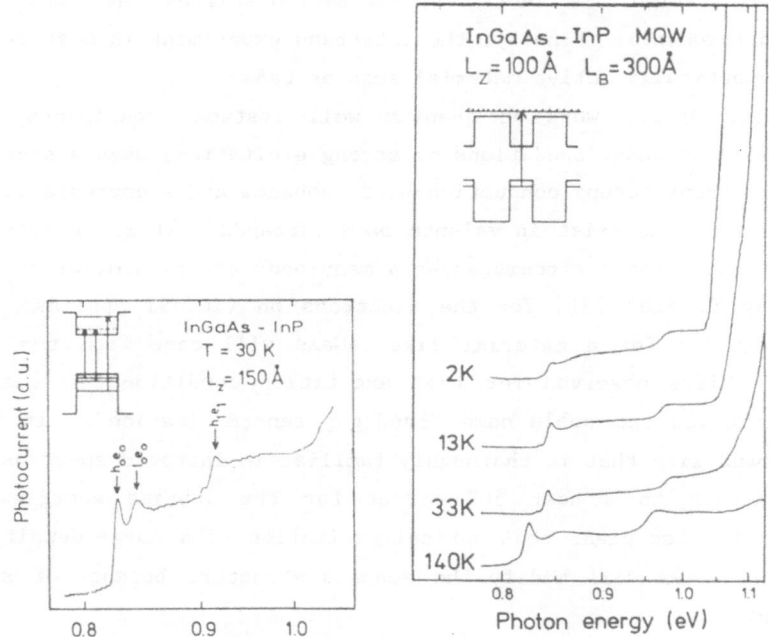

Fig. 11. Optical interband subband excitation observed as photoconductive response in an intrinsic InGaAs quantum well. The figure on the right shows onset of conductivity with excitation to a miniband extending into the InP. Localization of carriers in the well reduces the photoresponse dramatically between 140K and 2K. Multiplication factors of 1, 20, 40, and 400 apply to traces at 140K, 33K, 13K, and 2K respectively (after M. Zachau et al., Ref. [30]).

between subband levels when the 2D states are occupied is by now thoroughly established. Thus absorption, luminescence, and photoconductivity have all been observed when, because of doping or band bending by an external potential, the Fermi level lies in one or another of the bands. It is only necessary to exercise some care in the interpretation. Band-gap light will reduce the depletion layer potential and thus reshape a potential well like that in Fig. 6a) to look more like the accumulation case in 6b). It is then called quasi-accumulation. When optical transitions in a degenerate system are involved there are the expected Fermi energy effects. Because of screening the excitonic final-state effects may well be different.

As regards the photoconductivity experiment on degenerately doped layered structures, there is a basic loss of sensitivity for the lateral transport. The excitation will add only a few carriers to the existing many, and signal strength will be much reduced over the case where the system is an insulator prior to photoexcitation. There is, however, an interesting variant of the experiment where the transport is measured normal to the layer plane. Because of the potential barrier, carriers in the doped quantum well structure in Fig. 12 will not conduct. With the excitation of a carrier into one of the higher, unoccupied states in the well, there is a possibility that this electron will surmount the remaining barrier at finite temperature and be registered as a current. The two-step process is similar to photothermal ionization spectroscopy. This particular experiment can also be done advantageously at infrared energies where the radiation will excite carriers resonantly from the groundstate in the conduction band to the higher, thermally ionizable excited state. In such a way one can resonantly detect a given photon

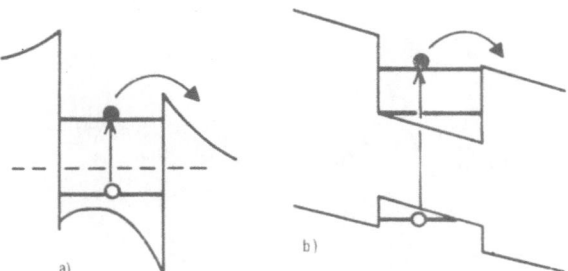

Fig. 12.   Photothermal ionization experiments that can be used to record
           subband excitation in doped (case a) and undoped (case b)
           quantum wells. Current flow is normal to the well in an
           applied perpendicular field.

energy. In this version the experiment is quite similar to the usual infrared absorption resonance.

Among the optical experiments there is one that directly compares with infrared spectroscopy of subbands. Resonant Raman scattering as originally conceived by G. Abstreiter and K. Ploog [31] is an experiment that applies for occupied 2D subbands in heterojunctions or quantum wells. The Raman excitation corresponds to a two-step process. An electron is first raised from a valence band state resonantly to an empty subband level. Simultaneously a carrier occupying a lower subband returns to the empty valence band state. The net excitation amounts to the transition of an electron between the subband levels and is registered in terms of a Stokes-shift of the scattered light. By now such Raman experiments have become a standard technique for sensitively looking at subband separations in a number of materials. The experiment requires a tunable dye-laser source at an energy that corresponds to a direct transition in the band structure. Since the pioneering experiment of G. Abstreiter on a voltage tuned $GaAs-Al_xGa_{1-x}As$ heterostructure interface, Raman scattering has been applied to study modulation-doped quantum wells in various III-V materials as well as subbands at the $Si-SiO_2$ interface.

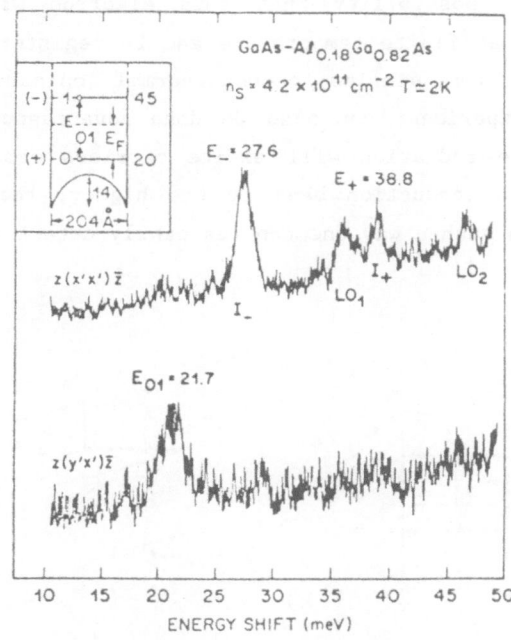

Fig. 13. Resonant Raman excitations in a modulation-doped GaAs quantum well with thickness 204 Å. Distinct excitation spectra are observed in the collective mode (upper trace) and the single particle, spin-flip mode (after A. Pinczuk et al., Ref. [32]).

The Raman experiments are so closely related to intersubband excitation that the considerations regarding depolarization shift and exciton-like corrections apply. The work of Pinczuk et al. [32] discovered that the Raman excitations occur also as doublet pairs of the type E_{01} and $\tilde{E}_{01}$. From this publication we show in Fig. 13 Raman spectra as they have come to be expected for this optical technique. What in infrared spectroscopy is the electrically active dipole mode, is called the collective excitation mode in Raman spectroscopy. The upper trace in the figure is observed with the back-scattered light polarized in the same direction as the normally incident exciting beam. The spectrum shows two electric modes of the $\tilde{E}_{01}$-dipole type. The peaks labeled E_- and E_+ result from a coupling of the $\tilde{E}_{01}$ electronic vibration to the optical phonons which have a comparable energy of excitation. The interaction effect is the same as has been observed for the infrared subband resonance on $Hg_{0.8}Cd_{0.2}Te$ by J. Scholz et al. [33]. The lower trace in Fig. 13 gives the signal in the crossed-polarization mode. This is known as the spinflip, single particle excitation and gives E_{01} in a manner related to the corresponding E_{01}-mode in infrared spectroscopy. The excitation exploits an internal degree of freedom leading to a vibration of the electron gas without a macroscopic electrical polarization. As such it is in principle similar to the parallel excitation in infrared absorption for nonparabolic systems or pairs of tilted ellipsoids, even though the microscopic mechanism is a different one and is special to optical excitation.

Our final sampling on optical subband experiments is one that represents a single heterojunction in a dynamical steady state but under nonequilibrium conditions. In the work by A. Rogachev and co-workers [4] it was first shown that for the case of a hole layer on Si(100) under illumination by an above band-gap light source, a new luminescence line appears. This so-called S-band (for surface) clearly represents the recombination of a surface-bound quantized hole state with an excited electron state subband level. The quasi Fermi levels φ_e and φ_h are indicated in Fig. 1f) for the case of an n-p dipole layer under non-equilibrium condition. Such an n-p configuration of layers has subsequently been found for Si(111) and Si(110) [34,35] to give rise to S-band signals. The luminescence peak reported by E. Koteles in [36] is most certainly of the same type. It is observed for a modulation-doped heterojunction interface of $Al_xGa_{1-x}As$ and GaAs. The S-band luminescence is by now a standard feature of many luminescence experiments and it clearly represents a contribution from surface bands under nonequilibrium conditions. A new variant of the Rogachev experiment has recently been reported by I. Kukushkin and V. Timofeev [37]. The latter is closely related to an experiment by M. Skolnick et al. [38].

Even though the luminescence line is certainly linked with quantized subbands it is not at all clear how its energy is related to the energy levels. There exists under illumination a dynamical state which is similar to the more familiar quantum well, only the electron and hole bands are somewhat displaced in real space (compare figure 1f)). In the papers of A. Rogachev the explanation for this potential well is sought in terms of his overscreening concept. Our own work [39] has attempted to account for the binding potential in terms of the fields and potential driving an ambi-polar diffusion current for carriers with different diffusion constants. A totally satisfactory explanation is still outstanding.

CHALLENGES IN SUBBAND SPECTROSCOPY. WHAT'S HAPPENING TODAY?

The investigation of energy levels in 2D systems is by no means a closed subject. Different and new layered structures are still being discovered and the more traditional ones are being reexamined in more detail. In these closing paragraphs the attempt is made to list some outstanding challenges, examples of current work in the field. The selection is necessarily a biased and personal view.

Looking around at new systems there are three that have attracted particular attention recently. One of these is the dynamical, non-equilibrium system of the type that we have discussed above. In addition to illumination, as for the Rogachev experiment, one can conceive of electrons and holes being injected by surface contacts into the gate-controlled surface region, thus providing tunable electroluminescence in Si. The second on my list is the δ-doping layer. Its subband structure has been explored for Si-donors in GaAs, as well as Sb in (100)Si [40]. Many interesting aspects such as the chemical shift question and multi-valley subband structure remain to be solved. There exist many variants of the δ-layer system such as a monolayer sheet of "compatible" atoms embedded in the MBE grown material. An Al layer in GaAs, or possibly a single Ge layer in Si may prove interesting. The idea is to examine systems which are no longer effective-mass type quantum wells but where atomic effects are essential. A third type of system is that involving high-quality and amorphous layers on crystalline substrates, such as the a-Si/c-Si structure. Also amorphous layer structures such as the $Si-Si_3N_4$ system that M. Hirose discusses in Ref. [41] deserve the attention of subband specialists. These are a few examples. There are many more.

Among the traditional heterojunction systems there is one case that merits additional subband attention. It is the $Hg_{0.8}Cd_{0.2}Te$ MIS structure involving either the anodic oxide, SiO_2 or CdTe as an insulating interface layer. The major point to be resolved here is how the subbands look in the presence of the strong Zener-tunneling between valence and conduction bands. The same system exhibits strong electric-potential induced spin-splitting that still needs to be looked at quantitatively.

A wide open field is the mechanism of intersubband relaxation and its relation to the linewidth in subband spectroscopy. A quite interesting experiment on this has recently been reported by A Seilmeier et al. [42]. More of this type are needed to resolve the question of how energy is dissipated in intersubband relaxation. It may be of interest to apply methods like photothermal deflection spectroscopy (PDS) to directly monitor heating effects caused by inelastic intersubband scattering processes after optical excitation. Resonant coupling of the intersubband excitation, such as the $\tilde{E}_{01}$ mode in Si, to optical phonons on both sides of the interface should be studied in more detail in order to understand the relaxation mechanism better.

Among the experimental techniques, there remains to study the very effective coupling to the subband excitation for the case of a strip line mode guided along the interface in a properly designed layer structure. One may expect that such a guided wave can be strongly modulated by the gate-voltage-controlled subbands. This geometry of excitation was briefly mentioned in connection with Fig. 9f).

A final thought, one that involves a wild flight of the imagination and is colored by all the spectacular developments in superconductivity recently, is the question of subbands in a thin epitaxial layer of this material, possibly interfacing a semiconductor. After all, subbands were observed in ordinary type 1 superconductors in the presence of a parallel magnetic field [43]. I look forward to the day when one has learned to embed the superconductor as a thin layer between semiconducting or insulating layers, being able to switch superconductivity on and off by a gate voltage controlling the electron density in the superconductor. Total madness - perhaps. But then again, what would everyone have said only a decade ago, if one had speculated about a mobility of 5,000,000 cm^2/Vsec in an interfacial layer?

REFERENCES

1. T. Ando, A. B. Fowler and F. Stern, Rev. Mod. Phys. 54:437 (1982).

2. A. B. Fowler, F. F. Fang, W. E. Howard and P. J. Stiles, Phys. Rev. Lett. 16:901 (1966).

3. A. Zrenner, H. Reisinger, F. Koch and K. Ploog, Proc. 17th Int. Conf. on the Physics of Semiconductors (San Francisco 1984), eds. J. P. Chadi and W. A. Harrison (Springer, New York, 1985) p. 325.

4. P. D. Altukhov, JETP Lett. 38:4 (1983); See also: P. D. Altukhov, A. M. Monakhov, A. A. Rogachev and V. E. Khartsiev, Sov. Phys. Solid State 27:359 (1985).

5. D. C. Tsui, Phys. Rev. Lett. 24:303 (1970).

6. U. Kunze, J. Phys. C 17:5677 (1984).

7. M. Zachau, F. Koch, K. Ploog, P. Roentgen and H. Beneking, Sol. State Commun. 59:591 (1986).

8. V. J. Goldman, D. C. Tsui and J. E. Cunningham, Phys. Rev. Lett. 58:1256 (1987).

9. I. Eisele, H. Gesch and G. Dorda, Surf. Sci. 58:169 (1976).

10. A. Därr, Ph.D. thesis, Tech. Univ. München; See also: Wen-Qin Zhao, F. Koch, J. Ziegler and H. Maier, Phys. Rev. B31:2416 (1985).

11. V. Mosser, R. Sizmann, F. Koch, J. Ziegler and H. Maier, Semicond. Sci. Technol. (to be published).

12. J. D. Beck, M. A. Kinch, E. J. Esposito and R. A. Chapman, J. Vac. Sci. Technol. 21:172 (1982).

13. D. C. Tsui and G. Kaminsky, Phys. Rev. Lett. 35:1468 (1975).

14. S. Uchida, G. Landwehr and E. Bangert, Sol. State Commun. 45:869 (1983).

15. A. Zrenner, F. Koch and K. Ploog, Proc. EP2DS VII (Santa Fe, 1987), p. 341, and to be published.

16. R. G. Wheeler and R. W. Ralston, Phys. Rev. Lett. 27:925 (1971).

17. F. Neppl, J. P. Kotthaus and F. Koch, Phys. Rev. B19:5240 (1979).

18. A. Kamgar, P. Kneschaurek, G. Dorda and F. Koch, Phys. Rev. Lett. 32:1251 (1974).

19. F. Schäffler and F. Koch, Sol. State Commun. 37:365 (1981).

20. W. P. Chen, Y. J. Chen and E. Burstein, Surf. Sci. 58:263 (1976).

21. T. Ando, Phys. Rev. B13:3468 (1976).

22. S. J. Allen, D. C. Tsui and B. Vinter, Sol. State Commun. 20:425 (1976).

23. T. Ando, Z. Phys. B26:263 (1977).

24. K. Weisinger, H. Reisinger and F. Koch, Surf. Sci. 113:102 (1982).

25. H. Reisinger, Ph.D. Thesis, Tech. Univ. München (1983).

26. W. Zawadski, J. Phys. C16:229 (1983).

27. Soe-Mie Nee, U. Claessen and F. Koch, Phys. Rev. B29:3449 (1984).

28. E. Gornik and D. C. Tsui, Phys. Rev. Lett. 37:1425 (1976).

29. R. Dingle, W. Wiegmann and C. H. Henry, Phys. Rev. Lett. 33:827 (1974).

30. M. Zachau, P. Helgesen, F. Koch, D. Grützmacher, H. Jürgensen and P. Balk, Sem. Sci. Technol. (to be published).

31. G. Abstreiter and K. Ploog, Phys. Rev. Lett. 42:1308 (1979).

32. A. Pinczuk, J. M. Worlock, H. L. Störmer, R. Dingle, W. Wiegmann and A. C. Gossard, Sol. State Commun. 36:43 (1980).

33. J. Scholz, F. Koch, H. Maier and J. Ziegler, Sol. State Commun. 45:39 (1983).

34. F. Martelli, Sol. State Commun. 55:905 (1985).

35. P. D. Altukhov, A. V. Ivanov, Yu. N. Lomasov and A. A. Rogachev, Sov. Phys. Solid State 27:1016 (1985).

36. E. S. Koteles, J. Y. Chi and R. P. Holmstrom, Proc. SPIE Conf. on Modern Optical Characterization Techniques for Semiconductors and Devices (1987) (to be published).

37. I. V. Kukushkin and V. B. Timofeev, JETP 92:258 (1987).

38. M. S. Skolnick, J. M. Rorison, K. J. Nash, D. J. Mowbray, P. R. Tapster, S. Bass and A. D. Pitt, Phys. Rev. Lett. 58:2130 (1987).

39. R. Küchler, A. Asenov, C. H. Perry and F. Koch, Sem. Sci. Technol. (to be published).

40. H. P. Zeindl, T. Wegehaupt, I. Eisele, H. Oppolzer, H. Reisinger, G. Tempel and F. Koch, Appl. Phys. Lett. 50:1164 (1987).

41. M. Hirose, S. Miyazaki, Proc. 3rd Int. Conf. on Modulated Semiconductor Structures (Montpellier 1987) (to be published).

42. A. Seilmeier, H.J. Hübner, G. Abstreiter, G. Weimann and W. Schlapp, Phys. Rev. Lett. 59:1345 (1987).

43. F. Koch and P. A. Pincus, Phys. Rev. Lett. 19:1044 (1967).

INFRARED EXCITATIONS IN ELECTRONIC SYSTEMS WITH REDUCED DIMENSIONALITY

Jörg P. Kotthaus

Institut für Angewandte Physik
Universität Hamburg, Jungiusstr. 11
2000 Hamburg 36, F. R. Germany

INTRODUCTION

Modern semiconductor devices are increasingly based on structure in which the electronically active channel is an electron or hole system with reduced dimensionality. The classical example is the metal-oxide-semiconductor-field-effect-transistor (MOSFET) in which electrons can be confined at the semiconductor-oxide interface to form a quasi-two-dimensional electron system (2DES). Another more recently developed device is the high-electron-mobility transistor (HEMT) in which a suitable doping profile induces a 2D electron channel at the heterojunction potential barrier. The extraordinary electronic properties of such two-dimensional electron systems at interfaces are summarized in part in an extended review by Ando, Fowler and Stern[1]. Fig. 1 schematically pictures the band diagram in the confinement direction perpendicular to the interface for both devices. In the MOS-system (Fig. 1a) a positive gate voltage V_g is applied between the p-type semiconductor and the metal gate and causes bending of the conduction and valence bands near the semiconductor-oxide interface. The electrons are confined in a narrow potential well bounded by the oxide barrier and the conduction band edge. In the modulation doped heterojunction (Fig. 1b) n-type doping of the large gap semiconductor (e.g. $Al_xGa_{1-x}As$) causes electron transfer to the undoped p-type lower gap semiconductor (e.g. GaAs) with a larger electron affinity. The charge separation at the interface thus again causes the formation of a potential well that confines the electrons in GaAs near the interface as illustrated in Fig. 1b.

In both devices the characteristic width of the confining potential is usually comparable to the de-Broglie wavelength of the electrons in the channel and, at least at low temperatures, much smaller than the elastic mean free path of the electrons. Thus the electronic motion in the potential well perpendicular to the interface becomes quantized. The possible electron energies may be written as:

$$E(k) = \frac{\hbar^2 k^2}{2m_x} + \frac{\hbar^2 k^2}{2m_y} + E_1 \, . \tag{1}$$

Here the first two terms reflect the kinetic energy of motion parallel to the interface, whereas $E_i (i=0,1,1,...)$ denotes the quantized energy states in the potential well. Such a system is called a 2DES if only few subbands E_i are occupied and subband spacings are larger than energy broadening induced by scattering processes. The extraordinary electronic properties of such two-dimensional electron systems have been investigated with steadily rising interest during the last 20 years, initially mainly in the MOS-System on Si and currently extensively on III-V hetero-junctions and related structures. In the course of these studies infra-red spectroscopy has become an extremely important and widely used tool[2-4] since it makes possible the direct observation of the relevant electronic excitations in such two-dimensional electron systems. In the first part of this review I will try to illustrate with some recent spectroscopic experiments on single layer heterojunctions how infrared spectroscopy can be used advantageously to study subband structure and electronic interactions in two-dimensional systems.

With steadily decreasing lateral dimensions of semiconductor devices, which in currently most advanced integrated circuits have

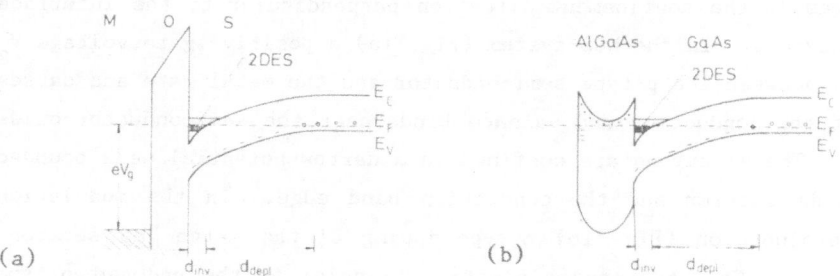

Fig. 1 Schematic band diagram in (a) a MOS-structure and (b) a AlGaAs-GaAs heterojunction with a 2D inversion layer (2DES) at the interface. E_c and E_v denote the conduction and valence band edges, respectively, E_F is the Fermi energy in the semiconductor. In both cases the inversion layer (d_{inv}) is separated by a depletion layer (d_{depl}) from the bulk semiconductor.

dropped to below 1μm, it also becomes increasingly important to investigate and understand the effects of lateral confinement on the electronic properties of electron channels at interfaces. In a number of laboratories recently channel widths of down to about 100nm have been realized. Along with such lateral confinement of inversion electrons to a length scale which is comparable to intrinsic lengths of the inversion electron system at low temperatures one now observes novel electronic phenomena that appear with the transition from 2D to 1D electronic behavior. The mesoscopic size of single narrow inversion channels has been observed to cause strong fluctuations of the channel conductance as a function of electron density and magnetic fields that are related to quantum interference and resonant tunneling phenomena. Recent publications[5-7] illustrate the activity in this field. There have also been several attempts to verify lateral quantization in such narrow channels. Conductance oscillations in multiwire MOS-devices on Si[8] as well as aperiodic Shubnikov-de Haas (SdH) oscillations in the magnetoconductance of narrow inversion channels on GaAs[9] have been linked to quantization into 1D subbands. In the second part of this review I want to summarize recent spectroscopic studies of the transition from two-dimensional to one-dimensional electronic behavior in laterally periodic MOS- and heterojunction microstructures. In such experiments it has now become possible to observe directly intersubband transitions between adjacent one-dimensional subbands[10,11]. This constitutes a very direct proof of lateral quantization in narrow inversion channels and allows one now to measure the relevant quantization energies.

SPECTROSCOPIC TECHNIQUES

The characteristic energies of the electronic excitations in electron systems with reduced dimensionality at semiconductor interfaces are typically between 1 meV and 100 meV. The lower limit is in most cases set by energy broadening of the electronic excitations caused by scattering. The excitation energies in confined electron systems thus correspond to photon frequencies that lie in the infrared with typical values $\bar{\nu}=\omega/(2\pi c)$ ranging from 10 cm^{-1} to 1000 cm^{-1}. It is therefore very natural that infrared spectroscopy is used extensively to study electronic excitations experimentally in such 2D or 1D systems. In many cases infrared spectroscopy is well complemented by Raman spectroscopy and, for quantum well structures, by interband optical studies[1].

Infrared spectroscopy is mostly done in simple transmission experiments. Radiation is either generated at discrete frequencies with far-infrared lasers[2,4] or a broadband lamp source is used in conjunction with a Fourier transform spectrometer to measure spectra in the frequency domain[3,12]. Fig. 2 schematically pictures a typical experimental apparatus for Fourier transform spectroscopy. The infrared radiation from the lamp source is transmitted through a Michelson interferometer with one moving mirror, then focussed onto the sample, and finally detected with a cryogenic bolometer. To achieve sufficiently large infrared signals the active sample area must be much larger than the infrared wavelengths and is typically several mm in diameter. Samples are mostly cooled to cryogenic temperatures and often placed in the center of a superconducting solenoid. With an appropriately chosen beam splitter and suitably selected mirror velocity in the interferometer the detector records an interferogram that is Fourier-transformed to give the intensity distribution in the frequency regime of interest. To determine the transmission coefficient T of a given sample it is necessary to take interferograms with and without the sample and then obtain T by normalization of the transformed spectra. For the low dimensional electron systems considered here it is often possible to measure a reference spectrum without removing the sample. This is done either by switching the areal charge density N_s with a gate voltage between 0 and a preselected value or by using a magnetic field to tune the spectral features of interest to sufficiently different frequency regions. One

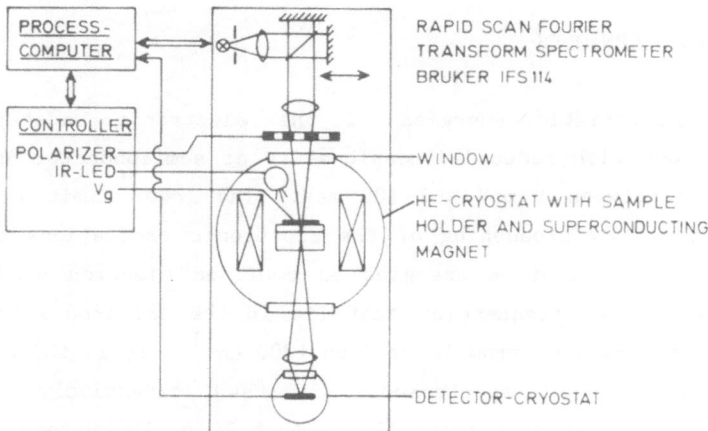

Fig. 2. Experimental arrangement for infrared transmission spectroscopy with a Fourier transform spectrometer on semiconductor samples at low temperatures (from Ref. [12]).

thus can measure the ratio $T(N_s)/T(N_s=0)$ or $T(B)/T(B=0)$ with high sensitivity.

The measured transmission coefficients T are often compared with calculated ones. The transmission coefficient T for linearly polarized infrared radiation incident normally onto a thin electron layer characterized by a complex dynamical sheet conductivity $\sigma(\omega)=\sigma_r+i\sigma_i$ normalized to $(\varepsilon_o/\mu_o)^{\frac{1}{2}}$, the reciprocal free space impedance, can easily be calculated in the following limits from Maxwell equations. The semi-conductor substrate is assumed to be nonabsorbing in the spectral regime of interest and have refractive index n. It is wedged on the back side to avoid interference effects within the substrate[13]. The distance between the electron layer and the vacuum-semiconductor interface is much smaller than the infrared wavelength in the semiconductor and thus can be neglected. Then one obtains:

$$\frac{T(N_s)}{T(0)} = \frac{(1+n)^2}{(1+n+\sigma_r)^2 + \sigma^2} \ . \tag{2}$$

If, as in the case of MOS structures or gated heterojunctions, the sample has a semitransparent gate with normalized sheet conductivity σ_g on top of the electron layer this is easily incorporated into eq. (2) by replacing n by $n+\sigma_g$. In most experimental situations one has, however, $\sigma_g \ll n$ and thus can neglect the effect of the semitransparent gate. If the dynamical conductivity of the electron layer is sufficiently small, i.e., if $\sigma_r,\sigma_i \ll (1+n)$, eq. (2) can be written in a very simple form, namely:

$$-\frac{\Delta T}{T} \equiv 1 - \frac{T(N_s)}{T(0)} \approx \frac{2}{1+n}\sigma_r \ . \tag{3}$$

In this small signal approximation the relative change in transmission $-\Delta T/T$ is directly proportional to the real part of the dynamical conductivity. Thus the relative change in transmission $\Delta T/T$ measured in spectroscopic experiments on electron systems confined to a thin layer rather directly reflects the dynamical conductivity $\sigma(\omega)$ of the electron system. Infrared active elementary excitations in the electron layer will then appear as resonances in the dynamical conductivity $\sigma(\omega)$.

EXCITATIONS OF 2D ELECTRON SYSTEMS IN HETEROJUNCTIONS

Two-dimensional electron systems in single layer heterojunctions have been studied extensively in recent years because of their extremely high mobility at low temperatures. This high mobility which in the GaAs-AlGaAs system has exceeded 2,000,000 cm^2/Vsec makes such hetero-junctions very attractive both from the technological as well as from the fundamental physics point of view. In the following I will use some recent spectroscopic experiments, mainly on 2D electron layers at the GaAs-AlGaAs interface, to illustrate where infrared spectroscopy can help us to increase our understanding of such systems. The fundamental excitations that will be considered are, namely, cyclotron resonance, intersubband resonance between two-dimensional subbands, and 2D plasmons.

Cyclotron Resonance

The first infrared spectroscopic experiment successfully done on a two-dimensional electron system was cyclotron resonance on MOS-structures on Si (100). One reason for that is that cyclotron resonance yields relatively strong signals in straightforward transmission experiments. The interest in cyclotron resonance arises from the fact that it can be used to determine directly the effective mass for motion parallel to the interface. The cyclotron frequency $\omega_c = eB/m$ is an accurate probe to study the effective mass m* for motion in the 2D plane and its dependence on electron density N_s and spacing $\hbar\omega_c$ between Landau levels. The width of the cyclotron resonance and its dependence on sample quality, N_s and magnetic field B can be used to get direct insight into scattering phenomena. However, the strong signals that result from cyclotron resonance limit the usability of cyclotron resonance to examine scattering phenomena in high mobility heterojunctions. Above certain mobilities cyclotron resonance line shapes are saturated and become insensitive to mobility, i.e. scattering time.

Fig. 3 compares a typical transmission spectrum caused by cyclotron resonance in a high mobility GaAs-AlGaAs heterojunction at temperature T=4.2K with a Gaussian and a Lorentzian line profile[14]. Obviously a Lorentzian profile describes the observed line well. Such a Lorentzian line is calculated from eq. (2) if we insert for $\sigma(\omega)$ the classical dynamical conductivity in a magnetic field. For the two opposing circular polarization directions the dynamical conductivity is

$$\sigma^{\pm}(\omega) = \frac{\sigma_0}{(\varepsilon_0/\mu_0)^{1/2}} \quad \frac{1}{1+i(\omega \pm \omega_c)\tau} \quad , \quad \sigma_0 = \frac{N_s e^2 \tau}{m^*} \quad . \tag{4}$$

Using eq. (2) for the calculation of $T^{\pm}$ we find that the cyclotron active

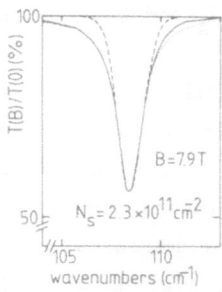

Fig. 3.   Cyclotron resonance signal measured on a GaAs-AlGaAs hetero-
           junction (solid line) at T=2K in comparison with a Gaussian
           (dashed) and a Lorentzian (dotted) line profile (from Ref.
           [14]).

mode σ^+ yields a Lorentzian line. For high mobility samples we obtain,
for the total relative transmission,

$$T(N_s)/T(0) = (1/2)(T^+(N_s)+T^-(N_s))/T(0) \simeq (1/2)(T^+(N_s)/T(0)+1)$$

and thus have the expression that is used to fit the spectrum in Fig. 3
with a Lorentzian line using N_s, m* and τ as adjustable parameters. Note
that in the spectral regime displayed in Fig. 3 T(B)/T(0) at fixed N_s is
identical to $T(N_s)/T(0)$ at fixed B.   Since the observed change in
transmission in Fig. 3 is close to 50% we can expect the cyclotron
resonance signal to be close to saturation, i.e. $T^+(N_s)/T(0)$ close to 0.
Note that in 2D layers, quite in contrast to 3D electron systems
saturation of the cyclotron resonance does not result in a deviation from
a Lorentzian profile. Thus one can only detect saturation from absolute
transmission values and not from the cyclotron resonance line shape. The
absolute transmission values, however, are only reliably compared to
theoretical fits in samples where the electron system is homogeneous and
completely covers the optical path.   In experimental situations where
this is not the case saturation effects may go unnoticed and lead to
misinterpretation of the cyclotron resonance linewidth.

To demonstrate that cyclotron resonance amplitude A and halfwidth b
(full width at half height) in high mobility heterojunctions become
rather insensitive to scattering time we have calculated[14] these
quantities for a single layer GaAs-AlGaAs heterojunction with eqs. (2)
and (4) using $\varepsilon=12.53$ and $m*=0.07m_0$ in their dependence on mobility μ and
density N_s. The result, displayed in Fig. 4, shows that at mobilities
$\mu>500,000$ cm^2/Vsec, which are often achieved in such heterojunctions,
both A and b become less sensitive to variations in μ. With increasing μ
the amplitude approaches 50% whereas the halfwidth approaches a constant

value that only depends on N_s and thus can no longer be used to determine scattering rates.

In heterostructures with intermediate mobility, analysis of the cyclotron resonance lineshape as a function of magnetic field can give valuable information, e.g., on the influence of screening on the scattering rate. This is demonstrated in Fig. 5 which shows a series of cyclotron resonance spectra measured on a single InAs quantum well sandwiched between two layers of GaSb[15]. Both, the cyclotron amplitude and linewidth oscillate strongly as a function of magnetic field. These oscillations reflect the importance of screening. Whenever Landau levels n are completely filled, as indicated by the arrows in Fig. 4., the cyclotron resonance linewidth exhibits maxima. In these cases the density of states at the Fermi energy is low and screening of scattering centers is ineffective. For half-filled Landau levels one has a high density of states at the Fermi energy and scattering centers are strongly screened resulting in a rather narrow cyclotron resonance width. Quantitative analysis of spectra as in Fig. 5 should help in understanding the nature of scattering processes in heterojunctions and quantum wells.

Even in high mobility heterojunctions where the cyclotron resonance linewidth no longer is a good measure of scattering rates cyclotron resonance can always be used to study changes in the effective mass caused, for example, by quantum phenomena, non-parabolicity, and interaction between electrons and optical phonons. To illustrate this

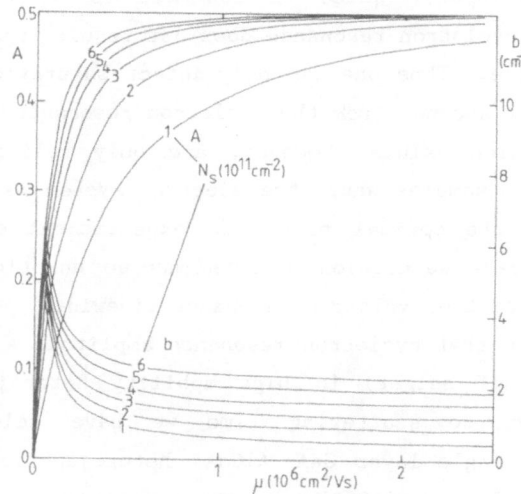

Fig. 4. Calculated resonance amplitude A and halfwidth b for cyclotron resonance excited by linearly polarized light in a single layer GaAs-AlGaAs heterojunction versus mobility μ at various electron densities N_s (from Ref. [14]).

Fig. 6 shows the magnetic field dependence of the cyclotron mass in two high mobility GaAs-AlGaAs heterojunctions[16]. The masses are extracted from the measured resonance frequency $\omega_c = eB/m^*$. At low magnetic field, i.e., high filling factors $\nu = n_s h/eB$, where several spin-split Landau levels are occupied, the cyclotron resonance mass shows oscillatory behavior. These oscillations arise from non-parabolicity and the fact, that only Landau transitions are possible for which the upper or lower Landau level is partially occupied. The experimental data are compared to the masses calculated from the energy spacing of discrete Landau levels. These calculations can semiquantitatively describe the observed oscillatory behavior of m^*. At sufficiently high magnetic fields, where only the lowest Landau level is occupied ($\nu < 2$) one observes a steady increase of m^* with B which is in fair agreement with the field dependence calculated for m^* from the $0 \rightarrow 1$ transition between the lowest two Landau levels. From such cyclotron resonance experiments one can also extract a mass-value for B=0 by extrapolation. The dependence of such a B=0 mass on electron density is derived from several samples with similar depletion charge N_{depl} and displayed in Fig. 7. The observed increase of m^*/m_0 with N_s is well described by non-parabolicity calculated from bulk band parameters of GaAs for a fixed depletion charge N_{depl}. However, band structure effects apparently predominant in the results discussed above, are not the only effects visible in cyclotron resonance.

Both, electron-phonon[17,18] and electron-electron[19] interactions have been shown to influence cyclotron resonance spectra in simple hetero-

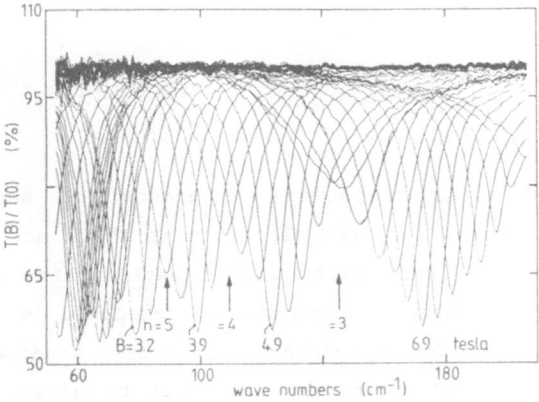

Fig. 5. Measured cyclotron resonance traces on a single InAs quantum well at T=1.8K for magnetic fields B between 2.1 and 8.2T. The arrows indicate where the nth Landau levels are completely filled (from Ref. [15]).

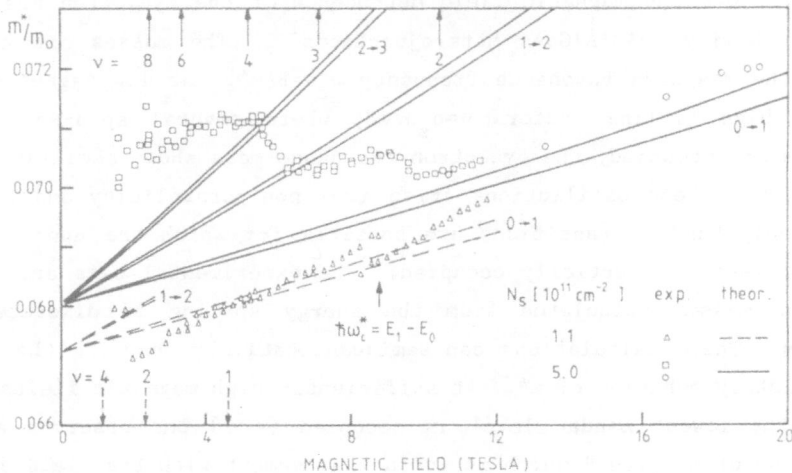

Fig. 6. Magnetic field dependence of measured and calculated cyclotron masses $m*/m_0$ for two GaAs-AlGaAs heterojunctions with electron densities N_s as indicated. The theoretical masses are calculated for transitions between spin-split Landau levels n (n+1) assuming a depletion charge density $N_{depl}=5\times10^{10}cm^{-2}$. Integer filling factors   are indicated at the top and bottom for the two samples, respectively (from Ref. [16]).

junctions on GaAs. At present, one still seems to be quite far from understanding all the details of cyclotron resonance in even the simplest two-dimensional systems.

Intersubband Resonance

Intersubband resonance between two-dimensional subbands tests the binding energies of the confined electrons in their motion perpendicular to the interface. It thus gives direct experimental insight into the confining potential. Whereas intersubband resonance in MOS-structures on Si and III-V compounds has been extensively studied with infrared spectroscopy[1,2,20,21] it is rather difficult to excite directly inter-subband transitions in single layer heterojunctions. This is because in nearly parabolic and isotropic conduction bands as in GaAs heterojunctions one needs the infrared exciting field to be polarized perpendicular to the interface to cause intersubband resonance. Such polarization is difficult to realize in frequency domain spectroscopy with a Fourier transform spectrometer. Excitation mechanisms with light polarized in the interface plane, that have been used successfully on Si-MOS structures[20,21], have not proven to be sufficiently strong in single layer heterojunctions.

However, intersubband resonances in heterojunctions can be detected in a normal transmission arrangement as in Fig. 2 via subband-Landau level coupling in magnetic fields tilted by an angle θ with respect to the interface normal. Such coupling can cause a splitting of the cyclotron resonance when the intersubband resonance energy between the ground and the first subband, E_{10}, equals the cyclotron energy or a multiple thereof. This was first observed by Schlesinger et al.[22] for $\hbar\omega_c = E_{10}$ and subsequently by Wieck et al.[23] also for $2\hbar\omega_c = E_{10}$. The principal mechanism for such coupling is illustrated in Fig. 8. Whenever a Landau level of the ground subband E_0 crosses the lowest Landau level in the first subband E_1 coupling between the two is mediated by a magnetic field component $B_{||}$ parallel to the interface[24]. Whereas the Landau level separation is essentially controlled by the perpendicular field component $B_\perp$, the parallel field component is responsible for the anticrossing at $p\hbar\omega_c = E_{10}$ (p=1,2) and also causes small diamagnetic shifts ΔE_0 and ΔE_1 of the subband energies. The splitting of cyclotron resonance that occurs at $p\hbar\omega_c = E_{10}$ is demonstrated in Fig. 9 for two tilt angles θ. At small tilt angles (e.g., $\theta=12°$) splitting is only observed at $\hbar\omega_c = E_{10}$. At larger tilt angles (e.g. $\theta=33°$) there also occurs splitting at $2\hbar\omega_c = E_{10}$, provided the second Landau level of the ground subband is occupied at this field. The p=2 coupling has the advantage of occurring at lower magnetic fields and thus can be used to measure intersubband energies of many samples with superconducting laboratory magnets.

Results of such a systematic study of intersubband energies in

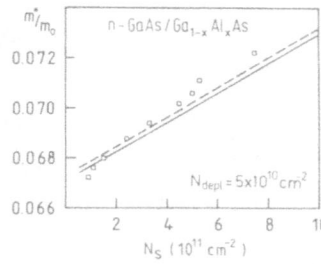

Fig. 7. Measured (squares) and calculated (solid line) cyclotron masses extrapolated for vanishing magnetic field $B \rightarrow 0$ vs electron density N_s. The dashed line is calculated at the Fermi energy from the B=0 dispersion $E(\vec{k})$ (from Ref. [16]).

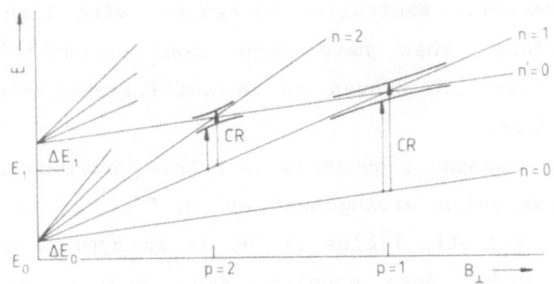

Fig. 8. Schematic diagram of the lowest Landau levels of a 2DES in a
tilted magnetic field. $B_\perp$ is the component of the magnetic
field perpendicular to the interface. The parallel B-component
causes diamagnetic shifts E_0 and E_1 of the subbands E_0 and E_1,
respectively. Anticrossing between originally energetically
degenerate Landau levels n of subband E_0 and n' of subband E_1
causes splitting of the cyclotron resonance (CR), whenever ph $_c$=
$E_1-E_0=E_{10}$ (from Ref. [24]).

single layer heterojunctions are summarized in Fig. 10. In these samples
the electron density has been raised from its original value (filled
symbols) by using the persistent photoeffect (open symbols). The
measured resonance energies are compared with calculations of the subband

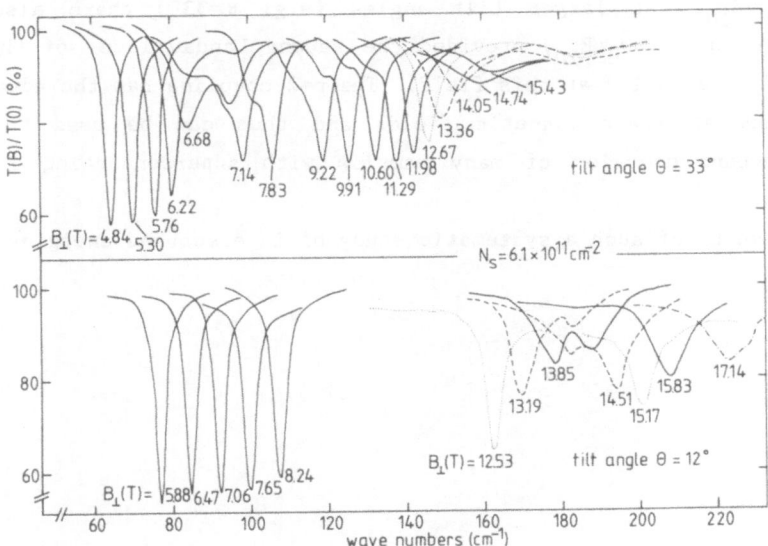

Fig. 9. Cyclotron resonance spectra of a GaAs-AlGaAs heterojunction with
density N_s at T=2K in a magnetic field tilted by an angle   from
the interface normal. The component of the magnetic field perpendicular to the interface, $B_\perp$, is indicated (from Ref. [23]).

splitting E_{10} for various depletion charge densities[25]. The calculated curves reasonably describe the observed N_s-dependence in a given sample. The sample-dependent energies at a given N_s can be attributed to different depletion charges. Such studies show that the confining potential in heterojunctions is rather well described by present theories. However, comparison with the experimental data shows that a precise knowledge of the depletion charge density is necessary to predict correct subband splittings.

2D Plasmons

Two-dimensional plasmons have, in contrast to 3D plasmons, a dispersion relation that strongly depends on wave vector q, even at long wavelengths. There it may be approximated by:[1]

$$\omega_p^2 = \frac{N_s e^2 q}{2\epsilon^* \epsilon_0 m^*} .$$

(5)

Here ϵ^* is an effective dielectric constant that describes the change of the dielectric constant across the interface as well as screening by a nearby metal gate. To excite 2D plasmons with infrared radiation at a well defined wave vector q one usually employs a periodic metal grating deposited, e.g., on the gate of a MOS-structure as shown in Fig. 11. Infrared radiation of wavelength λ incident normally onto such a grating with period $a \ll \lambda$ is spatially modulated in the near field of the grating and thus can couple to plasmons at wave vector $q = 2\pi m/a$ (m=1,2,3,...) and frequency ω_p, provided the grating is sufficiently close to the 2D electron system[3,4].

In Si-MOS structures the plasmon dispersion in 2D systems has been experimentally investigated in its dependence on N_s and q and has been found to be well described by eq. (5) as long as the wavevector q is not too large[1,3,4]. Recent studies of 2D plasmons in heterojunctions on GaAs at relatively large q, i.e. short plasmon wavelength λ_p, have revealed non-local effects. Such non-local effects on the plasmon dispersion are expected to become important when the plasmon wavelength becomes comparable to intrinsic lengths of the 2D system, namely the Fermi wavelength and the screening length[1]. Non-local behavior has a rather pronounced effect on the plasmon dispersion in high magnetic fields. In the presence of a magnetic field the dynamical spatial density modulation of the 2D system by the excitation of plasma waves can give rise to a non-local interaction between the plasmon and harmonics $n\omega_c$ (n≥2) of the cyclotron resonance. The strength of this non-local interaction is governed by the parameter $(\vec{q} \cdot \vec{v}_F / \omega_c)$, where $\vec{v}_F$ is the Fermi velocity. At sufficiently high q this interaction becomes visible in GaAs-AlGaAs heterojunctions by a splitting of the plasmon excitations at fields where $\omega_p \simeq 2\omega_c$[26,27]. This is shown in Fig. 12a where magnetoplasmon resonances are observed on the high field side of cyclotron resonance. In the magnetic field regime where $\omega_p(B)$ is close to $2\omega_c$ the plasmon dispersion, which outside this regime follows the classical expression $\omega_p^2(B) = \omega_p^2(B=0) + \omega_c^2$, splits into two branches ω_1 and ω_2. The strengths of the modes critically depend on their energetical distance from $\omega = 2\omega_c$. In

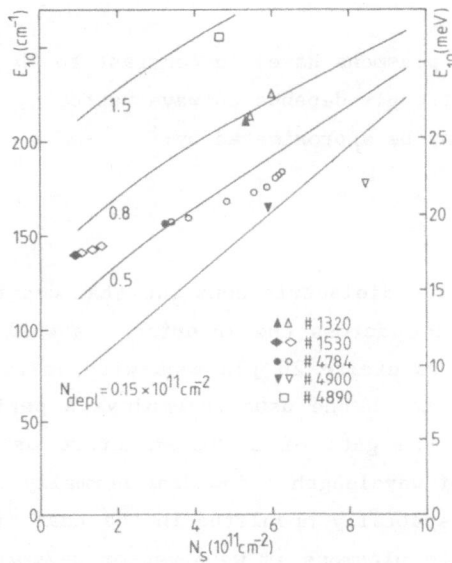

Fig. 10. Intersubband energies $E_1 - E_0 = E_{10}$ vs electron density N_s for various GaAs-AlGaAs single layer heterojunctions. The open symbols denote densities N_s achieved by the persistent photoeffect. The solid lines are subband spacings calculated by Stern and DasSarma[25] for various depletion charge densities N_{depl} (from Ref. [23]).

Fig. 12b the experimentally observed resonance positions and strengths are compared with calculations of the non-local behavior which include the effect of broadening by frequency-independent scattering[28]. The occurrence of a mode splitting in $\omega_p(B)$ and the quantitative agreement between theory and experiment clearly establishes that the non-local effects are as predicted by theory. In the experiment the interaction parameter is still relatively small. The observability of the effect is mainly due to the fact that here the non-local interaction is resonant and can be tuned by the magnetic field. Such an experiment demonstrates that plasmon spectroscopy can give access to lateral length scales in two-dimensional systems.

LATERALLY MICROSTRUCTURED DEVICES FOR SPECTROSCOPY

Since infrared spectroscopy has been used very successfully to study the electronic excitations in two-dimensional systems it is only natural that one wants to extend such studies to laterally confined two-dimensional and finally to quasi one-dimensional electron systems in semiconductors. This requires not only the preparatory techniques needed to confine electron systems to submicron dimensions. One also has to prepare such samples homogeneously on large active areas of typically 3mm in diameter to be able to sufficiently transmit infrared radiation. Both requirements can be satisfied by fabricating laterally periodic structures consisting, e.g., of many narrow and parallel inversion wires spaced at a submicron period evenly over the desired area. This is most easily started by using holographic lithography to expose gratings of submicron periodicity in a photoresist mask on the semiconductor substrate. The basic principle is illustrated in Fig. 13. The expanded beam of a laser operating at blue or ultraviolet wavelengths is symmetrically split into two beams. These are reflected from two symmetrically arranged mirrors and then interfere on the semiconductor substrate to form an intensity grating of period $a=\lambda/(2\sin\vartheta)$, where λ is the exposing wavelength and ϑ the angle of incidence. All structures discussed in the following are exposed with the 458nm line of an argon ion-laser which allows us easily to expose periodic structures down to $a=230$nm. The SEM picture of a typical photoresist grating after suitable exposure and development is shown in Fig. 14a. Such a photoresist

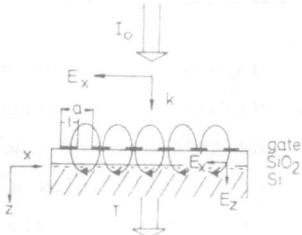

Fig. 11.   Schematic function of a grating coupler. Infrared radiation incident perpendicularly to the grating is spatially modulated in the near field and contains wavevector components $q=m(2\pi/a)$ $(m=1,2,...)$ (from Ref. [21]).

grating is used as a mask for the following preparatory steps. Depending on requirements, these include dry and wet etching processes, metal evaporation under a defined angle with respect to the surface normal ("shadowing") and often "lift-off" techniques to remove the photoresist with a metal cover. With these relatively straightforward techniques one can fabricate a variety of periodic structures at semiconductor interfaces on large areas with high homogeneity and reproducible feature definition with feature sizes down to about 1/10 of the exposing wavelength. A typical MOS-structure that can be fabricated in this manner is schematically shown in Fig. 14b. There the photoresist grating is used as a mask for a reactive ion etch of a thermal oxide of original thickness d_2 on Si. After etching grooves into the oxide such that the remaining thickness is d_1, the whole structure is evaporated with a thin NiCr gate to form a MOS-structure with periodic modulation of the oxide thickness. While geometric feature definition in such structures is not very difficult it is often tedious to find a combination of preparatory processes that assures good electronic properties in such laterally structured devices.

FROM 2D TO 1D ELECTRON SYSTEMS

In an effort to study the transition from 2D to 1D electronic systems we have prepared and studied various kinds of laterally periodic devices either as MOS-structures on Si[29-32] and InSb[10,11,33] or on GaAs-AlGaAs heterojunctions[10,11,30,32,34]. After discussing the effect of a laterally periodical electron density modulation on the plasmon dispersion I want to concentrate on intersubband resonance between 1D subbands observed in extremely narrow inversion channels.

Plasmons in Inversion Layers with Laterally Periodical Density Modulation

Plasmons in inversion layers with laterally periodical density modulation have first been studied on MOS-structures as introduced in Fig. 14b. In such devices the thickness of the gate oxide is periodically modulated with submicrometer periodicity a. With a positive gate voltage applied between the NiCr gate and the Si substrate one thus induces an electron inversion layer which has a high local density N_{s1} under the thin oxide (d_1) of width t_1 and a lower density N_{s2} under the thicker oxide (d_2). The infrared response of such an inversion layer with periodic N_s-modulation and sample parameters d_1=35nm, d_2=60nm,

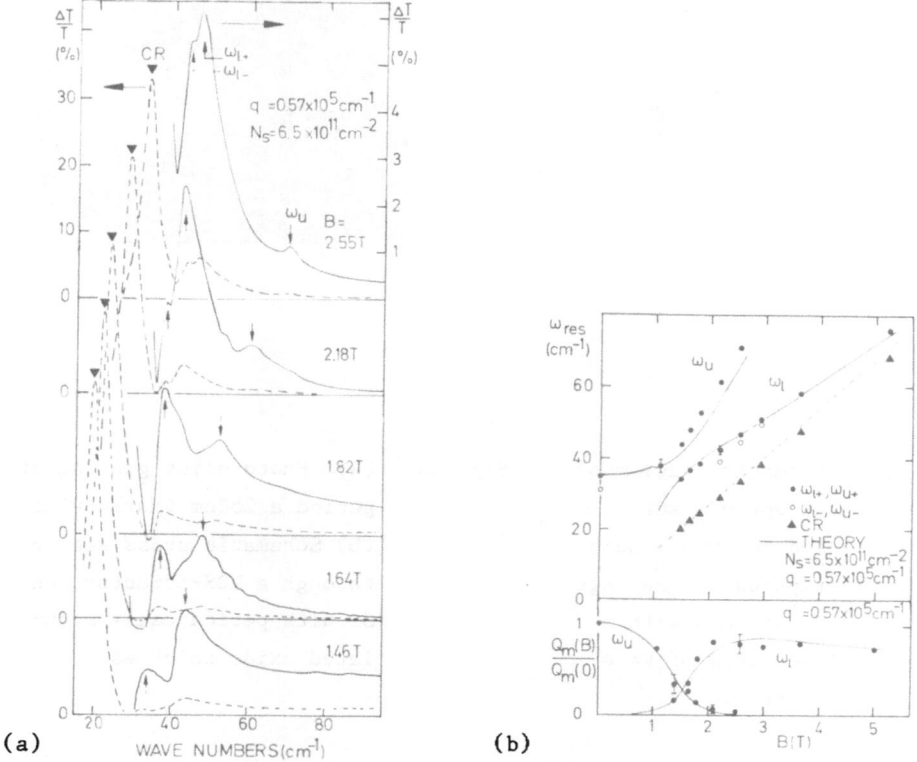

(a)

(b)

Fig. 12. Non-local behavior of magnetoplasmons in a GaAs-AlGaAs heterojunction with electron density $N_s = 6.5 \times 10^{11} cm^{-2}$. The spectra in
(a) show strong cyclotron resonances (CR, dashed line) and, at
higher frequency, (solid line, enlarged) magnetoplasmons at
wavevector q, that are split into modes ω_l and ω_u by non-local
interaction with the cyclotron resonance harmonic $2\omega_c$. The
resonance frequencies and strengths of the magnetoplasmons are
shown in (b) and are well described by the theory of Chaplik
and Heitmann[28] (solid line). Also shown are the cyclotron
resonance frequencies (triangles) which deviate from the
straight dashed line because of non-parabolicity (from Ref.
[26]).

a=650nm and $t_1/a \simeq 0.3$ measured at low temperatures (T$\simeq$10K) for various
gate voltages V_g above inversion threshold is shown in Fig. 15a[29]. For
light polarized perpendicular to the oxide grating (solid lines) one
observes strong plasmon excitations at wavevector q=$(2\pi/a)$m with m=1,2
which are not excited for parallel polarization (dashed line for the 4V
curve). Depending on sample parameters, namely t_1/t_2, these plasmons are

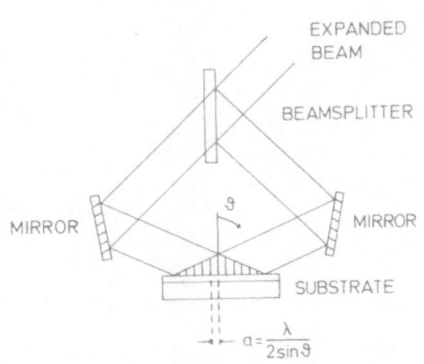

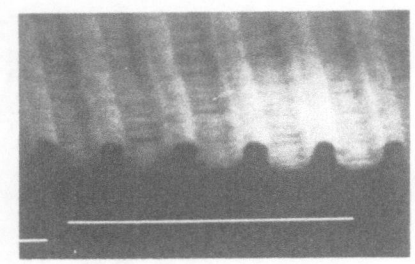

(a)

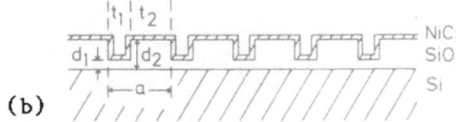

(b)

Fig. 13. Schematic diagram of
the optical setup
used to expose holo-
graphically gratings
of period a with a
laser beam of wave-
length λ.

Fig. 14. (a) Photoresist grating of
period a 250nm (Marker=1μm).
(b) Schematic cross section
through a MOS-structure on
Si with periodically modulated oxide thickness.

split into two branches ω^- and ω^+. Whereas the splitting of the m=1
plasmon is quite strong in Fig. 15a, where t_1/t_2 0.4, it is not observed
on samples with t_1/t_2 1.

The observed plasmon splittings find their natural explanation in
the effect of the lateral density "superlattices" on to the plasmon dispersion as schematically illustrated in Fig. 15b. The periodic density
modulation $N_s(x)$ of period a causes minigaps in the plasmon dispersion at
the artificial Brillouin zone boundaries $\pm n(\pi/a)$ (n=1,2,3,...), provided
that the nth Fourier expansion coefficient N_n of $N_s(x)$ is sufficiently
large that the plasmon splitting is larger than the plasmon width caused
by damping. In the sample geometry used here both the periodic oxide
modulation and the periodic density modulation can serve as "grating
coupler" to excite plasmons at q=m(2π/a). Thus one only sees even
minigaps, i.e. n=2 and n=4. In lowest order perturbation theory the
predicted plasmon splitting[35] is,

$$[\omega_+^2(n\pi/a)-\omega_-^2(n\pi/a)]/\omega_p^2(n\pi/a)= 2(N_n/\bar{N}_s) \ ,$$

where $N_s=(N_{s1}t_1+N_{s2}t_2)/a$ is the average charge density and ω_p the
unperturbed plasmon frequency that is obtained from eq. (5) by replacing
N_s with $\bar{N}_s$. This explains why we observe no splitting at m=1, i.e. n=2,

when t_1/t_2 1, i.e. N_2 0. Another point worth mentioning is that with sufficiently strong density modulation the electron system itself may dominate as "grating coupler". In this case zone-folding of the plasmon dispersion will make the modes at $q=2\pi/a$, $4\pi/a$,... zone-center plasmons

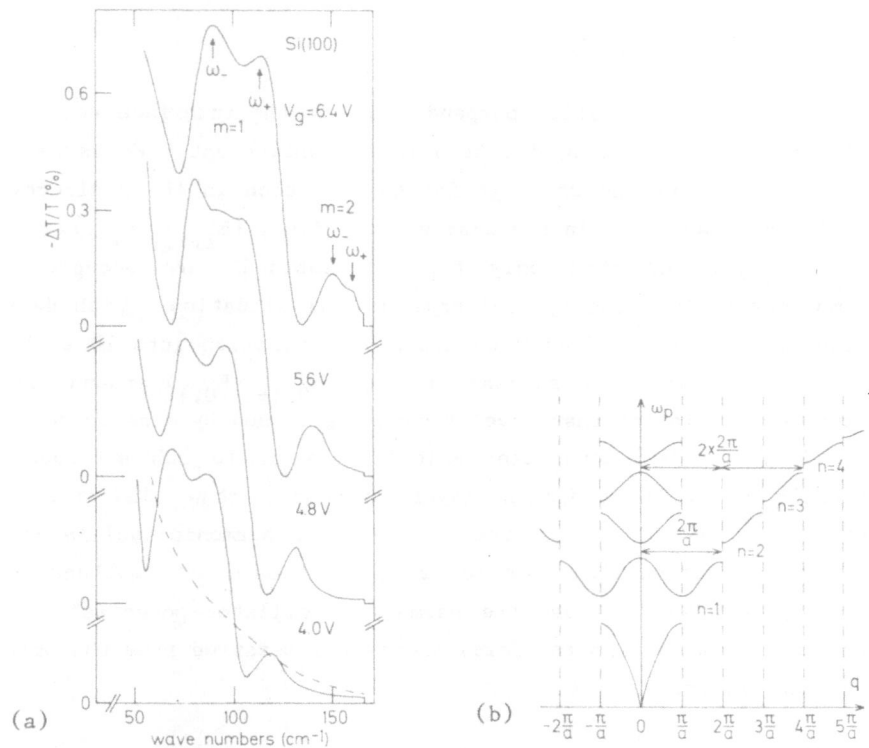

(a)   (b)

Fig. 15.   Plasmons in an inversion layer with laterally periodical
           density modulation of period a. The plasmon spectra (a) are
           measured on a MOS-capacitor with modulated oxide thickness at
           T=10K for various gate voltages V_g above inversion threshold
           and with light polarized perpendicularly to the oxide grating
           except for the dashed curve at 4.0V, where parallel
           polarization is used. The effect of the density modulation on
           the plasmon dispersion is schematically shown in (b) (from Ref.
           [29]).

which appear as q=0 modes. The above discussion shows that plasmons may serve as probes to study the Fourier-components of the density modulation in laterally periodic systems. Using additional grating couplers with period b≈2a such studies have recently been extended to the zone edge, i.e. coefficients N_n with odd n[36].

Intersubband resonance between 1D subbands

A transition from 2D to 1D electronic behavior will occur when lateral confinement causes quantization of the originally 2D subbands into 1D subbands with energies

$$E(k) = \frac{\hbar^2 k_y}{2m_y} + E_{ij} \; . \tag{6}$$

Here i denotes quantization perpendicular to the interface and j quantization in the x-direction due to lateral confinement. We assume in the following that subband spacings for quantization in the z-direction are much larger than those in the x-direction, i.e., $(E_{i+1,j} - E_{i,j}) \gg (E_{i,j+1} - E_{i,j})$, and that only the i=0 subbands are occupied. Both assumptions reflect the typical experimental situation. With decreasing lateral confinement widths W we expect quantization into 1D subbands to become visible when the subband spacing $(E_{0,j+1} - E_{0,j})$ around the Fermi energy becomes larger than level broadening caused by elastic scattering, i.e., $\Gamma = h/2\tau$. Self-consistent calculations of 1D subband spacings in laterally confined inversion layer systems[37] show that the lateral confining potential is somewhere between a harmonic-oscillator and a square-well potential. For a rough estimate of subband spacing $E_{0,j+1} - E_{0,j} = \hbar\Omega$ we thus use the harmonic oscillator potential. In this approximation we obtain the Fermi energy E_F, measured from the bottom of the potential well as

$$\frac{m^*}{8} \Omega^2 W^2 = E_F \; , \tag{7}$$

where W is the full width of the confining potential at the Fermi energy. If several 1D subbands are occupied we can assume that for a given electron density N_{sc} in the inversion channel the Fermi energy may be approximated by the relation valid for a 2D electron system, namely $E_F \approx \pi\hbar^2 N_{sc}/(m^* g_v)$, where g_v is the valley degeneracy. For not too low electron densities N_{sc}, i.e., sufficiently high E_F, we thus can estimate Ω as:

$$\Omega \simeq \frac{\hbar}{m^* W} \, (8\pi N_{sc}/g_v)^{1/2} \; . \tag{8}$$

It is worthwhile to note that, for a square well potential and subband indices j≥4, Ω approximately shows the same dependences on W, N_{sc} and m^* as eq. (8) but is numerically about a factor 1.6 larger. Eq. (8) is only valid at sufficiently high electron densities N_{sc} in the inversion channels. To have an estimate of the subband spacing $\hbar\Omega$ at very low N_{sc}, where only the ground subband is occupied, we can use for the harmonic oscillator potential $E_F \simeq \hbar\Omega/2$ and thus have

$$\Omega \simeq \frac{4\hbar}{W^2 \cdot m^*} \; . \tag{9}$$

For an infinite square well we obtain for $(E_{01}-E_{00})/\hbar$ a similar result which is a factor of $3\pi^2/8$ larger than eq. (9). To observe quantization into 1D subbands by intersubband spectroscopy one needs to have $\hbar\Omega > \Gamma$. One thus wants structures with relatively high mobility and low effective mass of the inversion electrons so that W needs not to be too small. This makes heterojunctions on GaAs particularly suited. In addition, one has to avoid large variations of W over the sample area to limit inhomogeneous broadening of the intersubband resonance. Quite in contrast to transport experiments[8,9] there are no stringent requirements on sample temperature since Fermi level smearing does not seriously affect the observability of intersubband resonances.

We have prepared in our laboratory a variety of periodic structures of inversion channels of submicron width of GaAs, Si and InSb. Most suited are devices in which the inversion electron density in the channel can be controlled by field effect. Fig. 16 shows schematically two such structures that have been successfully used to observe intersubband transitions between 1D subbands. In the GaAs device (Fig. 16a) we start with a relatively standard single layer of highly mobile inversion electrons at the GaAs-AlGaAs interface. On the surface we define a photoresist grating of periodicity a and cover the structure with a semitransparent NiCr gate. By applying a negative gate bias with respect to the inversion layer we can predominantly deplete the stripes in the 2D channel where the gate is closest to the channel. Below a certain threshold voltage V_d these are fully depleted and narrow inversion channels remain under the photoresist stripes. Decreasing the bias to more negative values decreases the local density N_{sc} in these channels as well as their effective width W. The Si-device, shown in Fig. 16b, is a rather standard MOS-capacitor with a typically 20nm thick thermal oxide and a semitransparent grating of period a and width t, serving as gate.

The GaAs device in Fig. 16a has proven very effective since it allows one to examine critically the transition from 2D to 1D electronic behavior[19]. To demonstrate this we discuss, in the following spectra measured on a heterojunction with a resist grating of period a=500nm and heights 120nm. At V_g=0V and T=4.2K there is a 2D inversion layer of density N_s=6x10^{11}cm^{-2} and mobility μ=80,000 cm^2/Vsec at the GaAs-AlGaAs interfaces, as derived from the analysis of the SdH oscillations. With decreasing V_g we find from SdH analysis that predominantly the stripes between the resist grating are depleted approximately linear with V_g. At V_g=-0.5V they are fully depleted and isolated inversion channels are left below the resist stripes. The change in the infrared excitations spectra is best demonstrated in a finite magnetic field with radiation polarized perpendicular to the grating, as shown in Fig. 17a. At V_g=0V, i.e. for the 2D inversion layer, we observe two kind of excitations, a strong cyclotron resonance at lower frequencies excited by the spatially uniform (q=0) infrared field and a 2D plasmon at higher frequencies excited at q=2π/a by the infrared field that is spatially modulated by the periodic grating gate structure. At the magnetic field used for Fig. 17a the magnetoplasmons lie at about twice the cyclotron frequencies and thus are split by non-local interaction with the cyclotron resonance harmonic as discussed before. At V_g=V_d=-0.5V, where isolated inversion channels are formed, the cyclotron resonance signal has disappeared. Instead we observe a single strong resonance that is cyclotron shifted with dispersion $\omega_{res}^2 = \omega_0^2 + \omega_c^2$, where ω_0 is its B=0 frequency. Note that this resonance is comparable in strength with cyclotron resonance and much stronger than the V_g=0V plasmons. If we further decrease V_g this resonance moves upward in frequency. In sufficiently high magnetic fields the single resonance at $V_g \leq V_d$ is excited with about equal strength for both polarization parallel and perpendicular to the grating and thus is a q=0 excitation. Fig. 17b shows the resonance signal at B=0 for gate voltage V_g above (lower part) and below (upper part) V_d. At B=0 the resonances are observed only for perpendicular polarization. The resonance at V_g=0V is the 2D plasmon at q=2π/a and is well described by eq. (5) if we use N_s=6x10^{11}cm^{-2}, the value derived from SdH analysis, m*=0.07m$_0$, and ε*=13.6, a value that reasonably corrects for screening of the metal gate. With V_g decreasing from 0V, at first the plasmon resonance shifts to lower frequencies and broadens. This shows that the average areal density $\bar{N}_s$ decreases. The broadening may be caused by the onset of an unresolved minigap in the plasmon dispersion at q=2π/a. At V_g=V_d the resonance suddenly starts to increase in frequency with decreasing V_g, i.e., decreasing average density $\bar{N}_s$. Simultaneously it becomes significantly stronger than the V_g=0V plasmon resonance as long as $\bar{N}_s$ is not too low. With decreasing V_g, i.e. decreasing $\bar{N}_s$, it approaches a finite frequency.

We have identified the resonance at $V_g \leq V_d$ as the intersubband resonance between 1D subbands in the narrow inversion channels[10]. The sudden change in character that occurs with the infrared excitations as

V_g is lowered below V_d, where isolated inversion channels are formed, is also documented by the observed resonance positions $\bar{\nu} = \omega/(2\pi c)$ that are shown in Fig. 18 for B=0 and two finite magnetic fields. At B=0 the gate voltage dependence of $\bar{\nu}^2$ exhibits a sudden change in slope. At finite B we find cyclotron resonance and plasmon excitations at $V_g > V_d$ but only a single resonance at $V_g = V_d$. Non-local coupling of the plasmon mode to the 2nd harmonic of the cyclotron frequency is only observed above V_d, as

(a)

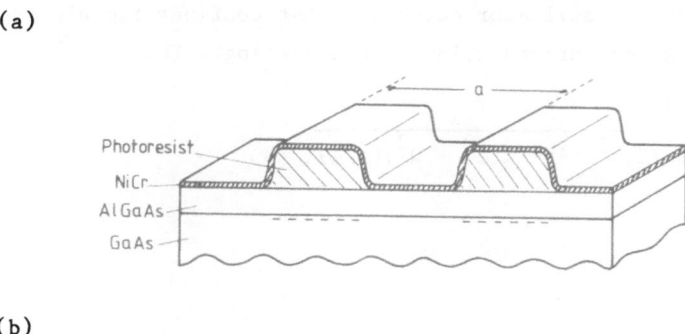

(b)

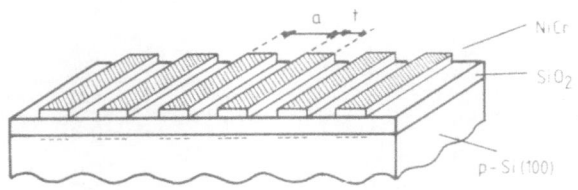

Fig. 16. (a) Schematic view of laterally microstructured field effect devices on a GaAs-AlGaAs heterostructure (a) and a Si-MOS structure (b). The location of the isolated electron channels is indicated by dashes (from Ref. [32]).

seen at the resonance positions for B=2.19T. At B=0 and $V_g > V_d$, the squared plasmon frequency decreases linearly with decreasing gate voltage. This implies that the average density $\bar{N}_s$ decreases in proportion to V_g. We can use this to determine $\bar{N}_s$ at $V_g = V_d$. At $V_g \leq V_d$ we can write $\bar{N}_s = N_{sc}(W/a)$ and thus can reliably estimate W at $V_g = V_d$ with the known value of $N_{sc} \simeq 5.4 \times 10^{11} \text{cm}^{-2}$ to be W≈160nm. The value of $\bar{N}_s(V_g = V_d)$ is found to be consistent with the one derived from the strength of the observed cyclotron resonance signal at $V_g > V_d$. With the above values of W and N_{sc} we estimate with eq. (8), at $V_g = V_d$, a subband spacing of

$\hbar\Omega\simeq2.5\,$meV which is remarkably close to the observed value of $\hbar\Omega\simeq3.7\,$meV. Also, the observed behavior of the resonance frequency with decreasing $V_g<V_d$ is qualitatively as expected for 1D intersubband resonance. Another convincing proof that we are observing 1D intersubband resonances is the fact that the resonance energy approaches a finite frequency as N_s approaches zero. If we use this frequency ($\bar{\nu}\simeq47\,$cm^{-1}) and eq. (9) we estimate that the effective channel width may become as narrow as 30nm.

At B=0 and $V_g\leq V_d$ we can describe the observed resonance signal with the classical expression for the normalized dynamical conductivity $\sigma_{xx}(\omega)$ of a simple harmonic oscillator potential that confines the electronic motion in x-direction perpendicular to the grating. This is:[34]

$$\sigma_{xx}(\omega) = \frac{\bar{\sigma}_0}{(\varepsilon_0/\mu_0)^{\frac{1}{2}}} \frac{1}{1 - i(\omega^2 - \Omega^2)(\tau/\omega)} \qquad (10)$$

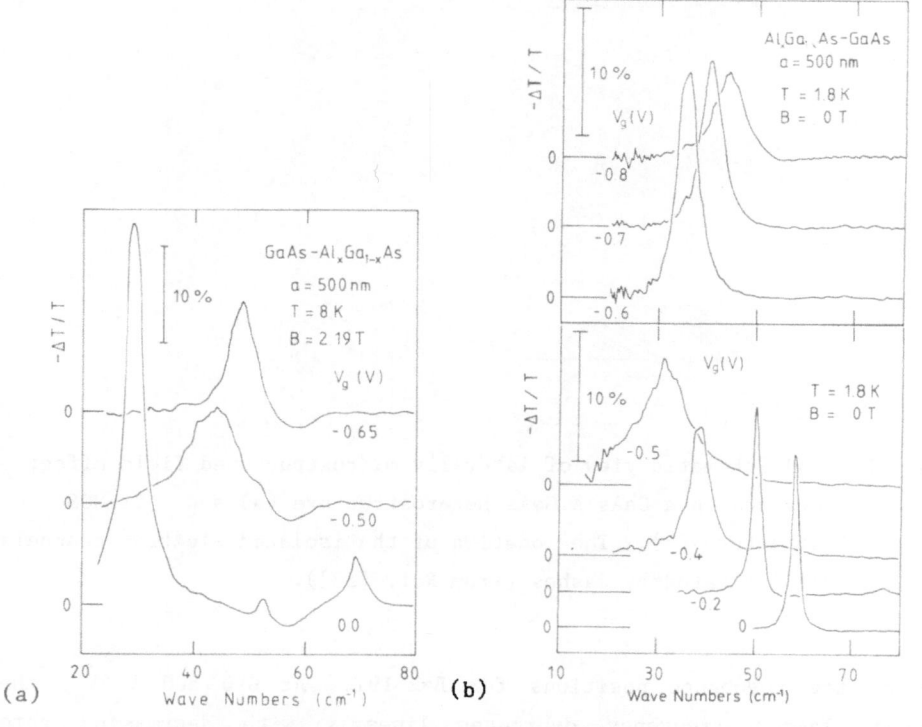

Fig. 17. Infrared excitations in a laterally microstructured GaAs-AlGaAs field effect device as sketched in Fig. 16 (a) for polarization perpendicular to the grating. Representative spectra at different gate voltages V_g are shown in (a) with a magnetic field B applied perpendicularly to the inversion layer, in (b) for B=0. The lower part in (b) represents spectra in the 2D, the uppper part in the 1D regime (from Ref. [32]).

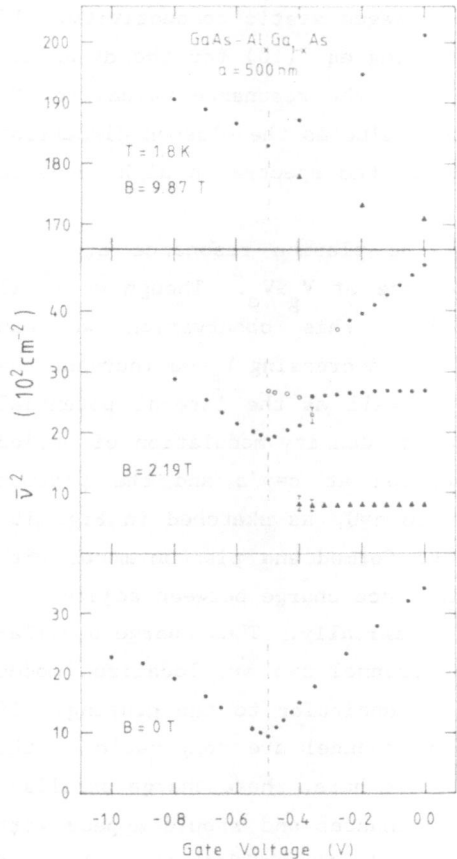

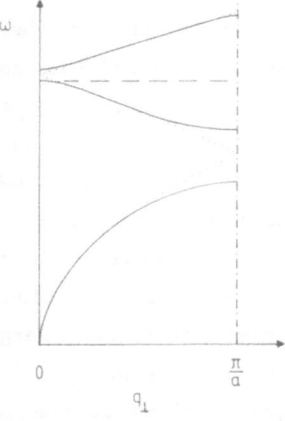

Fig. 18. Square of the resonance positions vs gate voltage with and without magnetic field B. The perpendicular dashed line indicates where crossover from 2D to 1D behavior occurs. In the 2D regime cyclotron resonances are marked by triangles, strong plasmon resonances by full circles, and weak ones by open circles (from Ref. [32]).

Fig. 19. Excitation frequencies in an electron layer with lateral N_s-modulation of period a vs. wavevector $q_\perp$ perpendicular to the direction of uniform density. The dotted line gives the zone-folded 2D plasmon dispersion. With increasing N_s-modulation mini-gaps appear at the zone boundary and in the zone center (solid line). Once isolated channels are formed plasmons at q=0 transform into 1D inter-subband resonances (dashed line).

where $\sigma_0 = N_s e^2 \tau / m^*$ is an appropriately averaged static conductivity. If we fit the B=0 resonances with eq. (2) using eq. (10) for the dynamical conductivity we obtain from the strength of the resonance values for N_s that at $V_g = V_d$ give approximately the same value as the plasmon dispersion [eq. (5)] and a cyclotron resonance fit to the spectra in high magnetic fields.

At first, it is surprising that the plasmon resonance at $V_g > V_d$ transforms into a 1D intersubband resonance at $V_g \leq V_d$. Though we still lack a quantitative theory to explain this observation we can qualitatively describe it as follows. With decreasing V_g we increase the laterally periodic density modulation as well as the lateral potential modulation. As discussed above, sinusoidal density modulation of period a causes a gap in the plasmon dispersion at $q = \pi/a$ and the plasmon observed at $q = 2\pi/a$ is zone-folded back to $q = 0$, as sketched in Fig. 19. At $V_g \leq V_d$ isolated inversion channels are formed and plasmon modes with arbitrary $0 < q < \pi/a$ are no longer possible since charge between adjacent channels can no longer be interchanged dynamically. Thus charge oscillations are only possible within a given channel and are localized modes with no dispersion for the direction perpendicular to the grating. If the lateral quantization energies in the channel are comparable to the observed transition energies, as is the case here, these charge oscillations are indeed the 1D intersubband resonances and should appear with frequency close to the zone-center plasmon in the folded zone scheme of Fig. 19, as indicated by the dashed horizontal line. If, at sufficiently large values of the lateral confinement length W, the subband spacings become negligibly small, the 1D intersubband resonance transforms into a classical depolarization resonance at frequency ω_d given by[34,38]

$$\omega_d^2 = \frac{2N_{sc} e^2}{\epsilon^* \epsilon_0 m^* W} \tag{11}$$

As in the case of 2D intersubband resonances[1] the 1D intersubband resonances observed here should be shifted from the subband spacing by depolarization and exciton-like effects. However, lateral quantization, screening by the metallic gate, as well as interaction effects between neighbouring channels will significantly reduce the real depolarization shift to a value which is much smaller than the classical value, eq. (11). Here, theoretical work is necessary to be able to predict realistically the difference between the 1D subband spacing and the energy of the 1D intersubband resonance. Comparison to what is known about such shifts in 2D systems[1] leads us to believe that also in 1D subband spectroscopy the difference between the measured transition energy and the subband spacing is moderate and grossly overestimated by eq. (11).

The laterally microstructured GaAs-AlGaAs heterojunction discussed above has proven nearly ideal to study the transition from 2D to 1D electronic behavior and to understand how 2D plasmon modes transform into 1D intersubband resonances. Laterally periodic MOS-structures on Si, as sketched in Fig. 16b, are well suited to study the influence of a depletion potential on the 1D intersubband resonance. To show this, I want in the following to present some recent results on such devices fabricated on p-Si (100) with a thermal oxide of 20nm thickness and a grating gate with period a=230nm and typical stripe width t=80nm. In such MOS-capacitors we apply the gate voltage V_g between the gate and a substrate contact and only can charge an electron inversion layer at low temperatures when we illuminate the sample with band gap radiation to generate minority carriers. After charging, we can switch off the band gap radiation to establish inversion conditions. In the dark we then can raise the gate voltage by an amount V_{SB} without noticeably changing the inversion charge. The voltage V_{SB} acts as substrate bias voltage and just increases the depletion field near the Si-SiO$_2$ interface. For laterally homogeneous systems this technique has been used previously and is discussed in more detail by Batke and Heitmann[12]. We characterize our laterally periodic MOS capacitors by quasi-static magnetotransport experiments at radio frequencies using capacitive source-drain contacts[31]. From these experiments we find that the inversion threshold is typically V_t=-0.6V and that the charge density in the channels is nearly proportional to (V_g-V_t) with $N_{sc} \simeq 1.1 \times 10^{12}(V_g-V_t)$cm^{-2}V^{-1}. Typical peak mobilities at low temperatures (T≤10K) are found to be 3000cm^2/Vsec and occur at $(V_g-V_t) \simeq 1V$.

Spectra of 1D intersubband resonances in these devices are shown in Fig. 20a at a fixed value of the substrate bias voltage V_{SB} and for various gate voltages above inversion threshold[32]. One significant difference in comparison to the GaAs-AlGaAs heterojunction discussed above is that here both the average density $\bar{N}_s = N_{sc}(W/a)$ and the amplitude of the lateral potential modulation decrease with decreasing (V_g-V_t). Therefore we find here that the B=0 resonance frequency ω_0 decreases with decreasing $\bar{N}_s$, whereas it increases with decreasing $\bar{N}_s$ in the GaAs-AlGaAs devices. Also, to achieve comparable resonance energies as in the GaAs-AlGaAs structures we have to use significantly smaller lateral dimensions to compensate for the larger effective mass m*=0.2m$_0$ (see eqs. (8) and (9)). All other features of the resonances as in Fig. 20a, such as the dependence on magnetic field and the polarization characteristics,

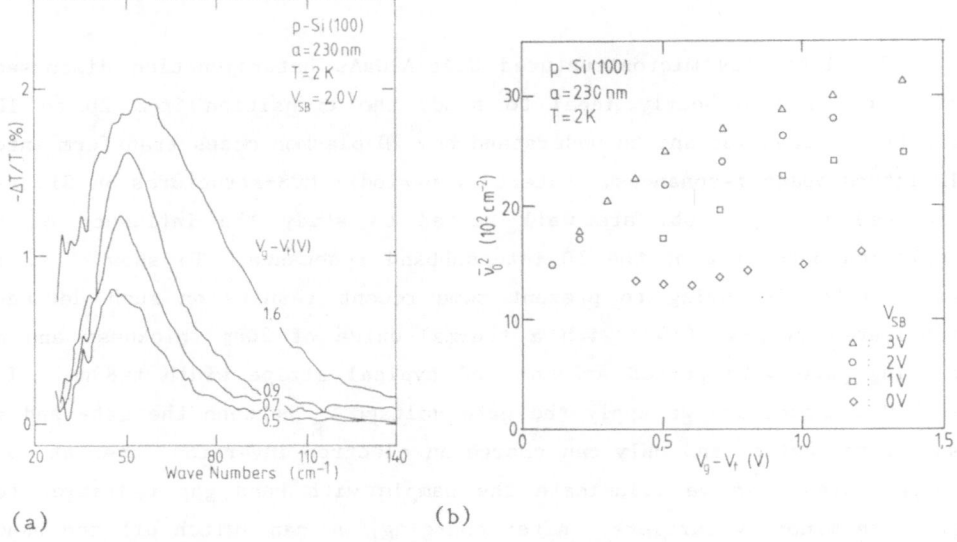

(a)　　　　　　　　　　　　　　　(b)

Fig. 20.　1D intersubband resonances in a laterally microstructured
MOS-device as sketched in Fig. 16 (b). Spectra in (a) are
taken at fixed substrate bias voltage V_{SB} and various gate
voltages V_g above inversion threshold V_t. Squared resonance
frequencies $\bar{\nu}_0$ are displayed in (b) vs V_g-V_t at various values
of V_{SB} (from Ref. [32]).

are essentially the same as discussed for the GaAs-AlGaAs heterojunctions. The dependence of the B=0 resonance frequency $\bar{\nu}_0$ on gate voltage
and substrate bias voltage is summarized in Fig. 20b. As expected here,
the resonance frequencies decrease with decreasing V_g, i.e. decreasing
N_{sc}, but extrapolate to a finite frequency. Application of a substrate
bias field apparently increases lateral confinement and shifts the 1D
intersubband resonance to higher frequencies, at least at not too low
values of $(V_g$-$V_t)$. Note that except at very low values of both V_{SB} and
$(V_g$-$V_t)$ (both < 1V) the strength of the resonance which is a measure of
$\bar{N}_s$ is not strongly affected by substrate bias. Since we lack a
quantitative description of the influence of the substrate bias field on
the resonance frequencies $\bar{\nu}_0$ we can only compare the observed frequencies
$\bar{\nu}_0$ at V_{SB}=0V with model calculations. The intersubband energies $\hbar\omega_0$
observed at V_{SB}=0V are comparable to the ones estimated from eq. (9) or
from self-consistent calculations on similar structures[37] to within a
factor of two. A more quantitative comparison is only reasonable if
self-consistent calculations are carried out for devices similar to the
ones used in our experiments.

From fits of the observed resonances with eqs. (2) and (10) we can extract a phenomenological scattering time τ which is typically 30% smaller than the value derived from magnetoconductance studies. Since the 1D intersubband resonances are most likely significantly inhomogeneously broadened, because of variations of the effective confinement width W over the sample area, we cannot derive any meaningful information on scattering processes in quasi-1D systems from such observations, except that we can use the observed linewidth to give a lower limit for the momentum scattering time in 1D systems.

We have also observed 1D intersubband resonances in laterally microstructured MOS-devices on InSb. Qualitatively, we observe the same features that we find in MOS-structures on Si but achieve higher subband spacings because of the relatively low effective mass in InSb[10,33]. There, additional complications arise because several 2D subbands are likely to be occupied. More details on the InSb structures can be found in a recent summary by Merkt et al.[33].

CONCLUSION

I have tried to illustrate how infrared spectroscopy can be used effectively to investigate the electronic excitations in semiconductor systems with reduced dimensionality. For two-dimensional systems I have only been able to show a very small fraction of the many interesting spectroscopic studies that have been carried out during the last 14 years and that have significantly increased our understanding of such systems. Now that the transition from 2D to 1D systems has been realized and proven by spectroscopic studies I expect that infrared spectroscopy will continue to give us a detailed insight into the electronic excitations of one-dimensional inversion channels and laterally periodic microstructures. Many exciting studies seem possible, ranging from dispersion of 1D plasmons[39] to the realization of the Bloch oscillator or related emission devices that make use of the laterally periodic microstructures. The preparatory control of laterally periodic structures now seems sufficient to start such new directions.

ACKNOWLEDGEMENTS

This summary has been based on many collaborative studies. I want to thank all my coauthors in these studies, as identified in the reference section, for their valuable contributions. I also wish to acknowledge continuous financial support by the Deutsche Forschungsgemeinschaft and the Stiftung Volkswagenwerk.

REFERENCES

1. T. Ando, A. B. Fowler, and F. Stern, Electronic properties of two-dimensional systems, Rev. Mod. Phys. 54:437 (1982).
2. J. F. Koch, Spectroscopy of surface space charge layers, Surf. Sci. 58:104 (1976).
3. D. C. Tsui, S. J. Allen, Jr., R. A. Logan, A. Kamgar and S. N. Coppersmith, High frequency conductivity in silicon inversion layers: Drude relaxation, 2D plasmons and minigaps in a surface superlattice, Surf. Sci. 73:419 (1978).
4. J. P. Kotthaus, High-frequency magnetoconductivity in space charge layers on semiconductors, Surf. Sci. 73:472 (1978).
5. A. B. Fowler, G. L. Timp, J. J. Wainer and R. A. Webb, Observation of resonant tunneling in silicon inversion layers, Phys. Rev. Lett. 57:138 (1986).
6. W. J. Skocpol, P. M. Mankiewich, R. E. Howard, L. D. Jackel, T. M. Tennant and A. D. Stone, Universal conductance fluctuations in silicon inversion-layer nanostructures, Phys. Rev. Lett. 56:2865 (1986).
7. S. B. Kaplan and A. Hartstein, Universal conductance fluctuations in narrow Si accumulation layers, Phys. Rev. Lett. 56:2403 (1986).
8. A. C. Warren, D. A. Antoniadis and H. I. Smith, Quasi one-dimensional conduction in multiple, parallel inversion lines, Phys. Rev. Lett. 56:1858 (1986).
9. K. F. Berggren, T. J. Thornton, D. J. Newson and M. Pepper, Magnetic depopulation of 1D subbands in a narrow 2D electron gas in a GaAs:AlGaAs heterojunction, Phys. Rev. Lett. 57:1769 (1986).
10. W. Hansen, M. Horst, J. P. Kotthaus, U. Merkt, C. Sikorski and K. Ploog, Intersubband resonance in quasi one-dimensional inversion channels, Phys. Rev. Lett. 58:2586 (1987).
11. J. P. Kotthaus, Infrared spectroscopy of lower dimensional electron systems, Physica Scripta, in press.
12. E. Batke and D. Heitmann, Rapid-Scan Fourier transform spectroscopy of 2D space charge layers in semiconductors, Infrared Phys. 24:189 (1984). 13. G. Abstreiter, J. P. Kotthaus, J. F. Koch and G. Dorda, Cyclotron resonance of electrons in surface space charge layers on Si, Phys. Rev. B14:2480 (1976).

14. A.D. Wieck, Ph. D. thesis, University of Hamburg 1987, unpublished.

15. D. Heitmann, M. Ziesmann and L.L. Chang, Cyclotron-resonance oscillations in InAs quantum wells, Phys. Rev. B34:7463 (1986).

16. F. Thiele, U. Merkt, J. P. Kotthaus, G. Lommer, L. Malcher, U. Rössler and G. Weimann, Cyclotron resonance masses in n-GaAs/Ga$_{1-x}$Al$_x$As heterojunctions, Solid State Commun. 62:841 (1987).

17. M. Horst, U. Merkt, W. Zawadzki, J. C. Maan and K. Ploog, Resonant polarons in a GaAs-GaAlAs heterostructure, Solid State Commun. 53:403 (1985).

18. M. A. Brummell, R. J. Nicholas, M. A. Hopkins, J. J. Harris and C. T. Foxon, Modification of the electron-phonon interactions in GaAs-GaAlAs heterojunctions, Phys. Rev. Lett. 58:77 (1987).

19. Z. Schlesinger, W. I. Wang and A. H. MacDonald, Dynamical conductivity of the GaAs two-dimensional electron gas at low temperature and carrier density, Phys. Rev. Lett. 58:73 (1986).

20. A. D. Wieck, E. Batke, D.Heitmann and J. P. Kotthaus, Parallel excitation of hole and electron intersubband resonances in space charge layers on silicon, Phys. Rev. B30:4553 (1984).

21. D. Heitmann and U. Mackens, Grating coupler induced intersubband resonances in electron inversion layers of silicon, Phys. Rev. B33:8269 (1986).

22. Z. Schlesinger, J. C. M. Hwang and S. J. Allen, Jr., Subband-Landau level coupling in a two-dimensional electron gas, Phys. Rev. Lett. 50:2098 (1983).

23. A. D. Wieck, J. C. Maan, U. Merkt, J. P. Kotthaus, K. Ploog and G.Weimann, Intersubband energies in GaAs-Ga$_{1-x}$Al$_x$As hetero-junctions, Phys. Rev. B35:4145 (1987).

24. U. Merkt, Spectroscopy of inversion electrons on III-V semiconduc-tors, in: "Festkörperprobleme: Advances in Solid State Physics", Vol. 27, P. Grosse, ed., Vieweg, Braunschweig (1987).

25. F. Stern and S. Das Sarma, Electron energy levels in GaAs-Ga$_{1-x}$Al$_x$As heterojunctions, Phys. Rev. B30:840 (1984).

26. E. Batke, D. Heitmann, J. P. Kotthaus and K. Ploog, Nonlocality in the two-dimensional plasmon dispersion, Phys. Rev. Lett. 54:2367 (1985). 27. E. Batke, D. Heitmann and C. W. Tu, Plasmon and magnetoplasmon excitation in two-dimensional electron space-charge layer on GaAs, Phys. Rev. B34:6951 (1986).

28. A. V. Chaplik and D. Heitmann, Geometric resonances of two-dimensional magnetoplasmons, J. Phys. C18:3357 (1985).

29. U. Mackens, D. Heitmann, K. Prager, J. P. Kotthaus and W.Beinvogl,

Minigaps in the plasmon dispersion of a two-dimensional electron gas with spatially modulated charge density, Phys. Rev. Lett. 53:1485 (1984).

30. E. Batke, W. Hansen, D. Heitmann, J. P. Kotthaus, U. Mackens, L. Prager and K. Ploog, Lateral structures of submicron periodicity on Si-MIS-systems and AlGaAs-GaAs heterojunctions, in: "Semiconductors, Quantum Well Structures, and Superlattices", K. Ploog and N. T. Linh., eds., Edition des Physique, Les Ulis (1986).

31. M. Wassermeier, H. Pohlmann and J. P. Kotthaus, Magnetoconductivity of inversion electrons in periodic MOS-microstructures on Si, in: "The Physics of Semiconductors", O. Engström, ed., World Scientific, Singapore (1987).

32. J. P. Kotthaus, W. Hansen, H. Pohlmann, M. Wassermeier and K. Ploog, Intersubband resonances in quasi-one-dimensional channels, Surf. Sci., to be published.

33. U. Merkt, C. Sikorski and J. P. Kotthaus, Spectroscopy of one-dimensional subbands on InSb, Superlattices and Microstructures, to be published.

34. W. Hansen, J. P. Kotthaus, A. Chaplik and K. Ploog, Electronic excitations in laterally microstructured AlGaAs-GaAs heterojunctions, in: "The Application of High Magnetic Fields in Semiconductor Physics", G. Landwehr, ed., Springer, Heidelberg (1987).

35. A. V. Chaplik, Absorption and emission of electromagnetic waves by two-dimensional plasmons, Surf. Sci. Rep. 5:289 (1985).

36. T. Zettler and J. P. Kotthaus, to be published.

37. S. E. Laux and F. Stern, Electron states in narrow gate-induced channels in Si, Appl. Phys. Lett. 49:91 (1986).

38. S. J. Allen, Jr., H. L. Störmer and J. C. M. Hwang, Dimensional resonance of the two-dimensional electron gas in selectively doped GaAs/AlGaAs heterostructures, Phys. Rev. B28:4875 (1983).

39. S. Das Sarma and W. Lai, Screening and elementary excitations in narrow-channel semiconductor microstructures, Phys. Rev. B32:1401 (1985).

ELECTRONIC STRUCTURE OF LATERALLY RESTRICTED SYSTEMS

Frank Stern

IBM T.J. Watson Research Center
Yorktown Heights, New York 10598, U.S.A.

INTRODUCTION

This lecture* is intended to describe some of the structures being built to achieve quasi-one-dimensional behavior of electrons in semi-conductors and some of the methods being used to calculate energy levels and charge densities for such structures. No attempt is made to describe the growing body of theoretical and experimental work on optical and transport properties. A basic knowledge of heterostructure physics, as presented in other lectures in this volume, is presupposed.

Rapid advances in processes for fabricating semiconductor samples have made possible structures in which carrier motion is restricted in one (or two or even three) dimensions, with carriers essentially free to move in the remaining dimensions. In this lecture we shall consider the case in which motion is restricted in a quantum sense, i.e. we have carriers confined in a "box" of some kind, whose size is less than or comparable to the Fermi length k_F^{-1}, the reciprocal of the diameter of the Fermi surface corresponding to the carrier density. There is another criterion for reduced dimensionality which involves a different physical length scale, namely the inelastic diffusion length $L_{in} = (D\tau_{in})^{1/2}$, where D is the diffusion constant and $1/\tau_{in}$ is the mean rate of inelastic--or more exactly, phase-breaking--collisions. The inelastic diffusion length can be relatively long at low temperatures, where phase-breaking transitions occur relatively rarely. Thus this form of reduced dimensionality, which I call reduced dimensionality in

----

* Two other lectures, on electrical transport properties, are based on lectures presented at the NATO Advanced Study Institute on Physics and Applications of Quantum Wells and Superlattices, Erice, 21 April - 1 May 1987. The manuscript prepared for those lectures will appear in the Proceedings of the Erice meeting, to be published by Plenum Press.

a transport sense, can be achieved rather easily at low temperatures
with structures whose dimensions are on the order of μm. Confinement
in the quantum sense, with which we are concerned here, generally
requires effective dimensions of order 0.1 μm or less. Such structures
have become available only in the last few years.

STRUCTURES

One can think of several ways to confine carriers. Perhaps the
simplest is geometrical confinement, as in a thin film or free-standing
wire. Alternatively, one can create a heterostructure in which carriers
are confined by the energy band discontinuity at a heterojunction or
semiconductor-insulator interface. Examples are the quantum wells and
heterojunction structures covered in other lectures of this series, as
well as silicon inversion layers and related structures, which have also
been actively studied.[1] Finally, one can confine carriers using potential profiles generated by external electrodes or by internal charge
distributions, by magnetic fields, or conceivably by nonuniform stress
distributions. Many of the structures that have been built or proposed
use more than one confinement scheme to limit the carrier motion and
reduce the dimensionality.

A special class of structures involves periodic variations of the
confining potential or geometry, as discussed for example in the lectures by Professor Kotthaus. The present lecture is concerned primarily
with isolated structures, but periodic structures in which there is
little or no overlap of charge between adjacent "wires" or "dots" can
be considered to be an array of independent regions for the purposes
of this lecture.

One of the earliest kinds of structures used to confine carriers
laterally is the so-called pinched accumulation layer used by Fowler
et al.[2] and by Dean and Pepper,[3] and illustrated in fig. 1. This
structure uses implanted p-type contacts to define a channel on the
surface of lightly doped n-type silicon. The channel can be controlled
by varying the (negative) voltage applied to these "pinching" contacts,
thereby varying the width of the p-n junction at each contact, as well
as the (positive) voltage on the gate electrode which controls the
electron density in the channel and also influences the channel width.
Electrical contact to the channel is made through implanted or diffused
contacts at each end. An important consideration is the fact that the
underlying lightly doped n-type silicon freezes out at low temperatures
and therefore does not add another conducting path in parallel with the
surface channel. This was perhaps the first demonstration of quasi-
one-dimensional behavior in a structure that allows the channel density
to be varied. A disadvantage of the structure is that the channel
confinement depends in a nontrivial way on the voltages applied to the

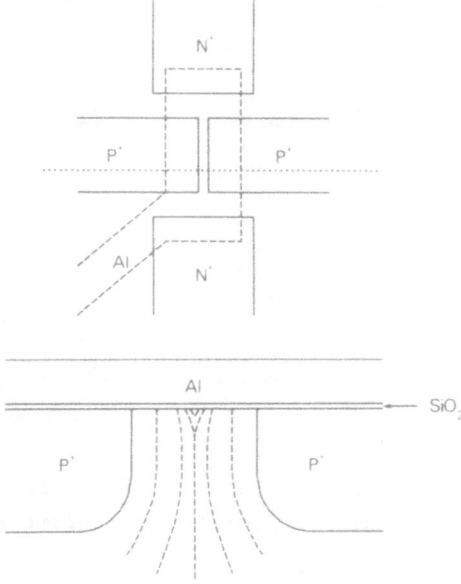

Fig. 1. Two schematic cross sections of the pinched accumulation layer
       structure. (After Fowler et al., ref. 2)

electrodes. Thus a detailed theory is required to extract the channel
parameters. An approximate calculation for this structure is described
briefly below.

A closely related structure used an external electrode, rather
than a diffused or implanted contact, to confine the channel. Such a
structure was used by Thornton et al.,[4] and others[5,6] on GaAs-based heterostructures. In this structure the density of carriers is primarily
controlled by the underlying heterostructure, most importantly by the
thickness of the undoped AlGaAs spacer layer, while the voltage applied
to the "pinching" electrodes primarily affects the channel width. As
we shall see later, the channel can be shut off completely if the gap
between the pinching electrodes is small enough. Near threshold, the
channel width and channel electron density cannot be easily controlled
independently in a given geometry.

A further refinement of this type of structure was first built by
Warren et al.,[7] who added a second insulator over the pinching electrode
and placed a second gate over the entire structure. They did this on
silicon devices, for which deposited silicon dioxide is available as
the second insulator. Their structure is periodic, as shown in
fig. 2, but can be used to form isolated channels. This structure can
be used in two modes: the inner gate can be biased positively,
attracting electrons under the gate, with the outer gate biased negatively to turn off the conduction under the gaps in the inner gate and
confine electrons under the inner gate. Alternatively, the inner gate

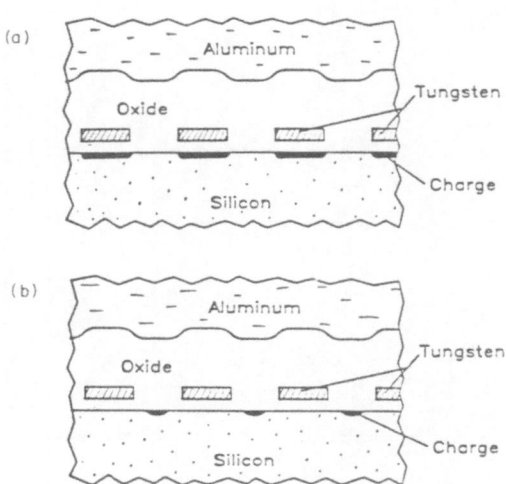

Fig. 2. Double-gate structure for lateral confinement, showing two
alternative modes of operation. (After Warren et al., ref. 7)

can be biased negatively, the outer gate positively, and electrons will
be induced under the gaps in the inner gate. This second mode appears
to give somewhat tighter lateral confinement of the carriers. Modeling
for this type of structure, both in a single-channel version and in the
periodic version, has been carried out and will be described below.

One of the simplest ways to induce a narrow channel is with a
narrow gate. Wheeler et al.[8] used photolithographic techniques to make
gates as narrow as 300 nm and Kwasnick et al.[9] used a shadow deposition
technique to make a gate 70 nm wide.

Geometrical confinement of carriers has been achieved by etching
mesas on silicon-silicon dioxide structures[10,11] as shown in fig. 3, and
on GaAs-based heterostructures.[12] This has the advantage that a physical
dimension is accurately known. It has the disadvantage that interface
charges at the edge of the channel influence the potential in the
channel and therefore lead to an effective width that may be consider-

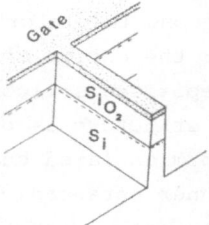

Fig. 3. Schematic illustration of an etched mesa structure. (After
Skocpol et al., ref. 10)

ably smaller than the geometrical width.[13] Etched structures have also been used to define isolated dots.[14,15] The dimensions continue to shrink, but the role of surface roughness and of surface states needs to be examined for these very small structures.

By slicing across the end of a heterostructure layer, one can use the heterostructure to provide lateral confinement and can use an external field to confine carriers near the interface. Such a structure was proposed by Sakaki,[16] as illustrated in fig. 4. A structure based on similar ideas, but using a wedge rather than a notch, was built by Petroff et al.[17] Calculations of the electronic states of a related structure in which both electrons and holes can be confined were carried out by Chang et al.,[18] who explored the ranges of layer thicknesses required for optimum one-dimensional behavior, i.e. the energy of the lowest one-dimensional subband well below that of higher-lying subbands.

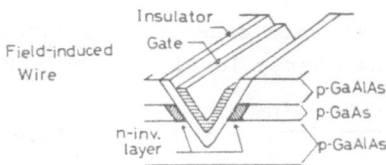

Fig. 4. Proposed structure for achieving quasi-one-dimensional electrons using heterojunction confinement. (After Sakaki, ref. 16)

Implantation can replace etching to provide an alternative method for lateral confinement, as shown by Cibert et al.[19] for GaAs-AlGaAs. Temkin et al.[20] used electron beam lithography and argon ion milling to define wires and "boxes" with feature sizes down to 30 nm on InGaAs/InP heterostructures.

Other kinds of structures have been proposed, in part in connection with the calculations described below, but the examples given here indicate the variety of structures that has already been implemented. It is likely that additional structures will be devised and built in the next few years, and that improving technology will allow smaller dimensions--and therefore tighter confinement.

MODELS, APPROXIMATIONS, RESULTS

This section lists some of the models that have been used to calculate electronic energy levels and charge densities in quasi-one-dimensional structures. The models generally involve one or more approximations and in most cases have been applied to simplified structures rather than to actual devices. This is a relatively young field, and additional advances can be expected in the near future.

The textbook examples for electronic states include a circular cylinder with an infinite potential barrier beyond a radius R, for which the radial wavefunctions are Bessel functions and the energy levels relative to the conduction band edge of the band from which the states are formed, here assumed to have an isotropic effective mass m, are

$$E = \frac{\hbar^2}{2m} (k_{nl}^2 + k^2), \qquad l = 0, 1, 2, \ldots, \qquad n = 1, 2, 3, \ldots, \qquad (1)$$

where $k_{nl}R$ is the n'th zero of the Bessel function of order l and k is the wave vector for motion along the wire. The levels with $l>0$ are doubly degenerate. If the barrier outside the wire is finite, boundary conditions must be matched at the interface and the energy levels are determined by an equation that must be solved numerically.

The other simple textbook example is a rectangular wire with sides L_1 and L_2, for which the energy levels in the infinite-barrier case are

$$E = \frac{\hbar^2}{2m} \left(\frac{\pi^2 n_1^2}{L_1^2} + \frac{\pi^2 n_2^2}{L_2^2} + k^2 \right), \qquad n_1, n_2 = 1, 2, 3 \ldots \qquad (2)$$

Somewhat unexpectedly, the energy levels can no longer be written down in a straightforward way for the rectangular wire when the barrier is finite, even when the potential everywhere outside the wire is a constant, because the problem is not separable--the potential is not a sum of a function of y and a function of z. (We choose axes such that the x is along the wire, the z is the direction of strong confinement, and y is the direction of lateral confinement.)

Perhaps the first attempt at understanding energy levels in channels under a narrow gate that went beyond the particle-in-a-box approximation was made by Maldague, in connection with unpublished measurements by F.F. Fang. Only a brief abstract of that calculation has been published.[21] Fang, like Wheeler[8] and Skocpol,[11] was looking for evidence of the laterally confined states by increasing the electron density and looking for structure in the conductance as the Fermi level passed through successive laterally confined states. Unfortunately, fluctuations in conductance associated with random scattering centers obscured the expected structure. We will return to this subject below.

The first paper to publish detailed results for narrow channels was that of Shik,[22] who treated a simplified version of the pinched accumulation layer device. He assumed that the side contacts were quasi-metallic, i.e. very heavily doped, and that they had infinite depth. He also ignored the nonzero thickness of the inversion layer and ignored quantization effects. Even with those approximations the problem is still not trivial. Shik obtained results for the threshold for forming a channel and for the channel density and channel width versus the voltages on the gate and on the pinching electrodes. In particular, he found that the channel width increases rapidly once the gate voltage increases beyond threshold.

Warren et al.[23] obtained a semi-classical solution for the charge densities and potentials in the grating structures shown in fig. 2. Their results agree qualitatively with those obtained subsequently in a full quantum-mechanical treatment, described below.

Quantum effects were considered explicitly by Lai and Das Sarma.[24] They used a simplified representation of the device geometry and used a variational trial function to find the energy and charge distribution in the lowest subband. They examined the adequacy of separation of variables, and found it to be a relatively good approximation in their example.

Brum et al.[25] studied the energy levels of electrons in a rectangular GaAs "wire" embedded in AlGaAs, with carriers supplied from a plane of dopants separated from the wire. They expanded the solutions in terms of the solutions for an infinite interface, with the wave functions for lateral confinement as expansion coefficients. A similar expansion had been used previously by Chang et al.[18] for a different structure. In this method, the envelope wave function of a particular state is written as a sum

$$\sum_n \alpha_n(y) \chi_n(z) \exp(ikx) \tag{3}$$

where the $\chi_n(z)$ are solutions of the usual equations for quantization normal to the interface and the $\alpha_n(y)$ describe the lateral motion. The equations to be solved then take the form

$$\left[-\frac{\hbar^2}{2m} \frac{d^2}{dy^2} + W_{nn} \right] \alpha_n(y) + \sum_{m \neq n} W_{nm}(y)\alpha_m(y) = \left[E - E_n - \frac{\hbar^2 k^2}{2m} \right] \alpha_n(y), \tag{4}$$

$$W_{nm}(y) = \int \chi_n(z) W(y,z) \chi_m(z) \, dz. \tag{5}$$

$W(y,z)$ is the difference between the actual two-dimensional potential and the one-dimensional z-dependent potential for the $\chi_n(z)$, and the E_n are the eigenvalues for the χ_n, which are taken to be real and normalized here.

The Poisson equation for the two-dimensional potential was solved analytically in the calculation by Brum et al.[25] The amount of charge transferred to the wire from the dopant plane is determined self-consistently in the course of the solution. A contour plot of potential profiles, not including the conduction band offset at the GaAs-AlGaAs interface, is shown in fig. 5.

The framework suggested by Chang et al.[8] and by Brum et al.[25] can be used for most structures in which the lateral confinement is a perturbation on the vertical confinement, and should be practicable even if the perturbation is not particularly weak. For most geometries, one would need to solve Poisson's equation numerically in two dimensions, solve a one-dimensional Schrödinger equation or several coupled equations, and iterate to self-consistency.

In a separate calculation, Brum et al.[26] also determined energy levels for holes in a quasi-one-dimensional heterostructure.

Laux and Stern[27] in their original calculation, for a silicon structure similar to a section of the two-gate structure of Warren et al.,[7] solved both the two-dimensional Poisson equation and the two-dimensional Schrödinger equation numerically. Apart from approximations made in determining the potential, including neglect of image effects and many-body effects in the original calculation, their calculation is a straightforward solution of the problem within the context of the effective mass approximation. They treated a case in which all the geometrical features of the structure are along cubic axes, so there are no complications associated with tilted conduction band valleys, and solved the Schrödinger equation in the form

$$-\frac{\hbar^2}{2}\left(\frac{\partial}{\partial y}\frac{1}{m_y^{(1)}}\frac{\partial\zeta_{i,j}^{(1)}}{\partial y}+\frac{\partial}{\partial z}\frac{1}{m_z^{(1)}}\frac{\partial\zeta_{i,j}^{(1)}}{\partial z}\right)+[V(y,z)-E_{i,j}^{(1)}]\zeta_{i,j}^{(1)}(y,z)=0, \quad (6)$$

where the superscript (1) identifies the valley orientation, and the subscripts i and j can be identified with the number of nodes in the z

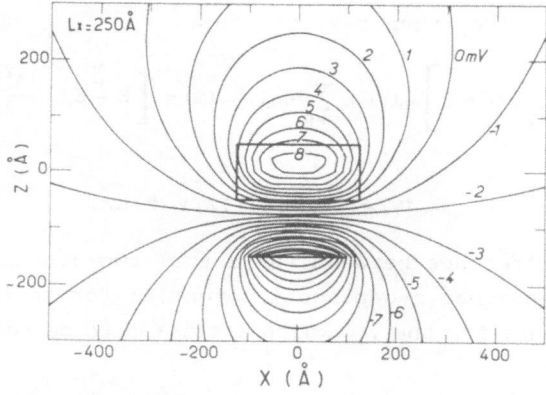

Fig. 5. Potential profiles from a self-consistent solution for electrons in a rectangular GaAs "wire" embedded in AlGaAs. Electrons are transferred to the GaAs from a plane of dopants outside the "wire." (After Brum et al., ref. 25)

dependence and y dependence of the wave function, respectively. The boundary condition for the solution is that it decay to zero far from the channel. The potential energy in the original Laux and Stern calculation is $V(y,z) = -e\phi(y,z) + \Delta E_c$, where ΔE_c is the variation of the bottom of the conduction band associated with variations in composition (i.e., the conduction band discontinuity at the Si-SiO$_2$ interface), and ϕ is the solution of Poisson's equation

$$\nabla \cdot [\varepsilon(y,z)\nabla\phi(y,z)] = -\rho(y,z) \tag{7}$$

with suitable boundary conditions determined by voltages applied at the contacts. The charge density $\rho(y,z)$ includes the charge density of the channel electrons,

$$\rho_{inv}(y,z) = -e \sum_{l,i,j} N_{i,j}^{(l)} [\zeta_{i,j}^{(l)}(y,z)]^2, \tag{8}$$

$$N_{i,j}^{(l)} = \frac{g_v^{(l)}}{\pi} \left(\frac{2m_x^{(l)} k_B T}{\hbar^2} \right)^{1/2} F_{-1/2}\left(\frac{E_F - E_{i,j}^{(l)}}{k_B T} \right), \tag{9}$$

where the sum is over all states i,j and ladders l, $m_x^{(l)}$ is the effective mass along the wire axis for the lth ladder, $g_v^{(l)}$ is the valley degeneracy, equal to 2 here, E_F is the Fermi energy, and F_k is the Fermi-Dirac integral of order k. The total charge density ρ in Eq. (7) also includes the contribution from ionized impurities and any other fixed charges and, at elevated temperatures, of holes in the substrate.

Each set of valleys, labeled by superscript (l), gives rise to a ladder of energy levels with nodal properties characterized by subscripts i and j. As in the usual two-dimensional case, the two silicon conduction band valleys whose heavy mass is perpendicular to the Si-SiO$_2$ interface lead to the ladder with lowest energy. However, the other two sets of valleys, which are degenerate in the usual two-dimensional case, have different energies because the symmetry of the x and y motion has been removed by the lateral confinement. The valleys whose heavy mass is in the y direction (we are somewhat imprecise in equating directions in real space and in k space, but there is no ambiguity in the present case) have somewhat lower energies than the valleys whose heavy mass is in the x direction. For the examples considered in refs. 27 and 28, the states associated with the higher valleys are not occupied by electrons.

The validity of separation of variables was examined qualitatively by Laux and Stern by comparing their numerical wave functions with approximate wave functions that are products of a function of y and a function of z. The difference is found to be very small for the lowest one-dimensional subband, as had also been found by Lai and Das Sarma,[24] but becomes significant for the higher subbands with several nodes both in the y and z directions.

The only self-consistent quantum mechanical calculation published so far that attempts to model an actual device geometry is that by Laux and Warren,[28] for the structure of ref. 7, shown in fig. 2b above. They calculated the voltage increments at which successive one-dimensional subbands pass through the Fermi level. The device parameters were taken from experiment, but the relative fraction of the 200 nm period taken up by the inner gate was not precisely measured and was varied in the calculation. The results, shown in fig. 6, indicate that there is a considerable discrepancy between the calculated voltage increments and the voltage increments at which Warren et al.[7] found structure in the conductance of their 250 parallel "wires", a configuration that reduces the sample-dependent fluctuations which had plagued early effort to look for lateral quantization, mentioned above. Inclusion of image effects and correlation effects in the potential did not change the calculated results appreciably. This problem requires further examination, both theoretically and experimentally.

The procedures used to solve the Poisson and Schrödinger equations have been briefly described in refs. 27 and 28, and recent improvements which reduce the computation time significantly are presented in ref. 29. A typical mesh to discretize the Poisson equation is of order 60×60 points, not uniformly spaced. The mesh for the Schrödinger equation covers only the area where channel electrons are present, but is large enough that the wave functions have decayed to very small values near the boundary. It also has about 60×60 points, again not evenly spaced. The Schrödinger equation for each ladder of levels thus

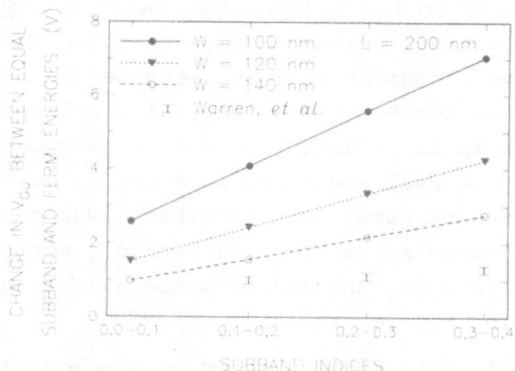

Fig. 6. Calculated voltage steps between crossing of successive one-dimensional subbands through the Fermi level for the two-gate periodic structure of Warren et al., ref. 7, for various gate widths. The barred points give the spacings seen in the conductance of the Warren et al. device. (After Laux and Warren, ref. 28)

has several thousand eigenvalues, of which Laux et al. need to find only
the few with significant carrier occupation to carry out a self-consistent calculation. In some cases additional eigenvalues are found
if more information on the energy level spectrum is needed. A single
solution--i.e. a fully self-consistent solution of the Poisson and
Schrödinger equations for a single set of voltages on the electrodes-
-typically takes 5-15 cpu minutes on an IBM 3090 vector machine.

Recently Laux et al.[30] examined the states in a split-gate GaAs-
AlGaAs heterostructure. The problem differs from the silicon case in
several ways: there is a doped layer to provide carriers to the heterojunction, and there is an exposed surface at which charged surface
states appear. A full solution of the problem would have to include a
way to determine current flow, since--unlike the case for silicon--
there is not an essentially ideal insulator to block current flow. The
surface charge was calculated at various temperatures near and somewhat
below room temperature using a program that treats currents in the
drift-diffusion approximation but ignores quantum effects. The interface charge calculated in that program was then simplified, by omitting
some edge effects, and was inserted into the quantum mechanical calculation as a frozen-in interface charge at the low temperatures of the
quantum mechanical calculation. Results for the energy levels associated with lateral confinement of the states are shown in fig. 7 for a
structure with a 400 nm gate opening. The level spacing varies from

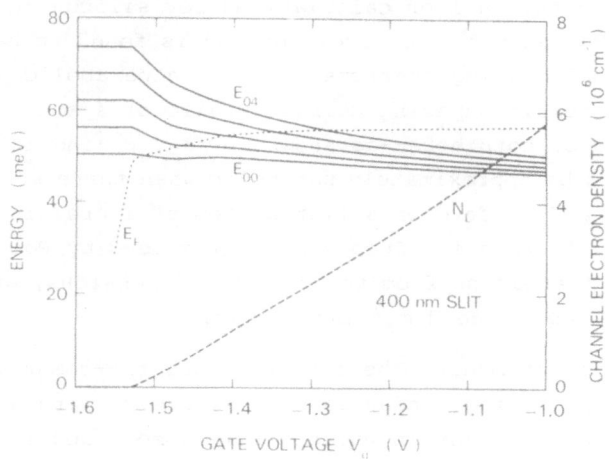

Fig. 7. Calculated energy levels associated with lateral confinement
of electrons under a 400 nm gate opening on a GaAs-AlGaAs heterostructure. The integrated density of channel electrons is
also shown, as is the position of the Fermi level. (After Laux
et al., ref. 30)

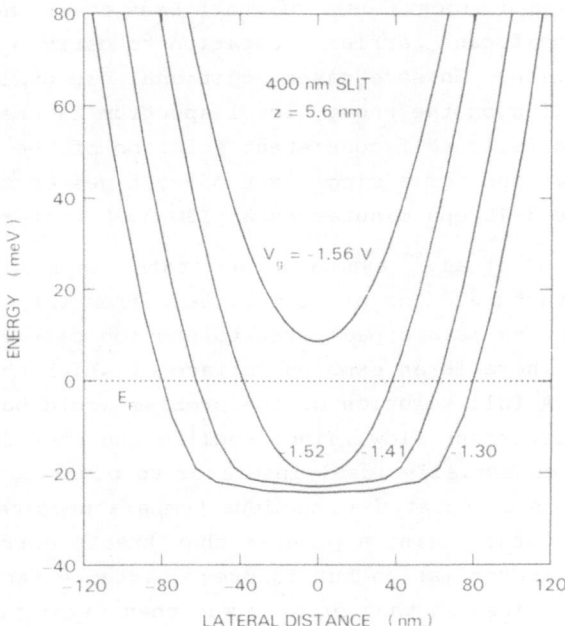

Fig. 8. Potential profiles along a line 5.6 nm inside the GaAs for the
structure whose lateral energy levels are shown in fig. 7.
(After Laux et al., ref. 30)

about 5 meV near threshold to about 1 meV when the channel has about
5×10^6 electrons per cm.

Here, as in the earlier calculations for silicon and in the very
early work by Maldague,[21] the level spacing is found to be intermediate
between the equal spacing characteristic of a parabolic potential and
the rapidly increasing spacing characteristic of a square well. The
actual potentials, both here--as shown in fig. 8--and in the silicon
case turn out to be approximately parabolic when there are few electrons
in the channel and to develop a flat bottom of increasing width as the
channel density increases. Such a model was used by Poole et al.[31] in
describing the transition from three to two dimensions, and by Berggren
et al.[32] in the quasi-one-dimensional case.

Magnetic fields modify the properties of two-dimensional systems
significantly, producing Landau levels for fields with a component
perpendicular to the plane in which carriers move and modifying the
states when there is a component parallel to the plane. The effects
in quasi-one-dimensional systems are different, because a strictly
one-dimensional system cannot support Landau levels. Some of the
effects of magnetic fields on quasi-one-dimensional states in a GaAs-
AlGaAs heterostructure with side electrodes have been explored by
Berggren et al.[32]

All the calculations described above used the effective mass approximation, albeit in a generalized form for the valence band calculations by Brum et al.[25] Microscopic calculations based on pseudopotentials have been carried out for GaAs wire embedded in AlGaAs by the group at Newcastle,[33] who found explicit charge distributions for several one-dimensional subbands.

CONCLUSIONS

A number of methods, with different levels of approximation, have been described for calculating the energy levels and charge distributions in quasi-one-dimensional systems. The calculations tend to be more extensive than the corresponding calculations for quasi-two-dimensional electron systems because of the greater geometrical complexity. For structures in which the vertical confinement is stronger than the lateral confinement, the method used by Chang et al.[18] and Brum et al.[25] should be both practicable and accurate. For very small structures and for structures in which there is no clear basis for partial separation of variables, the method used by Laux et al.[27-30] may be required because it solves the two-dimensional Schrödinger equation directly. Finally, for very small structures or for heterostructures involving states at different points in the Brillouin zone, microscopic calculations like those of the Newcastle group[33] may be required. There is now a fairly wide choice of methods, one or more of which should be adequate for most structures made or proposed so far.

Energy levels and quantum states provide only a framework for studying the properties of quasi-one-dimensional systems. The literature of quasi-one-dimensional physics is already quite large and includes many papers on screening, on states bound to impurities, on excitonic states, on dielectric response and plasmons, on optical properties and excitons, and on transport properties both for the low-temperature regime where localization and fluctuation effects are observed and for higher temperatures. While much of the interest in recent years has focused on metallic systems which are one-dimensional in the transport sense, recent development show that it is now possible to fabricate semiconductor systems which are quasi-one-dimensional even in the quantum sense. We can expect increasing opportunities to confront theory with experiment, and vice versa.

I am indebted to Steve Laux for a very fruitful collaboration on the calculations described here and to Jose Brum, David Frank, Milan Jaros, and Alan Warren for helpful discussions.

REFERENCES

1. T. Ando, A. B. Fowler, and F. Stern, Electronic properties of two-dimensional systems, Rev. Mod. Phys. 54:437 (1982).

2. A. B. Fowler, A. Hartstein and R. A. Webb, Conductance in restricted-dimensionality accumulation layers, Phys. Rev. Lett. 48:196 (1982).

3. C. C. Dean and M. Pepper, The transition from two- to one-dimensional electronic transport in narrow silicon accumulation layers, J. Phys. C 15:L1287 (1982).

4. T. J. Thornton, M. Pepper, H. Ahmed, D. Andrews and G. J. Davies, One-dimensional conduction in the 2D electron gas of a GaAs-AlGaAs heterojunction, Phys. Rev. Lett. 56:1198 (1986).

5. H. Z. Zheng, H. P. Wei, D. C. Tsui, and G. Weimann, Gate-controlled transport in narrow GaAs/$Al_xGa_{-x}As$ heterostructures, Phys. Rev. B 34:5635 (1986).

6. M. Frei, D. C. Tsui and G. Weimann, Gate-controlled one-dimensional transport in high-mobility GaAs/AlGaAs heterostructures, Bull. Am. Phys. Soc. 32:889 (1987).

7. A. C. Warren, D. A. Antoniadis and H. I. Smith, Quasi one-dimensional conduction in multiple, parallel inversion lines, Phys. Rev. Lett. 56:1858 (1986).

8. R. G. Wheeler, K. K. Choi and R. Wisnieff, Quasi-one-dimensional effects in submicron width silicon inversion layers, Surf. Sci. 142:19 (1984).

9. R. F. Kwasnick, M. A. Kastner, J. Melngailis and P. A. Lee, Nonmonotonic variations of the conductance with electron density in ~70-nm-wide inversion layers, Phys. Rev. Lett. 52:224 (1984).

10. W. J. Skocpol, L. D. Jackel, E. L. Hu, R. E. Howard and L. A. Fetter, One-dimensional localization and interaction effects in narrow (0.1-μm) silicon inversion layers, Phys. Rev. Lett. 49:951 (1982).

11. W. J. Skocpol, L. D. Jackel, R. E. Howard, H. G. Craighead, L. A. Fetter, P. M. Mankiewich, P. Grabbe and D. M. Tennant, Magnetoconductance and quantized confinement in narrow silicon inversion layers, Surf. Sci. 142:14 (1984).

12. H. van Houten, B. J. van Wees, M. G. J. Heijman and J. P. Andre, Submicron conducting channels defined by shallow mesa etch in GaAs-AlGaAs heterojunctions, Appl. Phys. Lett. 49:1781 (1986).

13. K. K. Choi, D. C. Tsui and K. Alavi, Experimental determination of the edge depletion width of the two-dimensional electron gas in GaAs/$Al_xGa_{1-x}As$, Appl. Phys. Lett. 50:110 (1987).

14. M. A. Reed, R. T. Bate, K. Bradshaw, W. M. Duncan, W. R. Frensley, J. W. Lee and H. D. Shih, Spatial quantization in GaAs-AlGaAs multiple quantum dots, J. Vac. Sci. Technol. B 4:358 (1986).

15. K. Kash, A. Scherer, J. M. Worlock, H. G. Craighead and M. C. Tamargo, Optical spectroscopy of ultrasmall structures etched from quantum wells, Appl. Phys. Lett. 49:1043 (1986).

16. H. Sakaki, Scattering suppression and high-mobility effect of size-quantized electrons in ultrafine semiconductor wire structures, Jpn. J. Appl. Phys. 19:L735 (1980).

17. P. M. Petroff, A. C. Gossard, R. A. Logan, and W. Wiegmann, Toward quantum well wires: Fabrication and optical properties, Appl. Phys. Lett. 49:1275 (1986).

18. Y. C. Chang, L. L. Chang, and L. Esaki, A new one-dimensional quantum well structure, Appl. Phys. Lett. 47:1324 (1985).

19. J. Cibert, P. M. Petroff, G. J. Dolan, S. J. Pearton, A. C. Gossard and J. H. English, Optically detected carrier confinement to one and zero dimensions in GaAs quantum well wires and boxes, Appl. Phys. Lett. 49:1275 (1986).

20. H. Temkin, G. J. Dolan, M. B. Panish and S. N. G. Chu, Low-temperature photoluminescence from InGaAs/InP quantum wires and boxes, Appl. Phys. Lett. 50:413 (1987).

21. P. F. Maldague, One-dimensional behavior in narrow-gate silicon MOS devices, Bull. Am. Phys. Soc. 26:787 (1981).

22. A. Ya. Shik, Calculations relating to a semiconductor structure with a quasione-dimensional electron gas, Fiz. Tekh. Poluprovodn. 19:1488 (1985) [Sov. Phys. Semicond. 19:915 (1985)].

23. A. C. Warren, D. A. Antoniadis, and H. I. Smith, Semi-classical calculation of charge distributions in ultra-narrow inversion lines, IEEE Electron Dev. Lett. EDL-7:413 (1986).

24. W. Y. Lai and S. Das Sarma, Ground-state variational wave function for the quasi-one-dimensional semiconductor wire, Phys. Rev. B 33:8874 (1986).

25. J. A. Brum, G. Bastard, L. L. Chang and L. Esaki, Energy levels in some quasi uni-dimensional semiconductor heterostructures, in: "Proceedings of the 18th International Conference on the Physics of Semiconductors," Vol. 1, p. 396, World Scientific, Singapore (1987).

26. J. A. Brum, G. Bastard, L. L. Chang and L. Esaki, Energy levels in quasi uni-dimensional semiconductor heterostructures, Superlattices and Microstructures 3:47 (1987).

27. S. E. Laux and F. Stern, Electron states in narrow gate-induced channels in Si, Appl. Phys. Lett. 49:91 (1986).

28. S. E. Laux and A. C. Warren, Self-consistent calculation of electron states in narrow channels, Technical Digest of International Electron Devices Meeting, Los Angeles, December, 1986, p. 567.

29. S. E. Laux, Numerical methods for calculating self-consistent sol-
    utions of electrons states in narrow channels, NASECODE V: The
    Fifth International Conference on the Numerical Analysis of
    Semiconductor Devices and Integrated Circuits, Dublin, June,
    1987 (to be published by Boole Press, Dublin).
30. S. E. Laux, D. J. Frank and F. Stern, Quasi-one-dimensional elec-
    tron states in a split-gate GaAs-AlGaAs heterostructure, Seventh
    International Conference on Electronic Properties of Two-Dimen-
    sional Systems, Santa Fe, July, 1987 (to be published in
    Surf.Sci.).
31. D. A. Poole, M. Pepper, K.-F. Berggren, G. Hill and H. W. Myron,
    Magneto-resistance oscillations and the transition from three-
    dimensional to two-dimensional conduction in a gallium arsenide
    field effect transistor at low temperatures, J. Phys. C 15:L21
    (1982).
32. K.-F. Berggren, T. J. Thornton, D. J. Newson and M. Pepper, Magnetic
    depopulation of 1D subbands in a narrow 2D electron gas in a
    GaAs:AlGaAs heterojunction, Phys. Rev. Lett. 57:1769 (1986).
33. K. B. Wong, M. Jaros and J. P. Hagon, Confined electron states in
    GaAs-Ga$_{1-x}$Al$_x$As quantum wires, Phys. Rev. B 35:2463 (1987).

ELECTRONS AND HOLES IN QUANTUM BOXES

Garnett W. Bryant

McDonnell Douglas Research Laboratories
P. O. Box 516
St. Louis, Missouri 63166

INTRODUCTION

Quasi-zero-dimensional quantum microstructures can now be made which exhibit quantum carrier confinement in all three dimensions.[1-8] Individual atoms are the microscopic limit for very small, confined-electron systems. Bulk systems bounded by surfaces are the macroscopic limit for very large, confined-electron systems. The intermediate regime, where the crossover from atomic to bulk behavior occurs, is called the mesoscopic regime. Studies[1-3,9-11] of semiconductor microcrystallites, with dimensions, L, from one to several tens of nanometers, extend the investigation of carrier confinement effects away from the atomic limit. In this size regime, quantum confinement effects, which scale as $1/L^2$, are expected to dominate interparticle Coulomb effects, which scale as $1/L$. With the recent advances in the art of microfabrication, ultrasmall (20 nm $\leq$ L $\leq$ 500 nm) quasi-zero-dimensional quantum-well boxes[4-8] can be made which exhibit carrier confinement, extending the investigations away from the macroscopic limit and into the submicrometer size regime. The transition from negligible confinement effects to complete confinement (no significant intersubband Coulomb mixing) occurs in the size regime (20 nm $\leq$ L) that is currently accessible by microfabrication.

The properties of ultrasmall structures are governed by the physics of the mesoscopic regime. As a consequence of the quantum confinement in all three dimensions, the energy levels are discrete. For that reason, quantum boxes have generated much interest as a new class of artificially structured materials with atomic-like discrete states ideal for use in laser structures[8,12] and interesting nonlinear optical properties.[13] Order-of-magnitude increases in maximum gain and lower threshold injection currents have been predicted[12] for semiconductor laser structures which use quantum boxes instead of two-dimensional wells or bulk material as the lasing medium. Large changes in the optical absorption and index of refraction, and optical bistability are predicted[13] for the nonlinear optical properties of quantum box structures.

The predictions about the optical properties of quantum boxes are made for boxes in the size regime (L $\leq$ 10 nm) in which Coulomb effects are a weak perturbation to the quantum confinement effects. However, significant

improvement in microfabrication capability is still needed to make structures in this size regime. In this article we consider quantum box structures that can be fabricated currently. These structures are within the
transition regime where quantum confinement effects are significant
(L ≤ 100 nm) but not yet complete (L ≥ 10 nm) and where Coulomb effects are
an important perturbation. This article presents a theory for electrons
and holes in quantum boxes to show how the electronic properties of quantum
boxes change as the box size decreases from the bulk macroscopic limit to
the regime of complete confinement. We will show when the transition from
the bulk limit begins, when the transition to quantum confinement is complete, and how different electronic and optical properties change in this
transition regime. Using these results, we will provide a qualitative
assessment of how complete the quantum confinement must be for quantum
boxes to have the novel properties predicted in the limit of complete
confinement.

APPROACH

The carriers in a bulk (L ≥ 1.0 μm) semiconductor structure form a
many-electron system of weakly interacting particles which can be modeled
by the single-particle effective mass equation. In ultrasmall structures
(L ≤ 0.1 μm) the effective-mass approach still provides a good description
of the motion through the lattice. However, the carriers cannot be assumed
to be a weakly correlated, many-particle system. Consider a quantum box
constructed such that the confinement at an interface defines one of the
confined dimensions. For a typical inversion-layer charge density of
10^{11} cm^{-2}, a two-dimensional uniform gas in a square, 100-nm-wide box would
contain 10 carriers; in a box 10 nm wide, less than one carrier. A similar
estimate holds if the source of carriers is the injection current for a
semiconductor laser structure. In that case a typical carrier density[12] is
10^{18} cm^{-3}. A cube 10 nm wide contains one carrier. As a consequence, in
the size regime of complete quantum confinement a quantum box will contain
only a few carriers. To study the transition to the limit of complete
confinement we consider only quantum boxes which contain a few carriers.
Since such a quantum box is an interacting, few-particle system with
discrete single-particle levels, a strong analogy to atoms can be made.

Microfabricated quantum boxes are constructed from narrow, two-
dimensional quantum wells by processing the wells to confine the two-
dimensional motion.[4-7] Typically, the width, w, of the two-dimensional
well is an order-of-magnitude less than the length, L, of the side of the
box (see the inset in Fig. 3). To simplify our theory, we model quantum
boxes as two-dimensional, rectangular boxes with w=0 and with infinite
barriers. Use of a finite, narrow-well width w weakens the Coulomb effects
slightly but would not change the results qualitatively. Carriers can
tunnel from boxes with finite barriers. However, this tunneling[4] does not
change the electronic properties qualitatively until L ≤ 5 nm. Actual
quantum boxes are typically circular, rather than square. However, shape
effects are minimal for structures with the same cross-sectional area.[15]

The few-particle systems have been studied[16-18] by solving the multiparticle effective-mass Schrodinger equation for two-dimensional interacting particles confined to the quantum box. The particle interaction is the
Coulomb interaction screened by the background dielectric constant. Polarization and image charge effects can be significant when there is a large
dielectric discontinuity between the quantum box and the surrounding
medium.[9] This is not the case for microfabricated boxes made, for example,
with GaAs wells and AlGaAs barriers.

The Hamiltonian is

$$H = \sum_i \left(-\frac{\hbar^2}{2m_i} \nabla^2 + V_i\right) + \sum_{i<j} \frac{e_i e_j}{\epsilon |\vec{r}_i - \vec{r}_j|} \tag{1}$$

$$V_i = 0 \quad \text{if } |x| < L_x/2 \text{ and } |y| < L_y/2 \tag{2}$$

$$\quad = \infty \quad \text{otherwise,}$$

m_i is the effective mass and e_i is the charge of particle i. L_x and L_y are the lengths of the sides of the box. Two approaches have been used to solve Eq (1). First, accurate ground states for a single exciton in a square box have been calculated by use of a variational approach. The exciton variational wave function, $\mathcal{I}$, is a product of the electron and hole ground-state subband states and a linear combination, s, of Gaussian functions of the electron-hole separation.

$$\mathcal{I}(\vec{r}_e, \vec{r}_h) = \cos(kx_e) \cos(ky_e) \cos(kx_h) \cos(ky_h) \, s(\vec{r}_e, \vec{r}_h) \tag{3a}$$

$$s(\vec{r}_e, \vec{r}_h) = \sum_n c_n \exp(-\alpha_n[(x_e - x_h)^2 + (y_e - y_h)^2]) \tag{3b}$$

with $k = \pi/L$. Accurate energies and wave functions are obtained by use of 5 to 10 Gaussians.

Second, a configuration-interaction (CI) approach has been used to calculate exciton excited states and to include correlations in the electronic structure of boxes with several electrons or with several electrons and holes (excitonic molecules). The multiparticle wave function is expanded in terms of Slater determinants constructed from the single-particle noninteracting eigenstates. Because the barriers are infinite, a basis set of wave functions which are separable in the two directions that define the well can be used. The single-particle, one-dimensional eigenstates (sines and cosines) are used as the basis functions. The kinetic energy and interaction matrix elements are found by use of the Slater determinant basis and the Hamiltonian is diagonalized to find the eigenstates.

In the infinite barrier model, all kinetic-energy matrix elements scale as $1/L^2$ and all interaction matrix elements scale as $1/L$ when the dimension L of the box is changed without changing the box shape. This scaling determines the nature of the multiparticle system. For small L, the Coulomb interactions are insignificant compared to the single-particle level spacings; the particles are independent and uncorrelated. As L increases, the interactions become significant and the multiparticle states become correlated. The multiparticle states evolve continuously, as L increases, from the exact, independent-particle states of the small L limit. In the infinite barrier model, the results for multielectron systems are independent of the electron mass m_e and dielectric constant ϵ if all energies are scaled by the effective Rydberg, $R_e = e^2 m_e/2a_o^2 \epsilon^2$, and the lengths are scaled by the effective Bohr $a_e = a_o \epsilon/m_e$. To illustrate the important size-scales for multielectron systems, we present results for GaAs wells, $m_e = 0.067 m_o$ and $\epsilon = 13.1$. The energies are scaled by R_e. Results are also presented for excitons in GaAs boxes with hole masses in the box $m_h = 0.090 m_o$ and $0.377 m_o$. Energies are also scaled by the effective electron Rydberg $R_e = 5.3$ meV.

CORRELATED CARRIERS IN QUANTUM BOXES

In multielectron systems the electron-electron repulsion separates particles, weakening the importance of Coulomb effects relative to the box confinement. In contrast, the electron-hole attraction provides additional confinement, thus increasing the importance of the Coulomb effects and the mixing of higher subband states in the exciton state. As a consequence, electron-electron correlations are easier to explain and are described more accurately by a CI calculation with a limited set of basis functions than are electron-hole correlations with the same basis set. We present results for the multielectron systems first.

The evolution of the energy levels of an interacting, multielectron confined system that occurs when the box size changes is shown in Figs. 1 and 2 for two simple systems: two interacting carriers in (a) long, narrow ($L \equiv L_y = 10\ L_x$) quantum-boxes (Fig. 1) and (b) square quantum-boxes (Fig. 2). The energies are scaled by a factor R_s which scales as $1/L_y^2$ (except for case (f) in Fig. 2) to account for the scaling of the kinetic energies. If the Coulomb interactions were unimportant, then the scaled energy levels would be independent of L. The increases in the scaled ground-state energy and the changes in the level spacings that occur as L increases show how important the electron-electron interaction becomes in large boxes.

The internal motion of electrons in long, narrow structures is quasi-one-dimensional because the carriers are confined to the lowest subband of the narrow (x) direction. Mixing of higher x-subbands is insignificant in the size regime covered in Fig. 1. The internal motion in the square box

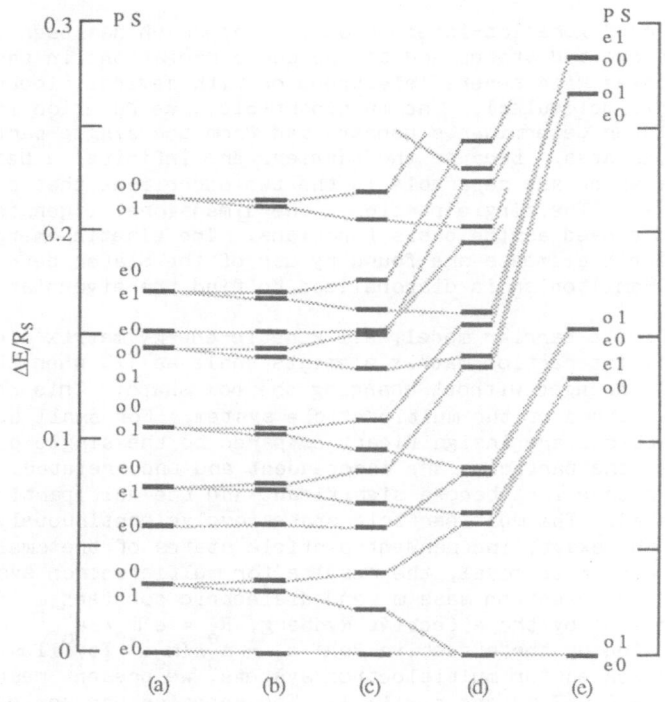

Fig. 1 Energy levels for two electrons in a long, narrow ($L_y = 10\ L_x$) GaAs quantum-well box. The spacings are measured relative to the ground-state energy E_g and scaled by R_s. (a) For noninteracting electrons, $L_y = 1$ nm, $E_g = 1.905\ R_s$ and $R_s = R_e \times 10^5$. For interacting electrons, L_y (nm), E_g/R_s and R_s/R_e are, respectively: (b) 1, 1.907, 10^5 ; (c) 10, 1.923, 10^3; (d) 100, 1.987, 10; and (e) 1000, 2.314, 0.1. The parity P (e = even, o = odd) and total spin, S, of each state are indicated. The energy levels with total spin S are independent of the z-component of spin, s_z.

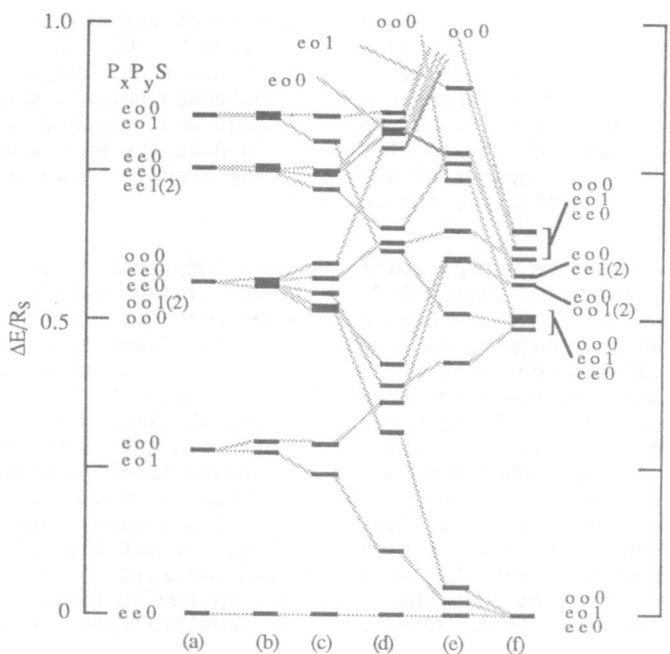

Fig. 2   Energy levels for two electrons in a square, GaAs quantum-well box.
L_y(nm) E_g/R_s and R_s/R_e are, respectively: (a) for noninteracting electrons,
1, 0.377, 10^4 ; and for interacting electrons (b) 1, 0.387, 10^4 ; (c) 10,
0.456, 10^2 ; (d) 100, 0.996, 1; (e) 10^3 , 3.925, 0.01; and (f) 2×10^3, 4.373,
0.0022. The x and y parity and total spin of each state are indicated.
States with odd-x, even-y parity are degenerate with even-x, odd-y parity
states and are not shown. The energy levels with total spin S are
independent of s_z. Additional degeneracies not due to parity or spin are
shown in parentheses.

is two-dimensional. The effective Coulomb interaction increases as the
dimensionality is lowered. The Coulomb contribution to the ground-state
energy, E_g, is larger, when measured on a common energy scale, in a long,
narrow box than in a square box with the same long (L_y) dimension.

For $L_y \lesssim 1$ nm, the scaled energy levels shift slightly from the non-
interacting levels, and exchange splitting occurs. Typically, states with
total spin S=1 have lower energy (Hund's rule) than S=0 states with the
same parity. When $L_y \approx 10$ nm, the scaled energy levels shift substantially
from the levels of the noninteracting system and begin to cross in long,
narrow boxes. Significant Coulomb contributions to E_g begin on this length
scale. When $L_y \approx 100$ nm, substantial reordering of the levels occurs both
in square boxes and in long, narrow boxes.

In the large-L limit, where interactions dominate the kinetic energy,
the system should become strongly correlated, with the electrons located to
minimize the repulsive interaction as in a Wigner lattice (WL) for unconfined systems. The signature of the WL states in a confined system is the
degeneracy of the levels. For example, in a long, narrow box there is one
way to put two particles with the same spin on the box axis to minimize the
direct Coulomb interaction. For two particles with opposite spin, there
are two configurations. In a square box the particles would sit on opposite ends of the same diagonal in the WL limit. The degeneracy would be
double the degeneracy for a long, narrow box since there are two equivalent

diagonals. In long, narrow boxes, the evolution of the states into levels with the degeneracies of the WL limit occurs when $L \geq 0.1$ μm and the splitting of levels that are degenerate in the WL limit is no longer apparent. The WL limit is reached at larger L for square boxes because the effective Coulomb interaction is weaker. Ceperley[19] found that a two-dimensional electron gas becomes a WL when $r_s > 33$ a_e. In GaAs the WL would occur for $r_s > 0.34$ μm. This is consistent with the length scale on which the confined square system approaches the WL limit.

The evolution of the energy levels is more complicated for systems with more than two particles. When $N > 2$, the level mixing is more complex, degeneracies in the WL limit are higher, and the transition to the WL limit occurs at larger L, requiring more accurate calculations to reach the WL limit. We calculated the cases of three and four particles in long, narrow boxes with accuracy adequate for tracing the evolution to the WL limit. For square boxes, the level degeneracies of the WL limit are not obvious when the number of particles is incommensurate with the symmetry of the structure; for example, when a square box contains three particles. Our results suggest that three spin-parallel electrons in a square box have four degenerate states in the WL limit. This is the degeneracy expected from the symmetry of the box. However, we have not yet been able to calculate with adequate accuracy the states for two parallel and one opposite spin particles to confirm that the degeneracy in the WL limit is twelve, as needed to be consistent with the results for parallel spin particles.

For boxes in which the confinement effects dominate, the carrier density, $\sigma(\vec{r})$, approaches the independent-carrier density. As L increases, the carriers move apart along the diagonals with $\sigma(\vec{r})$ peaking farther from the center, and the density on the diagonals increases relative to the density off the diagonals. In addition, as L increases, the particle positions become more strongly correlated. As L increases, the probability, $\sigma(\vec{r}, \vec{r}_o)$, for finding one particle at $\vec{r}$ if the second is on a diagonal at $\vec{r}_o$ increases for $\vec{r}$ at the opposite end of the same diagonal and decreases for other $\vec{r}$.

The multielectron wave functions evolve, as L increases, from the single-configuration Slater determinant states of the noninteracting system by mixing in other configurations with the same parity and spin. For $L \leq 10$ nm, the only important configuration (probability ≥ 0.95) in the interacting ground state is the noninteracting ground-state configuration. For $L \geq 100$ nm, the excited noninteracting configurations are mixed in with comparable or greater probability than the noninteracting ground-state.

In few-electron systems, the bulk limit (the Wigner lattice) is reached when $L \sim 1000$ nm. In an electron-hole system, the interparticle attraction reduces the interparticle separation, thus enhancing Coulomb effects and reducing confinement effects. The transition to the bulk limit should occur at smaller L in electron-hole systems. The evolution of the ground-state energy of an exciton in a square box as L decreases is shown in Fig. 3. At small L, the confinement energy increases as $1/L^2$ and the Coulomb energy increases as $1/L$. The confinement energy dominates and the ground-state energy increases monotonically with decreasing L. Similar results have been obtained in calculations for microcrystallites.[9-11] In boxes with finite barriers, tunneling would keep all energies finite as L decreases. However, tunneling does not qualitatively change the results until $L \leq 5$ nm. At large L (≥ 100 nm), the exciton ground-state energy approaches the energy for the unconfined, two-dimensional exciton ($4 R_\mu$ where R_μ is the exciton effective Rydberg). The shift from the limiting value is less than five per cent for $L \geq 100$ nm.

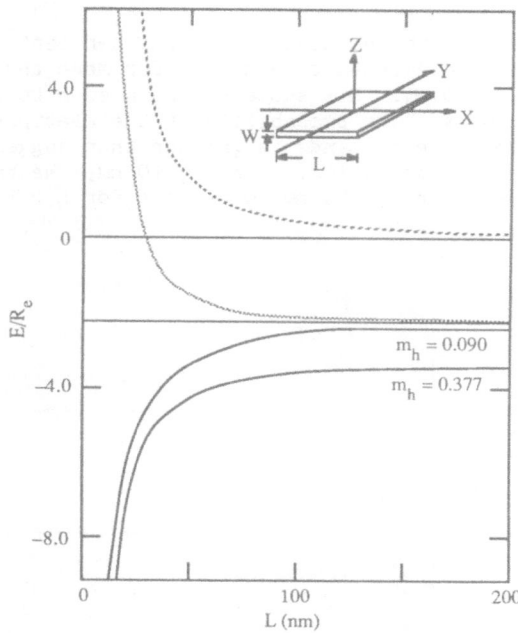

Fig. 3 Ground-state energy of an exciton confined in a square box of width L. The solid (upper dashed) curve is the energy of an (non)interacting electron-hole ($m_h = 0.09$) pair. The lower dashed curves are the exciton Coulomb energies for $m_h = 0.09$ and $m_h = 0.377$. The horizontal solid line gives the unconfined-exciton energy ($m_h = 0.09$). The inset shows the configuration of the box.

In the limit of complete quantum confinement, the exciton Coulomb energy should be independent of hole mass because the light and heavy holes will occupy only one subband and have the same wave function. The light-and heavy-hole Couloub energies become comparable when $L \lesssim 10$ nm. Two length scales should define the transition regime to complete confinement. The radius for a bulk, two-dimensional exciton, R_{2D}, is 11 nm (7.4 nm) for $m_h = 0.090$ (0.377). Confinement effects should occur when the exciton and the box have similar dimensions. Confinement effects should also dominate Coulomb effects when the energy separation between the first two noninter-acting pair levels is greater than the Coulomb energy. This occurs for $L \lesssim 55$ nm (35 nm) for $m_h = 0.090$ (0.377).

The electron-hole separation of the confined exciton, $R \equiv \langle (\vec{r}_e - \vec{r}_n)^2 \rangle^{1/2}$, is shown in Fig. 4. For $L \gtrsim 100$ nm, R is unaffected by the confinement. For $L \lesssim 10$ nm, R approaches the value for complete confinement. R has also been calculated for $m_h = 0.090$ using only the part, s, of the exciton wave function due to Coulomb correlations. For large L, the correlated part is the same as the bulk exciton wave function. As L decreases, the Coulomb-induced correlations become less important and the correlations due to confinement reduce R below the limit for free excitons. A similar compen-sation of confinement and Coulomb effects at large L keeps the ground-state exciton energy close to the bulk limit for $L \gtrsim 100$ nm, even though the Coulomb and confinement energies have changed from the bulk limit.

One indication of the extent of confinement is the number of electron and hole subbands needed in the CI calculation to accurately model the ex-citon. The one subband approximation is exact in the limit of complete confinement. For $L \lesssim 10$ nm, the one subband approximation accounts for more than ninety per cent of the exciton binding; for $L \gtrsim 60$ nm, this approximation accounts for no more than fifty per cent of the binding. Use of two electron and hole subbands accounts for more than ninety per cent of the binding for L ≤ 20 nm. Six subbands are needed to obtain the same accuracy for L ≤ 60 nm. The probability of the electron and hole being in the lowest electron-hole subbands is greater than suggested by the accuracy of the one subband approximation. For L ≤ 10 nm, the probability is 0.99; for L = 20 nm, 0.95; for L = 60 nm, 0.64; and for L ≥ 100 nm, less than 0.5.

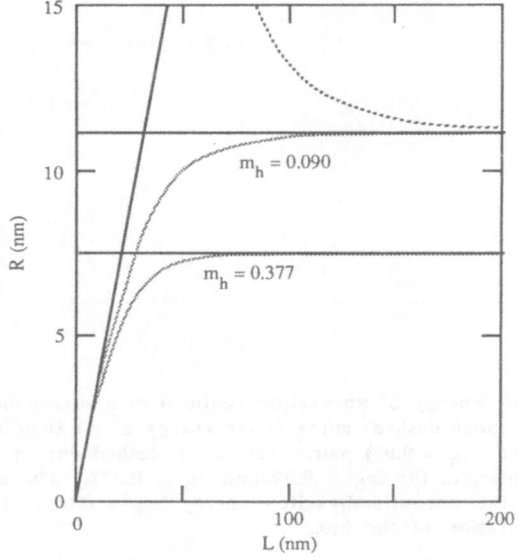

Fig. 4 Electron-hole separation, R, of an exciton confined in a square box of width L. The horizontal solid lines indicate the separation for unconfined light- and heavy-hole excitons. The other solid line is the separation of an uncorrelated electron-hole pair confined to the lowest pair of subband states. The dashed curves that interpolate between the solid lines are the R for correlated, confined electron-hole pairs. The upper dashed curve is R for m_h = 0.090, calculated using only the correlated part of the exciton wave function.

The effects of confinement in quantum boxes have been observed by photoluminescence.[4-7] Energy shifts of 2-5 meV, as shown in Fig. 3 for L ~ 50 nm, are typical of the observed shifts. However, as mentioned, shifts in Coulomb energy and confinement energy are comparable and compensating. Thus, any attempt to determine box sizes from the shifts of exciton photoluminescence peaks must be analyzed carefully. Experimental results suggest that exciton photoluminescence efficiency increases as L decreases. Figure 5 shows the exciton oscillator strength normalized to the box area as a function of L. The normalized exciton oscillator strength is proportional to the ratio, f/f(total), of the exciton oscillator strength to the total available oscillator strength. Rapid increase in f/L^2 for L ≤ 50 nm is indicative of the increased photoluminescence efficiency. This rapid rise in normalized oscillator strength is also indicative of the enhanced optical properties predicted for boxes in the regime of complete confinement. This enhancement begins for L ~ 50 nm, which is well above the limit of complete confinement, and increases rapidly for L ≤ 25 nm.

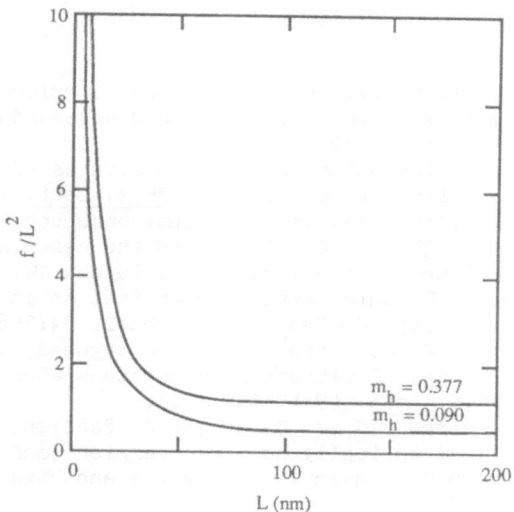

Fig. 5 The exciton ground-state oscillator-strength, f, normalized to box area for an exciton confined in a square box of width L. f/L^2 is shown in arbitrary units.

CONCLUSIONS

The electronic structure of quantum microstructures is determined by a competition between confinement effects and interparticle interactions. The electron-hole and electron-electron correlations can be important on the mesoscopic length scale in quantum boxes. The correlations are a few-particle rather than a many-particle effect. Electron-electron correlations are weaker because the interparticle repulsion separates the particles; consequently, the confinement effects are more significant at large L for few-electron systems. The transition to complete confinement occurs when $L \leq 10$ nm.

In small ($L \leq 10$ nm) infinite-barrier boxes the correlations are weak. However, the importance of Coulomb interactions in small boxes defined by finite barriers is underestimated by our model because the infinite-barrier model overestimates the competing kinetic energy effects. In boxes defined by finite barriers, correlation effects should extend to smaller size-scales, and Coulomb repulsion should inhibit the pairing of electrons. For example, in the smallest square box ($L = 1$ nm) considered in Fig. 2, the energy cost to pair two electrons is 0.01 R_s = 5.3 eV, more than any realistic finite barrier in a GaAs structure. Thus, correlation effects should be important even in the limit of complete quantum confinement.

Quantum boxes in the complete confinement limit are predicted to have enhanced optical properties, due to the discrete energy spectrum, that would be ideal for applications in semiconductor laser structures and non-linear optics. This enhancement of the optical properties should persist well above ($L \leq 50$ nm) the regime of complete confinement. Thus quantum box structures in the size regime that can presently be constructed should be able to provide some of these enhanced properties.

ACKNOWLEDGMENTS

This research was conducted under the McDonnell Douglas Independent Research and Development program.

REFERENCES

1. C. J. Sandroff, D. M. Hwang, and W. M. Chung, Carrier confinement and special crystallite dimensions in layered semiconductor colloids, Phys. Rev. B 33:5953 (1986).
2. J. Warnock and D. D. Awschalom, Picosecond studies of electron confinement in simple colored glasses, Appl. Phys. Lett. 48:425 (1986).
3. L. Brus, Zero-dimensional "exciton" in semiconductor clusters, IEEE J. Quantum Electron. QE-22:1909 (1986) and the references therein.
4. M. A. Reed, R. T. Bate, K. Bradshaw, W. M. Duncan, W. R. Frensley, J. W. Lee, and H. D. Shih, Spatial quantization in GaAs-AlGaAs multiple quantum dots. J. Vac. Sci. Technol. B4:358 (1986).
5. K. Kash, A. Scherer, J. M. Worlock, H. G. Craighead, and M. C. Tamargo, Optical spectroscopy of ultrasmall structures etched from quantum wells, Appl. Phys. Lett. 49:1043 (1986).
6. J. Cibert, P. M. Petroff, G. J. Dolan, S. J. Pearton, A. C. Gossard, and J. H. English, Optically detected carrier confinement to one and zero dimension in GaAs quantum well wires and boxes, Appl. Phys. Lett. 49:1275 (1986).
7. H. Temkin, G. J. Dolan, M. B. Panish, and S. N G. Chu, Low-temperature photoluminescence from InGaAs/InP quantum wires and boxes, Appl. Phys. Lett. 50:413 (1987).
8. Y. Miyamoto, M. Cao, Y. Shingai, K. Furuya, Y. Suematsu, K. G. Ravikumar, and S. Arai, Light emission from quantum-box structure by current injection, Jpn. J. Appl. Phys. 26:L225 (1987).
9. L. E. Brus, Electron-electron and electron-hole interactions in small semiconductor crystallites: The size dependence of the lowest excited electronic state, J. Chem. Phys. 80:4403 (1984).
10. Y. Kayanuma, Wannier excitons in microcrystals, Solid State Commun. 59:405 (1986).
11. H. M. Schmidt and H. Weller, Quantum size effects in semiconductor crystallites: Calculation of the energy spectrum for the confined exciton, Chem. Phys. Lett. 129:615 (1986).
12. M. Asada, Y. Miyamoto, and Y. Seumatsu, Gain and the threshold of three-dimensional quantum-box lasers, IEEE J. Quantum Electron. QE-22:1915 (1986).
13. S. Schmitt-Rink, D. A. B. Miller, and D. S. Chemla, Theory of the linear and nonlinear optical properties of semiconductor micro-crystallites, Phys. Rev. B 35:8113 (1987).
14. G. W. Bryant, Hydrogenic impurity states in quantum-well wires. Phys. Rev. B 29:6632 (1984).
15. G. W. Bryant, Hydrogenic impurity states in quantum-well wires: Shape effects, Phys. Rev. B 31:7812 (1985).
16. G. W. Bryant, D. B. Murray, and A. H. MacDonald, Electronic structure of single ultrasmall electron devices and device arrays, Super-lattices and Microstructures (in press).
17. G. W. Bryant, The electronic structure of ultrasmall quantum-well boxes, Phys. Rev. Lett. (submitted).
18. G. W. Bryant, Excitons in quantum boxes, Proc. of the Seventh Int. Conf. on the Electr. Props. of Two-Dimensional Systems (in press).
19. D. Ceperley, Ground state of the fermion one-component plasma: A Monte Carlo study in two and three dimensions, Phys. Rev. B 18:3126 (1978).

ELECTRONIC EXCITATIONS OF MULTI-QUANTUM-WELL STRUCTURES

Godfrey Gumbs*

Institut für Festkörperforschung der Kernforschungsanlage Jülich
D-5170 Jülich, West Germany

Department of Physics, Universtity of Lethbridge
Lethbridge, Alberta T1K 3M4 Canada

ABSTRACT

The dispersion relation of plasmons in a film of dielectric containing two-dimensional charged layers is calculated in the RPA. Two types of layered material are investigated corresponding to two-dimensional layers of electrons and alternating electron and hole layers, respectively. It is shown that when the substrate dielectric is included in the screened Coulomb potential energy between two point charges within the film the plasmon excitation spectrum for a layered electron gas system has the features recently observed in the Raman scattering measurements of modulation-doped GaAs-Al$_x$Ga$_{1-x}$As. An alternative method of probing the plasmon excitations is by means of high-resolution electron energy loss spectroscopy. We have calculated the linear response function $g(q_{||}, \omega)$ which determines the electron energy - loss spectra.

INTRODUCTION

Raman scattering experiments on modulation-doped GaAs-(Al$_{0.3}$Ga$_{0.7}$)As quantum wells have recently been reported of the intraband particle-hole and plasmon excitations[1]. These plasmons were predicted by Jain and Allen[2] (JA) for a model consisting of a finite number of two-dimensional electron gas (2DEG) layers. The plasmon dispersion relation of JA was calculated for a freely suspended film of 2DEG layers embedded in a medium of local dielectric constant ϵ whose role is to screen the Coulomb interaction between electrons within the quantum wells. In this paper, we calculate the effect of a substrate on this screening and obtain the resulting plasmon excitation spectrum. For this arrangement, we find that the symmetric and antisymmetric coupled surface plasmon modes of JA which protrude from the bulk plasmon band are altered so that one branch rises above the band while another branch drops below the band. This is, of course, due to the breakdown in the geometrical symmetry of the plasmon excitation spectrum about the mid-plane of the film, as was the case in the experimental results in Ref. 1.

For GaAs-(Al$_x$Ga$_{1-x}$)As, the band gap of GaAs is smaller than, and contained within, that of Al$_x$Ga$_{1-x}$As, giving rise to band gap discontinuities in both the valence and conduction bands of the resulting heterostructure. In Fig.

---

*Alexander von Humboldt Fellow

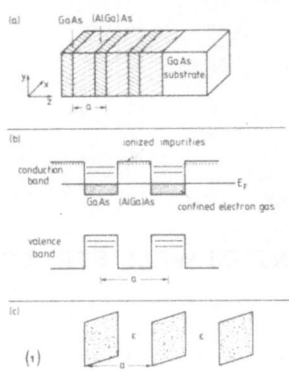
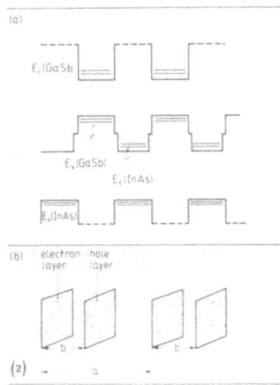

Fig. 1   (a) Structure of the GaAs-(Al_xGa_{1-x})As heterostructure. (b) Schematic diagram of the conduction and valence subbands of modulation-doped GaAs-(Al_xGa_{1-x})As. (c) The two-dimensional electron gas layers used to model the system.

Fig. 2   (a) Schematic diagram of the quantum-well structure of InAs-GaSb.
(b) The two-dimensional electron and hole plasma bi-layers which model the quantum-wells of InAs-GaSb.

1, a schematic representation of the electronic energy bands of the GaAs-(Al_xGa_{1-x})As quantum-well structure is given. The model we use to simulate this system consists of a rigid film of local dielectric constant ϵ and thickness L in which are embedded N equally spaced layers of 2DEG located at z_n = na (n = 0,1,...,N-1), where L = Na. (We do not take any account of phonon excitations). The film sits on a substrate of dielectric constant ϵ_o(z > L). The 2DEG is translationally invariant in the xy-plane where the electrons are free to move but are not allowed to hop from plane to plane. We also assume that the electron density is low enough that only the lowest subband in the conduction band is occupied. Intersubband excitations are not being considered in this paper since we are assuming that the quantum well is so narrow and the barrier height so high that we could treat it as two-dimensional (2D).

Another layered compound semiconductor of interest is InAs-GaSb. In this heterostructure, shown schematically in Fig. 2, the conduction band edge of InAs lies below the top of the valence band edge of GaSb, resulting in a transfer of electrons from GaSb to InAs and, subsequently, spatially separated electrons and holes in adjacent quantum wells.[3] We therefore model this system by alternating electron and hole layers.[4-7]

We have calculated the plasmon excitation energies for these two systems in the random-phase-approximation (RPA). The excitation energies are a few meV, as verified for GaAs-(Al_xGa_{1-x})As in the Raman measurements in Ref. 1. An alternative method of studying heterostructure plasmons is via electron energy loss spectroscopy (EELS). In an EELS experiment, an almost monochromatic beam of electrons is incident upon the surface under study. The probability that an incident electron of wave vector $\underset{\sim}{k}$ is scattered inelastically into a state of wave vector $\underset{\sim}{k}'$ into the range of energy losses between $\hbar\omega$ and $\hbar(\omega+d\omega)$ and into the solid angle $d\Omega_{\underset{\sim}{k}}$ around the direction of $\underset{\sim}{k}'$ is given by $P(\underset{\sim}{k},\underset{\sim}{k}')\hbar d\omega d\Omega_{\underset{\sim}{k}}$, where[8]

$$P(\underset{\sim}{k},\underset{\sim}{k}') = \frac{2}{(\pi e a_o)^2} \sec\theta \, \frac{k'}{k} \, \frac{q}{(q_{\|}^2 + q_z^2)^2} \, Im \, g(q_{\|},\omega). \tag{1}$$

Here $q_{\parallel} = k_{\parallel} - k_{\parallel}'$ is the momentum transfer parallel to the surface (z is the growth direction of the heterostructure), $q_z = k_z - k_z'$, a_o is the Bohr radius and θ is the angle between the incoming electron beam and the normal to the surface.

For thin films, we find that the loss intensity peaks in Im g at the surface plasmon frequencies are about two orders of magnitude greater than for metals, illustrating that the integrated EEL spectra should have strong signals. However, since we are dealing with small energy excitations, high resolution measurements are necessary. In this paper, we calculate the heterostructure plasmon dispersion relations together with the surface response function $g(q_{\perp},\omega)$, using the RPA.

GENERAL FORMULATION OF THE PROBLEM

EELS has proved to be a powerful spectroscopic tool for studying dynamical processes at surfaces.[8-11] The scattering probability in Eq. (1) is based on the assumption that the scattering mechanism of the incident electrons is long-range dipole scattering.[8] Thus the electric field from the external current density in the region z < 0 can be expressed as $E_{ext} = -\nabla\phi_{ext}$. Since $\nabla^2\phi_{ext} = 0$ in the region z > 0, the field in this region from the incident electrons can be written as a superposition of plane waves. The external potential induces a current density in the semiconductor which gives rise to an induced potential. Assuming that the heterostructure couples linearly to ϕ_{ext}, we obtain the total potential outside the medium in the vacuum region where z < 0 to be

$$\phi(\underline{r},t) = e^{i\underline{q}_{\parallel} \cdot \underline{r}_{\parallel} - i\omega t} [e^{-q_{\parallel} z} - g(q_{\parallel},\omega)e^{q_{\parallel} z}]. \tag{2}$$

We now turn to a calculation of the surface response function g for the periodic array of 2DEG layers (type - I) and biplanar system of electron and holes (type - II).

(a) $g(q_{\parallel},\omega)$ for Type-I Heterostructure

In calculating g, we imagine that there is an infinitesimal vacuum region between the 2DEG and the dielectric medium filling the space between adjacent planes. In this region, we assume that the electrostatic potential takes the form at $z_l = la$ ($l = 1,2,...,N-1$)

$$\phi_l(z) = A_l e^{-q_{\parallel} (z-z_l)} + B_l e^{q_{\parallel} (z-z_l)} \tag{3}$$

[$\exp(iq_{\parallel} \cdot \underline{r}_{\parallel} - i\omega t)$ factors out of the potential]. ϕ and $\epsilon d\phi/dz$ must be continuous at the vacuum - medium interface. Also the electric field is discontinuous across the 2DEG layer on which the charge density is

$$\sigma_l(q_{\parallel},\omega) = e^2\chi^o(q_{\parallel},\omega)\,\phi_l(z_l) = e^2\chi^o(q_{\perp},\omega)\,(A_l + B_l), \tag{4}$$

where χ^o is the single-particle density-density response function. After some straightforward algebra, we obtain

$$\begin{pmatrix} A_{l+1} \\ B_{l+1} \end{pmatrix} = \underline{T}_a(\alpha) \begin{pmatrix} A_l \\ B_l \end{pmatrix} = \underline{T}_a(\alpha)^{l+1} \begin{pmatrix} 1 \\ -g \end{pmatrix} \equiv \underline{M}_l \begin{pmatrix} 1 \\ -g \end{pmatrix}. \tag{5}$$

Here the elements of the T-matrix are given by

$$\left[\mathcal{T}_d(\alpha) \right]_{11} = \frac{1}{4\epsilon} \left[(1+\epsilon)(\epsilon+1+2\epsilon\alpha)e^{-q_{\parallel} d} + (1-\epsilon)(\epsilon-1-2\epsilon\alpha)e^{q_{\parallel} d} \right] \quad (6a)$$

$$\left[\mathcal{T}_d(\alpha) \right]_{12} = \frac{1}{4\epsilon} \left[(1+\epsilon)(\epsilon-1+2\epsilon\alpha)e^{-q_{\parallel} d} + (1-\epsilon)(\epsilon+1-2\epsilon\alpha)e^{q_{\parallel} d} \right] \quad (6b)$$

$$\left[\mathcal{T}_d(\alpha) \right]_{21} = \frac{1}{4\epsilon} \left[(1-\epsilon)(\epsilon+1+2\epsilon\alpha)e^{-q_{\parallel} d} + (1+\epsilon)(\epsilon-1-2\epsilon\alpha)e^{q_{\parallel} d} \right] \quad (6c)$$

$$\left[\mathcal{T}_d(\alpha) \right]_{22} = \frac{1}{4\epsilon} \left[(1-\epsilon)(\epsilon-1+2\epsilon\alpha)e^{-q_{\parallel} d} + (1+\epsilon)(\epsilon+1-2\epsilon\alpha)e^{q_{\parallel} d} \right] \quad (6d)$$

where $\alpha = (2\pi e^2/\epsilon q_{\parallel})\chi^o$ is the layer polarizability. Matching the electrostatic potential inside the semiconductor to the potential outside, we obtain in a straightforward way

$$g(q_{\parallel},\omega) = \frac{1}{D_I} \left[(1-\epsilon_o+2\epsilon\alpha)M_{11} - (1+\epsilon_o-2\epsilon\alpha)M_{21} \right] \quad (7a)$$

where

$$D_I \equiv (1-\epsilon_o+2\epsilon\alpha)M_{12} - (1+\epsilon_o-2\epsilon\alpha)M_{22}. \quad (7b)$$

M_{ij} are the elements of the matrix obtained by multiplying N-1 of the T-matrices defined in Eq. (5). The plasmon dispersion relation follows by setting D_I in Eq. (7) to zero. Equation (7) therefore provides us with a very convenient way of calculating the normal mode spectrum of a finite number of layers of 2DEG, since the dispersion relation involves the multiplication of a finite number of 2 x 2 matrices which is amenable to numerical computation. This dispersion relation has a simpler structure than the result obtained by Fourier transforming the RPA equation for the density-density response function and solving the resulting equation.[2] Also, it is straightforward to include the effect due to the substrate on the electron-electron interaction via the transfer-matrix method by choosing a value of ϵ_o not equal to unity. Alternatively, one could directly include the effect of the substrate on the Coulomb interaction and then use the result to calculate g. This method of calculation is also useful since it provides a convenient way of deriving the first-moment sum-rule for Im g. In this formalism, we calculate the inverse dielectric function which, in general, enters the expression for the energy loss of a charged particle to a material medium.[12-14]

First we calculate the surface response of a slab of dielectric before the 2DEG layers are embedded in the medium. By matching $\epsilon d\phi/dz$ and ϕ at the interfaces, calculation shows that the electric potential inside the slab due to the applied field in Eq. (2) is [we suppress the exponential factor depending on $q_{\parallel}$ and ω]

$$\phi_{ext}^{>}(z) = t_o e^{-q_{\parallel} z} + r_o e^{q_{\parallel} z}, \quad (8a)$$

where

$$t_o \equiv [(\epsilon+1) - (\epsilon-1)g_{slab}]/2\epsilon \quad (8b)$$

$$r_o \equiv [(\epsilon-1) - (\epsilon+1)g_{slab}]/2\epsilon. \quad (8c)$$

156

Here the surface response function for the dielectric slab is

$$g_{slab} \equiv \frac{\bar{\epsilon}(L) - 1}{\bar{\epsilon}(L) + 1} \tag{9a}$$

where

$$\bar{\epsilon}(L) \equiv \epsilon \left[\frac{1 - g_0 e^{-2q_\parallel L}}{1 + g_0 e^{-2q_\parallel L}} \right], \tag{9b}$$

with $g_0 \equiv (\epsilon - \epsilon_0)/(\epsilon + \epsilon_0)$. When the 2DEG layers are embedded in the slab, the surface response function is obtained by matching the total potential $\phi_{ext}^{>}(z) + \delta\phi_{ind}^{>}(z)$ inside the film to the external applied potential in Eq. (2). The induced potential $\delta\phi_{ind}^{>}$ is calculated in the RPA, assuming linear response coupling to the external field. After a little algebra, we obtain

$$g(q_\parallel, \omega) = g_{slab}(L) - \sum_{n,n'=0}^{N-1} v_{e,e}(0, z_n) \chi(z_n, z_{n'}) \phi_{ext}^{>}(z_{n'}) \tag{10}$$

for an equally-spaced layered electron gas (LEG). The density-density response function is given by

$$\chi(z_m, z_n) = \chi^0 \delta_{mn} + \chi^0 \sum_{n'=0}^{N-1} v_{e,e}(z_m, z_{n'}) \chi(z_{n'}, z_n). \tag{11}$$

For a semi-infinite geometry, the Coulomb interaction energy between two point charges is the sum of a bulk term and an image term depending on the distance between the source z and the image of z'. In the case of a film, infinitely many image points $\mathbf{r}_j'$ located at $(\mathbf{r}_\parallel', -z')$ and $(\mathbf{r}_\parallel', \pm z' + 2jL)$, $j = \pm 1, \pm 2,...$ must be introduced to satisfy the boundary conditions. This is illustrated in Fig. 3. A straightforward calculation shows that

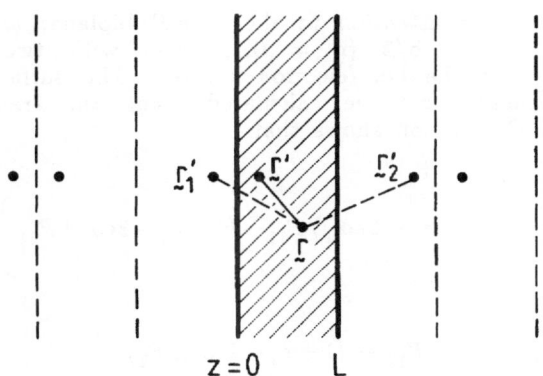

$$z = 0 \qquad L$$

Fig. 3 The slab geometry and the image points $\mathbf{r}_j'$ of $\mathbf{r}'$. The dashed vertical lines are inserted to identify the positions of the image points.

$$v_{e,e}(z,z') = \frac{2\pi e^2}{\epsilon q_{\parallel}} \left\{ A(z,z')\, e^{-q_{\parallel}|z-z'|} + \beta_{\infty} g_{\infty}\, e^{-q_{\parallel}|z+z'|} \right.$$

$$\left. + B(z,z')\, e^{-2q_{\parallel}L}\, e^{q_{\parallel}|z-z'|} + \beta_o g_o\, e^{-2q_{\parallel}L}\, e^{q_{\parallel}|z+z'|} \right\}, \qquad (12)$$

where $A = \beta_{\infty}$ and $B = \beta_o g_o^2$ for $z > z'$, and $A = \beta_o$ and $B = \beta_{\infty} g_{\infty}^2$ for $z < z'$. We also define $g_{\infty} \equiv (\epsilon - 1)/(\epsilon + 1)$ and $\beta_{o(\infty)} \equiv (1 - g_{o(\infty)}^2 e^{-2q_{\parallel}L})^{-1}$.

In the high frequency limit $\omega \gg g_{\parallel} v_F$, where v_F is the Fermi velocity, $\chi(z_n, z_{n'})$ can be approximated by its single-particle value $(n_s q^2 / m^* \omega^2)\delta_{nn'}$, where m^* is the electron effective mass and n_s is the areal density of the mobile carriers in the lowest band in the quantum wells. This asymptotic behavior allows us to apply the Kramers-Kronig relations to the function $g - g_{slab}(L)$ from which we obtain the sum-rule

$$2 \int_0^{\infty} \frac{d\omega}{\pi}\, \omega\, Im\, g(q_{\parallel}, \omega) = \omega_0^2 \qquad (13)$$

where

$$\omega_0^2 \equiv \frac{n_s q_{\parallel}^2}{m^*} \sum_{n=0}^{N-1} v_{e,e}(0, z_n)\, \phi_{ext}^{>}(z_n). \qquad (14)$$

For a semi-infinite equally-spaced LEG, ω_o is given by a very simple result since the lattice sum in Eq. (14) involves a geometric series. We obtain

$$\omega_o^{SI} \equiv \frac{\omega_p}{\epsilon + 1} \left[\frac{q_{\parallel} a\, e^{q_{\parallel} a}}{sinh(q_{\parallel} a)} \right]^{1/2}, \qquad (15)$$

where $\omega_p^2 \equiv 4\pi n_s e^2 / a m^*$.

(b) $g(q_{\parallel}, \omega)$ for Type-II Heterostructure

We now turn our attention to the type-II biplanar array having biplanes located at $z = na \pm b/2$ ($n = 0, 1, \ldots, N-1$) with two-dimensional planar polarizabilities $\alpha_{\pm}$ for the hole/electron sheets. The surface response function for this heterostructure can be calculated using the transfer-matrix method described above. Calculation shows that

$$g(q_{\parallel}, \omega) = \frac{1}{D_{II}} \left[(1 - \epsilon_o + 2\epsilon\alpha_+) P_{11} - (1 + \epsilon_o - 2\epsilon\alpha_+) P_{21} \right], \qquad (16a)$$

where

$$D_{II} \equiv (1 - \epsilon_o + 2\epsilon\alpha_+) P_{12} - (1 + \epsilon_o - 2\epsilon\alpha_+) P_{22} \qquad (16b)$$

and P_{ij} are the matrix elements of the 2 x 2 matrix $\underline{P}$ defined by

$$P = T_b\,(\alpha_-)\,[T_{a-b}\,(\alpha_+)\,T_b\,(\alpha_-)]^{N-1},\tag{17}$$

The plasmon dispersion relation for a type-II heterostructure follows immediately by setting D_{II} in Eq. (16) to zero. This simple form for the plasmon dispersion is very convenient for carrying out numerical calculations. We have also calculated the surface response function by separating out the explicit contribution from the dielectric slab. The result is

$$g(q_{||},\omega) = g_{slab}(L) - \sum_{n,n'=0}^{N-1} \sum_{\pm,\pm''} v_{e,e}\left(-\frac{b}{2},na \pm \frac{b}{2}\right)$$
$$\chi\left(na \pm \frac{b}{2}, n'\,a \pm ''\frac{b}{2}\right)\,\phi_{ext}^{>}\left(n'\,a \pm ''\frac{b}{2}\right).\tag{18}$$

Here the density-density response function χ is calculated in the RPA. g in Eq. (18) reduces to g_{slab} when the 2D layers are removed by setting the layer polarizability to zero. The surface response function for a type-II heterostructure calculated by Hawrylak, Wu and Quinn[15] does not satisfy this requirement.

Im g for a type-II LEG satisfies the sum rule in Eq. (13) where

$$\omega_0^2 \equiv q_{||}^2 \sum_{n=0}^{N-1} \sum_{\pm} v_{e,e}\left(-\frac{b}{2},na \pm \frac{b}{2}\right)\,(n_{\pm}^{(s)}/m_{\pm}^{*})\,\phi_{ext}^{>}\left(na \pm \frac{b}{2}\right)\tag{19}$$

where $n_{\pm}^{(s)}$ and $m_{\pm}^{*}$ are hole (+) and electron (-) areal density and effective mass, respectively. In the semi-infinite limit, we obtain

$$\omega_0^2 \equiv \frac{1}{(\epsilon+1)^2}\left[e^{-q_{||}b}\,\omega_p^{(+)2} + e^{q_{||}b}\,\omega_p^{(-)2}\right]\frac{q_{||}\,a\,e^{q_{||}(a-b)}}{sinh\,(q_{||}\,a)},\tag{20}$$

where $\omega_p^{(\pm)2} \equiv 4\pi n_{\pm}^{(s)}\,e^2/am_{\pm}^{*}$.

NUMERICAL RESULTS

In Fig. 4, we give a numerical example of the plasmon dispersion relation of a five layered type-I heterostructure. (All numerical calculations were performed at T = 0K.) To simulate GaAs-(AlGa)As, we chose the length of the unit cell as a = 890Å, ϵ = 13.1, the substrate dielectric constant ϵ_o = 20, the electron effective mass m* = 0.07 m_o (m_o = free electron mass) and the areal density n_s = 7.3 x 10^{11} cm^{-2}. The bulk plasmon band, corresponding to a LEG embedded in an infinite slab of dielectric constant ϵ, is also plotted. Note that increasing the substrate dielectric constant with respect to that of the doped AlAs layers gives plasmon modes lying slightly below the bulk band, as exhibited in the Raman scattering experiments of Fasol et al.[1] A value of ϵ_o = 1, as assumed by JA, on the other hand, gives a surface plasmon doublet which rises above the bulk band.[2] To further analyse the nature of the plasmon modes in Fig. 4, we have obtained the self-consistent solutions of Poisson's equation for the self-sustaining charge density oscillations on the layers. In the RPA, these are given by the eigenvalue equations

$$\rho(z_l) = \chi^o \sum_{n=0}^{N-1} v_{e,e}\,(z_l, z_n)\,\rho(z_n).\tag{21}$$

Solution of Eq. (21) shows that the highest mode in Fig. 4 has its largest amplitude on the surface at $z = 0$ whereas the lowest mode has its largest amplitude in the middle of the slab. Since the electric potential decays exponentially into the medium, we expect this mode to couple weakly to the external electric field.

For a fixed value of $q_{||}$, we have calculated Im g as a function of ω. The result when $q_{||} = 10^5 \text{cm}^{-1}$ in Fig. 4 is presented in Fig. 5. The single layer response function χ^0 was taken to be of the form suggested by Mermin[16] which conserves the local electron number by requiring electron-electron scattering to cause relaxation to a local equilibrium. The sample mobility was taken to be $\mu = 5 \times 10^4 \text{cm}^2/\text{V.s}$ corresponding to a collision time $\tau = \hbar \, m^* \mu/e = 2.2 \times 10^{-12} \text{sec}$. For $q_{||} = 10^5 \text{cm}^{-1}$, the five modes in Fig. 4 range from 5.35 meV to 12.20 meV. The coupling strengths follow from the charge density patterns of the modes, with weak coupling to modes with little weight at the surface.

In Fig. 6, we plot Im g for $q_{||} = 10^5 \text{cm}^{-1}$ with the parameter values for the effective mass, areal density and dielectric constant ϵ the same as in Fig. 5, but with $\epsilon_o = 1$. In this case, the highest plasmon mode is symmetric and the second highest branch is antisymmetric. As more and more layers are added, Im g for all modes drops relative to the highest mode.

A numerical example of the normal mode spectrum of a type-II heterostructure such as InAs-GaSb is shown in Fig. 7. There are five biplanes and we chose $b = a/2$. The bulk plasmon energy of electrons is $\omega_p^{(e)} = 1.1 \times 10^{14} \text{ sec}^{-1}$ and the Fermi velocity of the electrons is $v_F^{(e)} = 3.3 \times 10^7 \text{ cm sec}^{-1}$. The areal density of electrons is taken equal to the density of holes and the effective mass of a hole is assumed equal to half of the effective mass of an electron. As more and more layers are added, the bulk band is filled up, as illustrated in Fig. 8 for $q_{||}a = 2$. We note that the lowest plasmon mode splits off from the particle-hole spectrum when $q_{||} \sim a^{-1}$ as Fig. 7 shows. That is, in the strong coupling limit, $q_{||}a << 1$, the lowest mode behaves like two independent simultaneously excited particle-hole excitations, in which the oscillating charge densities are in phase on the two planes within a unit cell.

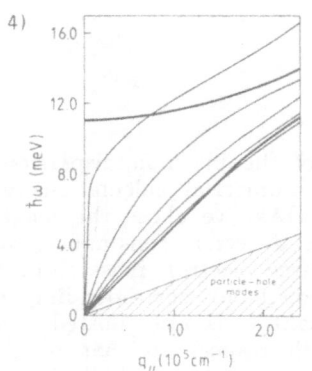
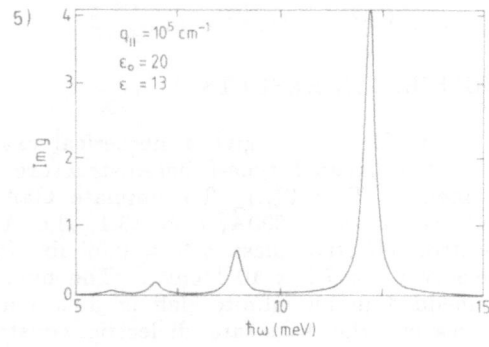

Fig. 4   Dispersion relation of a slab of dielectric containing five layers of 2DEG. The bold lines are the boundaries of the bulk plasmon-band. The particle - hole modes are also shown.

Fig. 5   Im $g(q_{||},\omega)$ is plotted as a function of energy transfer when the substrate dielectric constant is chosen as $\epsilon_o = 20$. The highest peak is at the surface plasmon energy at 12.2 meV.

In Fig. 9, Im g is plotted as a function of frequency for $q_{\|}a = 2$ when there are five electron-hole biplanes. In this example, the lowest frequency mode has strong coupling to the electric field (for $q_{\|}a \geq 1$) since the charge density oscillation of this mode has the largest weight at the surface. Two of the modes have such weak coupling that they do not show up in this plot.

ACKNOWLEDGEMENTS

The author acknowledges useful discussions with J. Harris and B.N.J. Persson. I am also thankful to J. Harris, B.N.J. Persson, K. Sturm and A. Liebsch for a critical reading of this paper. This work was supported in part by a grant from the Natural Sciences and Engineering Research Council of Canada.

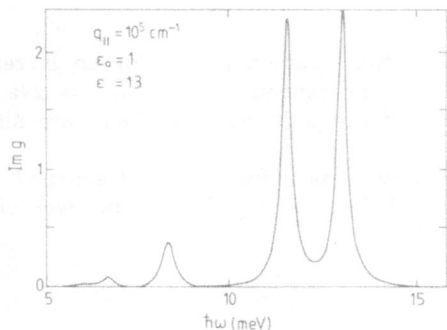

Fig. 6   Im $g(q_{\|},\omega)$ is shown as a function of the energy loss when vacuum is assumed to be on both sides of the dielectric medium ($\epsilon_0 = 1$). The two highest peaks are at the surface plasmon doublet.

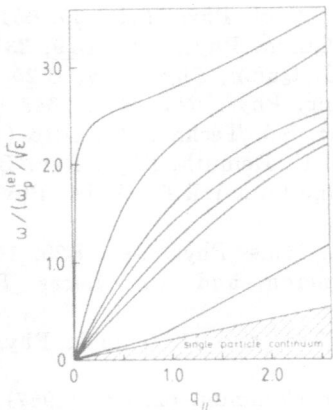

Fig. 7   Frequency of plasmon modes in a slab containing five biplanes of electrons and holes. The single particle-hole modes are also shown. The sample parameters used in the calculation are presented in the text.

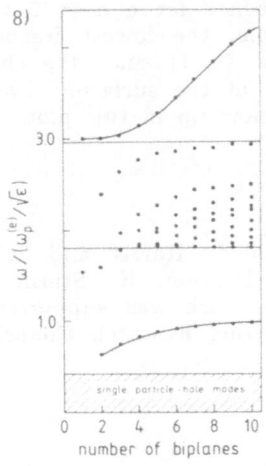

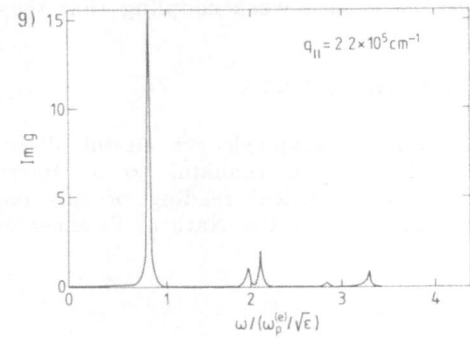

Fig. 8   The frequency of the plasmon modes for an increasing number of bi-planes for a fixed momentum transfer $q_{\parallel} = 2/a \approx 2.2 \times 10^5 cm^{-1}$. The boundaries of the bulk plasmon band are also shown.

Fig. 9   Im $g(q_{\parallel}, \omega)$ is plotted as a function of frequency for five biplanes of electrons and holes. The collision time was chosen so that $\hbar/\tau = 0.3$ meV.

REFERENCES

1. G. Fasol, N. Mestres, H.P. Hughes, A. Fischer, and K. Ploog, Phys. Rev. Lett. 56, 2517 (1986).
2. J.K. Jain and P.B. Allen, Phys. Rev. Lett. 54, 2437 (1985); Phys. Rev. B32, 997 (1985).
3. Type-II heterostructures were first discussed by G.A. Sai-Halasz, R. Tsu, and L. Esaki, Appl. Phys. Lett. 30, 651 (1977).
4. A.C. Tselis and J.J. Quinn, Phys. Rev. B29, 3318 (1984).
5. S. Das Sarma and J.J. Quinn, Phys. Rev. B.25, 7603 (1982).
6. N. Tzoar and C. Zhang, Phys. Rev. B33, 2642 (1986).
7. W.A. Harrison, J. Vac. Sci. Technol. 14, 1016 (1977).
8. B.N.J. Persson and J. E. Demuth, Phys. Rev. B30, 5968 (1984).
9. R.E. Palmer, J.F. Annett, and R.F. Willis, Phys. Rev. Lett. 58, 2490 (1987).
10. R.E. Camley and D.L. Mills, Phys. Rev. B29, 1695 (1984).
11. P. Lambin, J.P. Vigneron, and A.A. Lucas, Phys. Rev. B32, 8203 (1985).
12. N.J.M. Horing, T.C. Tso, and G. Gumbs, Phys. Rev. B36 (July 15, 1987).
13. G. Gumbs, Solid State Commun. 62, 365 (1987).
14. G. Gumbs, Solid State Commun. (accepted for publication).
15. P. Hawrylak, J.-W. Wu, and J.J. Quinn, Phys. Rev. B.32, 4272 (1985).
16. N.D. Mermin, Phys. Rev. B1, 2362 (1970).

MAGNETOPLASMA MODES OF THE TWO DIMENSIONAL ELECTRON GAS

C. Kallin

Department of Physics
McMaster University
Hamilton ON L8S 4M1
Canada

I. INTRODUCTION

A two-dimensional electron gas, such as is formed in the electron inversion layer of a GaAs/AlGaAs heterojunction or a silicon metal-oxide-semiconductor field-effect transistor (MOSFET), supports collective charge density wave oscillations at long wavelengths at the plasma frequency,

$$\omega_p(q) = \left(2\pi n_e e^2 q/\epsilon m^*\right)^{\frac{1}{2}}, \tag{1.1}$$

where n_e is the electron density and ϵ is the dielectric constant of the surrounding medium. In a strong perpendicular magnetic field, because of the Landau quantization of the kinetic energy, the long wavelength density oscillations occur near multiples of the cyclotron frequency $\omega_c = eB/m^*c$. These neutral excitations may be described as magnetoplasma modes or, equivalently, as "magnetic excitons," since they involve the promotion of an electron from an occupied state in one Landau level to an unoccupied state in another level. In an ideal, non-interacting two-dimensional electron gas, the energy of these excitations is just equal to the kinetic energy difference of the two Landau levels, $(n' - n)\hbar\omega_c$ and their lifetime is infinite. Both electron-electron interactions and disorder or impurity scattering will, in general, shift the energy of these modes and also give them a finite lifetime.

In this chapter, we focus on the effect of electron-electron interactions on the magnetoplasmon dispersion. Correlation effects can be very important in these systems because the reduced dimensionality enhances the importance of potential energy relative to kinetic energy. The presence of a strong magnetic field can further enhance correlation effects, and in fact, it gives rise to the fractional quantum Hall effect[1,2] at low electron densities and is believed to lead to Wigner crystallization at even lower densities.[3] Of course, disorder is always present in the experimental systems, but in the best devices (i.e. the ones with the highest mobilities) the scattering due to impurities and disorder can be very small. Thus, it is useful in the first instance to ignore the effect of disorder and concentrate on the effect of electronic correlations. The effect of disorder on these modes, and in particular, on the cyclotron mode, will be discussed in a later chapter.[4]

Anomalies have been observed in the cyclotron resonance which have been attributed to correlation effects.[5] Although in an ideal two-dimensional electron gas, electron-electron interactions can have no effect on the cyclotron resonance (this fact is known as Kohn's theorem[6]) the presence of impurities breaks the translational invariance and allows coupling to magnetoplasma modes at finite wavevector which are affected by electron-electron interactions. Thus, many body effects enter into the cyclotron resonance indirectly and the first step to understanding their effect is to calculate the magnetoplasma modes at all wavevectors.

In principle, one can measure the magnetoplasmon dispersion directly at all wavevectors experimentally, for example, by phonon absorption or reflection,[7] or by using gratings to generate electromagnetic absorption at finite wave vectors.[8] However, in practice it is very difficult to probe the interesting wavevectors which are of order the inverse magnetic length, $\ell_o^{-1} = (eB/\hbar c)^{\frac{1}{2}}$, where typically $\ell_o \sim 50\text{Å}$. Therefore, cyclotron resonance remains the simplest, although indirect, probe of the magnetoplasmon dispersion and for this reason we will focus on the excitations which occur near the cyclotron frequency. However, most of the formalism is directly applicable to the study of excitations near multiples of ω_c, as well as to spin excitations which occur near multiples of the cyclotron frequency shifted by the Zeeman energy.[9]

The energy spectra of the collective charge density wave oscillations are determined by locating the poles of the density response function, $\chi_\rho(\mathbf{q}, \omega)$. For each wave vector $\mathbf{q}$, one obtains a set of dispersion curves

$$E_m(\mathbf{q}) = m\hbar\omega_c + \Delta E_m(\mathbf{q}), \tag{1.2}$$

where $m = n' - n$ and the energy shift $\Delta E_m(\mathbf{q})$ depends on m and also on the filling factor, $\nu = 2\pi\ell_o^2 \cdot n_e$. There may be several branches for each value of m, which can be distinguished by an additional index μ when necessary.

As will be seen below, for the special case of integral filling and strong magnetic fields, the collective excitations have inverse lifetimes which are of the order of $(e^2/\epsilon\ell_o)^2/\hbar\omega_c$, and hence can be neglected if the magnetic field is sufficiently strong. However, in general, even in the absence of disorder, the collective excitations will have a finite lifetime which is given by the inverse of the imaginary part of ΔE_m. Lifetimes due to electron-electron scattering are difficult to calculate, and, for the two-dimensional electron gas in a magnetic field, they have only been estimated in the low density, long wavelength limit for temperatures sufficiently high that the dominant scattering is thermal.[10] The calculations which will be described here can only be used to determine energy shifts, not lifetimes. Therefore, throughout our analysis, ΔE_m is real.

If the magnetic field is sufficiently strong, the energy shift may be expanded in powers of $(e^2/\epsilon\ell_o)/\hbar\omega_c$. In general, we will only calculate ΔE_m to lowest order in which case it is proportional to $e^2/\epsilon\ell_o$. Also, as mentioned above, we restrict our analysis to the case $m = 1$, which is relevant to cyclotron resonance.

The simplest approximation in which to calculate the density response function, and one that is often used, is the random phase approximation (RPA).[11] However, as we shall see below, the RPA is unreliable at intermediate and large wave vectors because it ignores important exchange contributions to the energy. When an integral number of Landau levels are initially occupied, one can reliably treat the interaction in a self-consistent perturbation theory if the magnetic field is sufficiently strong.[9] This perturbative calculation, which is essentially a restricted Hartree-Fock approximation,[12] is exact to lowest order; that is, it gives $\Delta E_1(q)$ exactly to order $e^2/\epsilon\ell_o$ and neglects contributions of order $(e^2/\epsilon\ell_o)^2/\hbar\omega_c$. For larger values of m

When the filling factor is nonintegral, the perturbative approximation is unreliable. In this case, one can use a variational approach which is called the (generalized) single mode approximation (SMA). This approximation was first applied to the two dimensional electron gas in a magnetic field by Girvin et. al.[13,14] to study the low-lying neutral excitations which are relevant to the fractional quantum Hall effect. It was later generalized by MacDonald et. al.[15] to study other collective modes of the system. In the SMA, the excitation energies are related to the ground state pair correlation function. Thus, the success of the approximation is largely determined by the accuracy to which the pair correlation function is known. In fact, if the exact pair correlation function is used, the SMA gives the exact average excitation energy (for each wave vector) weighted by the corresponding oscillator strengths. Thus the approximation can give meaningful results for the energy shifts provided the starting assumption that the mode is a collective oscillation is satisified. It is important to note that one can retrieve the Hartree Fock approximation from the SMA by using the uncorrelated pair correlation function for the partially occupied Landau level.[15] However, one can do much better for the special fillings where the fractional quantum Hall effect is seen by using the pair correlation function determined by Laughlin's wavefunction.[2] This is expected to be a very good approximation to the exact correlation function at these densities,[16] and, hence, the calculated excitation energies should also be reasonably accurate.

The remainder of this chapter is organized as follows. The details of the model used to calculate the magnetoplasma modes, as well as the general formalism which uses the density response function, are explained in Sec.II. The case of integral filling is treated in Sec. III and that of nonintegral filling in Sec. IV. A brief discussion of the qualitative features of the results is given in Sec. V.

II. MODEL AND DENSITY RESPONSE FUNCTION

The electrons in the inversion layer of a GaAs or silicon device, at low temperatures, behave dynamically as a two dimensional electron gas and in a perpendicular magnetic field may be modelled by the following Hamiltonian:

$$\mathcal{H} = \mathcal{H}_0 + \mathcal{H}_{e-e}, \tag{2.1}$$

where

$$\mathcal{H}_0 = \frac{1}{2m^*} \sum_i (\mathbf{p}_i - e\mathbf{A}_i/c)^2 + g\mu_B \sum_i \mathbf{B} \cdot \mathbf{S}_i, \tag{2.2}$$

$$\mathcal{H}_{e-e} = \sum_{ij} v(\mathbf{r}_i - \mathbf{r}_j) - \frac{Nn_e}{2} \int d\mathbf{r}v(\mathbf{r}). \tag{2.3}$$

and $v(\mathbf{r})$ is an effective electron-electron interaction which is modified from the bare Coulomb interaction $e^2/\epsilon r$ because of the finite thickness of the inversion layer. The effective interaction which we will use in our calculations is more simply written in Fourier space as[17]

$$v(q) = \frac{2\pi e^2}{\epsilon q} F(q), \tag{2.4}$$

where

$$F(q) = \frac{1}{8} \left(1 + \frac{q}{b}\right)^{-3} \left(8 + 9\frac{q}{b} + 3\left(\frac{q}{b}\right)^2\right). \tag{2.5}$$

We have neglected the difference in dielectric constants on the two sides of the interface, as is appropriate for GaAs heterojunctions. Here $b^{-1} = Z_0/3$, where Z_0 is the average distance of the inversion layer electrons from the interface and which we for infinite b (the ideal two dimensional case) to $b\ell_o = 3$ in order to illustrate the dependence of the results on the inversion layer thickness.

We are interested in the low temperature regime where correlation effects are most noticeable and most of our analysis is restricted to zero temperature. In the case of integral filling, the lowest lying excitations are near the cyclotron energy and so for temperatures much smaller than this energy, thermal effects will be completely negligible. However, for fractional fillings there are many low-lying excitations, and it would be of interest to study the temperature dependence. This can be done, at least within the Hartree Fock approximation,[18] although it is not discussed here.

The magnetoplasmon dispersion curves are determined by calculating the density response function, which is related to the density correlation function by

$$\chi_\rho(\mathbf{q}, \omega) = -i \int_0^\infty dt\, e^{i\omega t} \langle [\rho_{\mathbf{q}}(t), \rho_{-\mathbf{q}}(0)] \rangle. \qquad (2.6)$$

The electron density operator is

$$\rho_{\mathbf{q}}(t) = e^{i\mathcal{H}t} \sum_j e^{i\mathbf{q}\cdot\mathbf{r}_j} e^{-i\mathcal{H}t}. \qquad (2.7)$$

This response function has poles at the frequencies $\omega(\mathbf{q})$ corresponding to charge density wave excitations or magnetoplasma modes of the system. The imaginary part of $\omega(\mathbf{q})$ is the decay rate of the excitation. In the following section, the density response function is calculated perturbatively in the electron-electron interactions.

III. INTEGRAL FILLING – HARTREE FOCK APPROXIMATION

The case where an integral number of Landau levels are initially occupied and the magnetic field is very large is particularly simple and is closely related to the problem of two particles of opposite charge in a magnetic field.[19] In fact, in the simplest cases, except for single particle self energy corrections, the excitation energy is simply the sum of the energy in the random phase approximation and the binding energy of the two particles.

What simplifies the case of integral filled Landau levels is the absence of any low-lying excitations. Since there are no allowed intra-Landau level excitations because of the Pauli exclusion principle, the lowest lying excitations are at energies near the cyclotron frequency. Therefore, one may treat electron-electron interactions perturbatively, with the small parameter being the typical interaction energy, $e^2/\epsilon\ell_o$, divided by the cyclotron energy, $\hbar\omega_c$. The shift to the magnetoplasma modes can be calculated exactly to lowest order in this parameter. This calculation is described in papers by Kallin and Halperin,[9] using a diagrammatic approach, and by MacDonald,[12] using time-dependent perturbation theory within the Hartree Fock approximation. These formalisms, which are equivalent and which we will refer to as the Hartree Fock approximation (HFA), contain some contributions from all orders in $(e^2/\epsilon\ell_o)/\hbar\omega_c$, but are only exact to lowest order.

The HFA gives a self-consistent integral equation to solve for the density response function which is shown diagrammatically in Fig. 1. This integral equation can be diagonalized leaving only a matrix equation to be solved. This calculation is described in detail in Ref. 9 and here we simply state the resulting secular equation for the excitation energies:

$$\sum_{nn'\sigma} \left[[\hbar\omega - (n' - n)\hbar\omega_c]\delta_{n',m'}\delta_{n,m}\delta_{\sigma\sigma'} - E_{\sigma'\sigma}^{\mathrm{HF}}(m', m, n', n; \mathbf{q}) \right] B_{n'n}^\sigma(\mathbf{q}, \omega) = 0, \qquad (3.1)$$

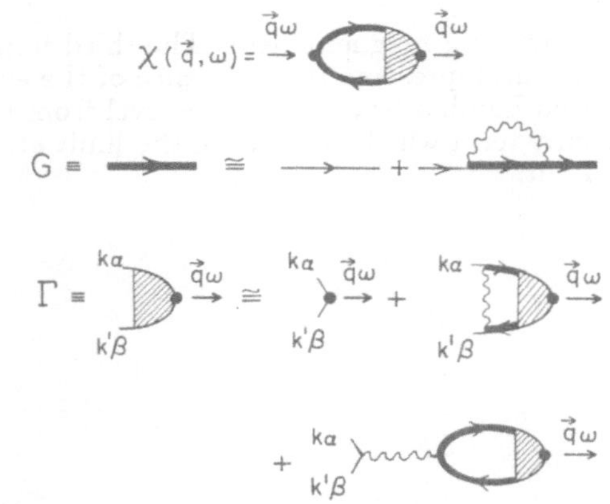

Fig. 1. Hartree Fock approximation for the response function $\chi(\mathbf{q},\omega)$ is shown. $G_\alpha(\omega)$ (thick line) is the single-particle Green's function and $\Gamma_{\alpha\beta}(k,k';\mathbf{q},\omega)$ is the vertex part. Wiggly lines represent the unscreened electron-electron interaction.

where

$$E_{\sigma'\sigma}^{\mathrm{HF}}(m',m,n',n;\mathbf{q}) = (\Sigma_{n'\sigma'} - \Sigma_{n\sigma})\delta_{n',m'}\delta_{n,m}\delta_{\sigma'\sigma} + (f_{m\sigma} - f_{m'\sigma})$$
$$\times \,[H(m',m,n',n;\mathbf{q}) - \delta_{\sigma',\sigma}X(m',m,n',n;\mathbf{q})]. \quad (3.2)$$

Here, $f_{m\sigma}$ is the occupation of the state with Landau level index m and spin σ,

$$H(m',m,n',n;\mathbf{q}) = v(q)F_{mm'}^{*}(\mathbf{q})F_{nn'}(\mathbf{q}), \quad (3.3)$$

$$X(m',m,n',n;\mathbf{q}) = \int \frac{d\mathbf{q}}{2\pi} v(q)e^{i(\mathbf{k}\times\mathbf{q})\cdot\mathbf{z}} F_{mn}(\mathbf{q})F_{n'm'}^{*}(\mathbf{q}), \quad (3.4)$$

$$\Sigma_{n\sigma} = -\sum_m f_{m\sigma}X(m,m,n,n;0), \quad (3.5)$$

$$F_{nm}(\mathbf{q}) = \left(\frac{m!}{n!}\right)^{\frac{1}{2}}\left(\frac{iq_x - q_y}{\sqrt{2}}\right)^{n-m} L_m^{n-m}\left(\frac{q^2}{2}\right)e^{-q^2/4}, \quad (3.6)$$

where L_m^{n-m} is a Laguerre polymonial.

The shifts in the magnetoplasma modes, $\omega(\mathbf{q}) - (n'-n)\hbar\omega_c$, are given by the eigenvalues of $E_{\sigma'\sigma}^{\mathrm{HF}}$. There are three different contributions to $E_{\sigma'\sigma}^{\mathrm{HF}}$: (i) The direct or Hartree contributions $H(m',m,n',n;\mathbf{q})$, which correspond to the bubble diagrams in Fig. 1 and are the only terms kept in the random phase approximation; (ii) the exchange or Fock contributions $X(m',m,n',n;\mathbf{q})$, which correspond to the ladder diagrams in Fig. 1; and (iii) the exchange self-energy corrections $\Sigma_{n'\sigma'} - \Sigma_{n\sigma}$. In terms of electron and hole propagators, the second term is the direct Coulomb interaction between the particle and the hole, and is related to the binding energy of two particles of opposite charge in a magnetic field. The third term is a constant, independent of wave vector, and represents the difference of the exchange self-energy of an electron in the excited Landau level and in the level from which the electron is removed. This is the only term which survives in the limit $q\ell_o \to \infty$, since both H and X vanish in this limit.

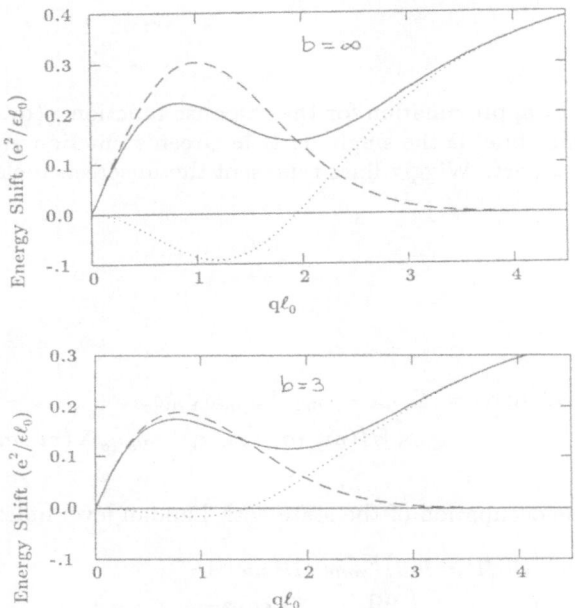

Fig. 2. The magnetoplasmon dispersion curves are shown in the Hartree Fock approximation for filling $\nu = 1$, in the limit as the inversion layer thickness goes to zero ($b \to \infty$) and for a finite thickness, $b = 3$.

The strong field approximation corresponds to restricting the sum over n and n' in Eq. (3.1) to keep only the terms of order $e^2/\epsilon\ell_o$. For example, for filling $\nu = 1$ and $m = n' - n = 1$, the only term which survives in the strong field approximation is $m' = n' = 1$, $m = n = 0$. If we neglect the width of the electron inversion layer so that $v(q)$ is simply the Coulomb potential, then analytic expressions (not involving integrals) can be found for all terms in $E^{\mathrm{HF}}_{\sigma'\sigma}$, and we find the energy shift is[9,20]

$$\Delta E_1(\mathbf{q}) = \frac{e^2}{\epsilon\ell_o} \frac{\pi^{\frac{1}{2}}}{8} \left(1 - e^{-q^2/4}\left[(1 + q^2/2)I_0(q^2/4) - \frac{q^2}{2}I_1(q^2/4)\right] + \left(\frac{2}{\pi}\right)^{\frac{1}{2}} qe^{-q^2/2}\right),$$

(3.7)

where I_n is a modified Bessel function of the first kind.

For finite inversion layer width, i.e. finite b in Eq. (2.5), the integrals in $E^{\mathrm{IIF}}_{\sigma'\sigma}$

must be done numerically. The results for $\nu = 1$ and $b \to \infty$ and $b = 1$ are shown in Fig. 2. Similar curves are obtained for $\nu = 2$.[9] For $\nu = 3$ an electron with minority spin can make a transition from the lowest Landau level to the second Landau level or an electron with majority spin can make a transition from the second Landau level to the third Landau level. These two transitions are coupled by the Coulomb interaction and Eq. (3.1) gives a 2×2 matrix to diagonalize. The results are shown in Fig. 3.

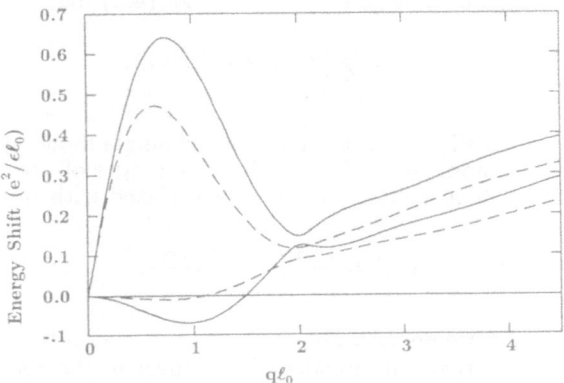

Fig. 3. The magnetoplasmon dispersion curves in the Hartree Fock approximation are shown for filling $\nu = 3$ and $b \to \infty$ (solid curves) and $b = 3$ (dashed curve).

IV. SINGLE MODE APPROXIMATION FOR PARTIALLY FILLED LANDAU LEVELS

The Hartree-Fock approximation discussed above may be extended to non-integral filling factors,[18] but it is not reliable for these densities because it neglects important correlations between electrons in the partially filled level. An approximation which does much better in treating these correlations is the single mode approximation (SMA). The SMA was originally applied to the two-dimensional electron gas by Girvin et. al.[13] to study the intra-Landau level modes relevant to the fractional quantum Hall effect. It was later generalized by MacDonald et. al.[15] to study inter-Landau level modes, such as the ones relevant to cyclotron resonance. The SMA is expected to be valid for any excitation which is collective in nature.

The essential feature of the SMA is the assumption that the excited states can be constructed by forming density waves in the ground state. In zero magnetic field, the SMA starts from the ansatz that an excited state with wave vector $\mathbf{q}$ may be written as

$$|\psi_{\mathbf{q}}\rangle = \rho_{\mathbf{q}}|\psi_0\rangle, \tag{4.1}$$

where $|\psi_0\rangle$ is the ground state and $\rho_{\mathbf{q}}$ is the density operator. In a magnetic field all the excitations of the non-interacting system occur at integer multiples of the cyclotron frequency (possibly shifted by the Zeeman energy, if one also considers excitations in which the spin of the electron is reversed). In the interacting system, in a strong magnetic field, the excitations will still occur near integer multiples of the cyclotron frequency and they may be grouped according to the Landau levels involved. Therefore, it is useful to separate the density operator into contributions corresponding to transitions between different pairs of Landau levels, i.e.

$$\rho_{\mathbf{q}\sigma} = \sum_{n'n} \rho_{\mathbf{q}\sigma}^{n'n},$$ (4.2)

where $\rho_{\mathbf{q}\sigma}^{n'n}$ is the part of the density operator which transfers an electron with spin σ from the nth to the n'th Landau level. (See Ref. 15.) Then the ansatz excited state wave function which is analogous to that of Eq. (4.2) may be written as

$$|\psi_{\mathbf{q}}\rangle = \sum_{n'n\sigma} C_{n'n}^{\sigma}(\mathbf{q})\rho_{\mathbf{q}\sigma}^{n'n}|\psi_0\rangle.$$ (4.3)

This approximation is usually called the generalized single mode approximation (GSMA), since it can be applied to all the collective modes of the system, including spin density excitations. The excitation energy associated with this state is

$$\Delta(\mathbf{q}) = \frac{\langle \psi_{\mathbf{q}}|\mathcal{H} - E_0|\psi_{\mathbf{q}}\rangle}{\langle \psi_{\mathbf{q}}|\psi_{\mathbf{q}}\rangle},$$ (4.4)

where E_0 is the ground state energy. By using the expression for $|\psi_{\mathbf{q}}\rangle$ in terms of the density operators and then minimizing with respect to the coefficients $C_{n'n}^{\sigma}(\mathbf{q})$, one finds the following secular equation for the excitation energies:

$$\sum_{n'n\sigma} [E_{\sigma'\sigma}(m',m,n',n;\mathbf{q}) - \Delta(\mathbf{q})S_{\sigma'\sigma}(m',m,n',n;\mathbf{q})]C_{n'n}^{\sigma}(\mathbf{q}) = 0.$$ (4.5)

Here, the matrix elements are defined in terms of the following ground state expectation values:

$$E_{\sigma'\sigma}(m',m,n',n;\mathbf{q}) = \langle \psi_0|\rho_{-\mathbf{q}\sigma'}^{mm'}[\mathcal{H}, \rho_{\mathbf{q}\sigma}^{n'n}]|\psi_0\rangle,$$ (4.6)

$$S_{\sigma'\sigma}(m',m,n',n;\mathbf{q}) = \langle \psi_0|\rho_{-\mathbf{q}\sigma'}^{mm'}\rho_{\mathbf{q}\sigma}^{n'n}|\psi_0\rangle.$$ (4.7)

Since the commutator of $\rho_{\mathbf{q}\sigma}^{n'n}$ with the kinetic energy part of $\mathcal{H}$ is given by

$$[T, \rho_{\mathbf{q}\sigma}^{n'n}] = \hbar\omega_c(n' - n)\rho_{\mathbf{q}\sigma}^{n'n},$$ (4.8)

it follows that

$$E_{\sigma'\sigma}(m',m,n',n;\mathbf{q}) = \hbar\omega_c(n' - n)S_{\sigma'\sigma}(m',m,n',n;\mathbf{q})$$
$$+ \frac{1}{2}\int \frac{d^2q}{(2\pi)^2}v(q)\langle \psi_0|\rho_{-\mathbf{q}\sigma'}^{mm'}[\rho_{-\mathbf{k}}\rho_{\mathbf{k}}, \rho_{\mathbf{q}\sigma}^{n'n}]|\psi_0\rangle.$$ (4.9)

The correlations all appear in the last term in Eq. (4.9). It can be shown that $E_{\sigma'\sigma}$ reduces to $E_{\sigma'\sigma}^{\mathrm{HF}}$ if the Hartree Fock approximation is used for the ground state $|\psi_0\rangle$.[21] By using the definition of the partial density operators, $E_{\sigma'\sigma}$ can be written in terms of the electron static structure factor in the ground state, $s(q)$, although the actual expression becomes increasingly complex for larger values of the Landau indices. The Hartree Fock approximation then corresponds to using the uncorrelated structure factor, $s(q) = 1 - \nu_o e^{-q^2/2}$, where ν_o is the filling factor of the partially occupied Landau level.

If we restrict the sum over Landau indices in Eq. (4.5) to $n'=n=0$, we obtain the expression derived by Girvin. et. al.[13] for the low-lying excitations relevant to the fractional quantum Hall effect. The modes near the cyclotron frequency are obtained by considering $n'-n=1$, where the values of n are determined by the filling factor. The case where all the electrons are in the lowest Landau level, $\nu < 2$, corresponds to $n'=1$, $n=0$ and has been treated by Oji and MacDonald.[18] In that case, if both spins are present, $\nu > 1$, one obtains a 2×2 matrix to diagonalize, since electrons of either spin can be excited to the next Landau levels. The comparison between the HFA and the GSMA is shown in Fig. 4 for the magnetoplasma modes at filling $\nu_\downarrow=1$ and $\nu_\uparrow=1/3$. The two modes are neither pure charge nor pure spin excitations in that they correspond to poles both in the charge density response function and in the spin density response function. However, the upper mode is mainly chargelike in that the residue of the corresponding pole in the charge density response function is larger than that in the spin response function and, in fact, at zero wave vector this mode is a pure charge density excitation. Similarly, the lower mode is mainly spinlike. One sees from Fig. 4 that the modes are stiffer in the GSMA than in the HFA. This is because the HFA, by neglecting the strong correlations in the partially filled Landau level, underestimates the energy cost of promoting an electron from the partially filled level to an empty level. On the other hand, these correlations are put into the GSMA, by assuming the partially filled level is well described by Laughlin's highly correlated wavefunction for 1/3 filling.[2]

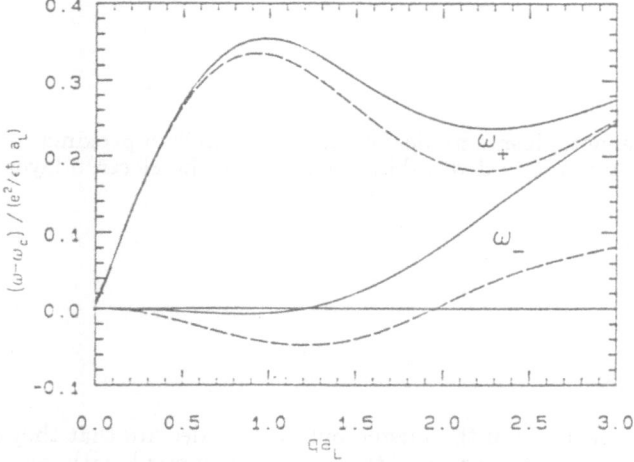

Fig. 4. The Hartree Fock (dashed curves) and single mode approximations (solid curves) for the magnetoplasma modes are compared at filling $\nu = 4/3$ and vanishing thickness of the electron layer.

The case of $2 < \nu < 3$ is of particular interest because anomalies have been seen in the cyclotron resonance at these densities. In addition, these are the lowest filling factors where one obtains three modes, corresponding to excitations from the lowest Landau level with both spins and from the second Landau level with majority spin. Therefore, there is a 3×3 matrix to be diagonalized in this case. It is not clear whether a Laughlin-like state is the ground state for fractional fillings with more than two Landau levels occupied. Although structure has been seen in the resistivity suggesting the existence of a fractional quantum Hall state, no plateaus have been observed in the Hall resistivity which would confirm the existence of such a state. Calculations suggest that a Laughlin-like state may be marginally stable at filling

$\nu = 7/3$ and somewhat more stable at $\nu = 11/5$.[20] We have calculated the excitations at various fractional fillings between 2 and 3, assuming both a Laughlin state in the partially filled level and a Hartree Fock state. The calculation is very similar to those detailed in Refs. 15 and 18, and the resulting magnetoplasmon excitation curves are shown in Fig. 5 for $\nu = 7/3$. In all cases, the modes are softer in the SMA than in the HFA.

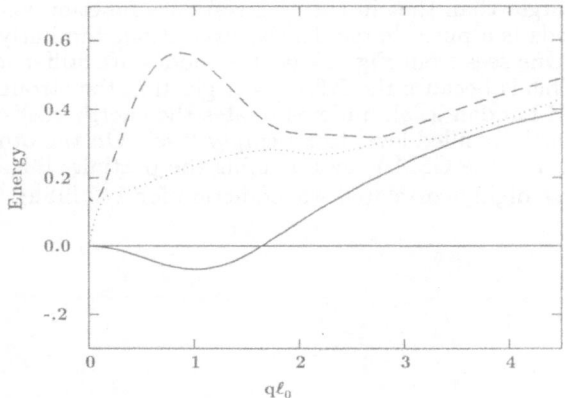

Fig. 5. The magnetoplasma modes in the single mode approximation are shown for density $\nu = 7/3$ and vanishing thickness of the electron layer.

V. DISCUSSION

The general features of the magnetoplasma modes are that they exhibit structure at wavelengths of the order of the inverse magnetic length, and at larger wave vectors they are shifted up in energy from the cyclotron frequency because of self-energy exchange effects. This structure will affect the cyclotron resonance, as will be discussed in the next chapter.[4] The random phase approximation describes the cyclotron mode well at small wave vectors, but breaks down at intermediate and large wave vectors and, in addition, misses other modes which also have a charge density component at finite wave vectors. The Hartree Fock approximation does well near integral fillings if the magnetic field is strong, but this approximation overestimates the energy shifts at fractional fillings. At fractional densities, the single mode approximation is expected to give a very good estimate of the energy shifts.

ACKNOWLEDGEMENTS

The material presented here is based on work done in collaboration with B.I. Halperin and with A.H. MacDonald. The author acknowledges the support of the Natural Sciences and Engineering Research Council of Canada and of the Alfred P. Sloan Foundation.

REFERENCES

1. H.L. Stormer, A. Chang, D.C. Tsui, J.C.M. Hwang, A.C. Gossard, and W. Wiegmann, Fractional quantization of the Hall effect, Phys. Rev. Lett. 50:1953 (1983).
2. R.B. Laughlin, Anomalous quantum Hall effect: An incompressible quantum fluid with fractionally charged excitations, Phys. Rev. Lett. 50:1395 (1983).
3. D.J. Yoshioka and P. A. Lee, Ground-state energy of a two-dimensional charge-density-wave state in a strong magnetic field, Phys. Rev. B 27:4986 (1983).
4. C. Kallin, Many body effects on the cyclotron resonance in electron inversion layers, in: "Interfaces, Quantum Wells and Superlattices,"

5. Z. Schlesinger, S.J. Allen, J.C.M. Hwang, P.M. Platzman, and N. Tzoar, Cyclotron resonance in two dimensions, Phys. Rev. B 30:435 (1984).
6. W. Kohn, Cyclotron resonance and de Hass-van Alphen oscillations of an interacting electron gas, Phys. Rev. 123:1242 (1961).
7. J.C. Hensel, B.I. Halperin, and R.C. Dynes, Dynamical model for the absorption and scattering of ballistic phonons by the electron inversion layer in silicon, Phys. Rev. Lett. 51:2302 (1983).
8. S.J. Allen, Jr., D.C. Tsui, and R.A. Logan, Observation of the two-dimensional plasmon in silicon inversion layers, Phys. Rev. Lett. 38:980 (1977).
9. C. Kallin and B.I. Halperin, Excitations from a filled Landau level in the two-dimensional electron gas, Phys. Rev. B 30:5655 (1984).
10. H. Fukuyama, Y. Kuramoto, and P.M. Platzman, Many-body effects on level broadening and cyclotron resonance in two-dimensional systems under strong magnetic fields, Phys. Rev. B 19:4980 (1979).
11. K.W. Chiu and J.J. Quinn, Plasma oscillations of a two-dimensional electron gas in a strong magnetic field, Phys. Rev. B 9:4724 (1974).
12. A.H. MacDonald, Hartree-Fock approximation for response functions and collective excitations in a two-dimensional electron gas with filled Landau levels, J. Phys. C 18:1003 (1985).
13. S.M. Girvin, A.H. MacDonald, and P.M. Platzman, Magneto-roton theory of collective excitations in the fractional quantum Hall effect, Phys. Rev. B 33:2481 (1986).
14. S.M. Girvin, Phonons, rotons and fractionally charged vortices in the quantum Hall effect, in: "Interfaces, Quantum Wells and Superlattices,"
15. A.H. MacDonald, H.C.A. Oji, and S.M. Girvin, Magnetoplasmon excitations from partially filled Landau levels in two dimensions, Phys. Rev. Lett. 55:2208 (1985).
16. F.D.M. Haldane and E. Rezayi, Finite size studies of the incompressible state of the fractionally quantized Hall effect and its excitations, Phys. Rev. Lett. 54:237 (1985).
17. T.Ando, A.B. Fowler, and F. Stern, Electronic properties of two-dimensional systems, Rev. Mod. Phys. 54:437 (1982).
18. H.C.A. Oji and A.H. MacDonald, Magnetoplasma modes of the two-dimensional electron gas at nonintegral filling factors, Phys. Rev. B 33:3810 (1986).
19. Yu.A. Bychkov, S.V. Iordanskii, and G.M. Eĺiashberg, Two-dimensional electrons in a strong magnetic field, Pis'ma Zh. Eksp. Teor. Fiz. 33:152 (1981) JETP Lett. 33:143 (1981)].
20. A.H. MacDonald and S.M. Girvin, Collective excitations of fractional Hall states and Wigner crystallization in higher Landau levels, Phys. Rev. B 33:4009 (1986).

MANY BODY EFFECTS ON THE CYCLOTRON RESONANCE IN ELECTRON INVERSION LAYERS

C. Kallin

Department of Physics
McMaster University
Hamilton ON L8S 4M1
Canada

I. INTRODUCTION

At low temperatures, the electrons in the inversion layer of a silicon metal-oxide-semiconductor field-effect-transistor or a GaAs/AlGaAs heterojunction behave dynamically as a two dimensional electron gas.[1] This leads to interesting correlation effects for a number of reasons. The reduced dimensionality typically enhances the importance of potential or interaction energies relative to the kinetic energy. Also, it is possible to vary the two-dimensional density of electrons and, hence, to vary the relative strength of the Coulomb potential. Finally, in the best devices, the scattering due to impurities is very small and can, in fact, be weaker than electron-electron scattering, making it possible to observe correlation effects experimentally. The presence of a strong perpendicular magnetic field can further enhance correlation effects because of the Landau quantization of the kinetic energy. For example, at low densities and strong magnetic fields, correlation effects give rise to the fractional quantum Hall effect.[2,3]

In this chapter we study the cyclotron resonance lineshape of an interacting two-dimensional electron gas at low temperatures. In an ideal (i.e. translationally invariant) system, electron-electron interactions have no effect on cyclotron resonance. This result is known as Kohn's theorem[4] and follows from the fact that cyclotron resonance, which probes a collective excitation at zero wavevector, is a center-of-mass probe, i.e., the perturbing potential is proportional to the total momentum. On the other hand, electron-electron interactions only affect the relative degrees of freedom, which do not couple to the center-of-mass motion in a translationally invariant system. However, the presence of disorder or impurities breaks the translational symmetry and allows coupling to magnetoplasma modes at nonzero wavevectors where correlation effects are important. Structure in the magnetoplasma modes of the two dimensional electron gas will then lead to structure in the cyclotron resonance or infrared absorption. Although correlation effects only enter in this indirect way, through impurity scattering, we will see that their influence on the cyclotron resonance can be very large in the limit of weak impurity scattering.

A number of anomalies have been seen in the cyclotron resonance of electron inversion layers which can be attributed to correlation effects. For example, the existence of higher harmonics, i.e. peaks in the optical absorption at integer multiples of the cyclotron frequency, has been explained by interparticle interactions.[5-7] In addition, it has been suggested that the line width broadening[8-10] or line-splitting,[11] which is observed at certain filling factors or electron densities, can also be explained by correlation effects.[11]

We will see that, in the purer (high mobility) samples, correlations actually affect the entire cyclotron resonance, that is its position and its shape. It is not just the anomalies which can only be understood by introducing correlation effects. To understand cyclotron resonance lineshapes in high mobility samples, one needs to introduce electron-electron interactions from the start. This can be understood qualitatively by comparing how disorder and impurity scattering affect the magneto-plasma modes of the non-interacting system to how scattering affects the magnetoplasma modes of the interacting system, which we studied in an earlier chapter.[12] In general, impurity scattering will broaden and shift these modes. This is shown schematically in Fig. 1. Modes which have many states nearby in energy (and non-vanishing matrix elements between these states) will be broadened the most. In the non-interacting system, the modes occur at the cyclotron frequency for all wavevectors, and, hence, impurities will be very effective in broadening the cyclotron mode. However, in the interacting system, the modes at finite wavevector are shifted away from the cyclotron energy and so one would expect impurities to be less effective in broadening the cyclotron mode in this case. Therefore, for a given impurity potential, at low temperatures, one expects a narrower cyclotron resonance line in the interacting system than in the non-interacting case. Also, because there is no longer any symmetry about the cyclotron frequency in the interacting system, even if one neglects nonparabolicity effects and Landau level mixing, one would expect the cyclotron resonance to be shifted away from the bare cyclotron frequency.

When impurity scattering is very strong, the broadening of the modes at long wavelengths will be larger than the shift away from the cyclotron frequency due to electron-electron interactions and the effect of the shift will be negligible. Thus, in lower mobility samples, correlation effects will be "washed out" by impurity scattering. (Another example where correlation effects can be observed only in a disordered sample, but where too much disorder will wash out the effect, is the fractional quantum Hall effect.[2,3] In that case, it follows from Lorentz invariance, rather than Kohn's theorem, that electron-electron interactions have no effect on the Hall conductivity of the ideal, pure system. Also, in the very dirty limit, disorder will wash out correlation effects. However, for weak but nonzero disorder, correlations can have a very substantial effect.)

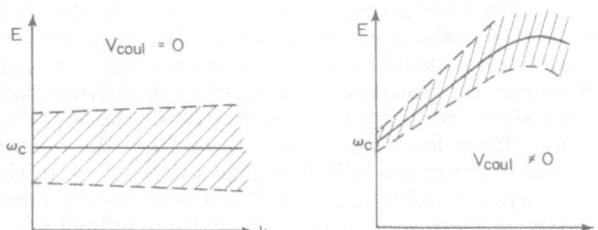

Fig. 1. Qualitative comparison of the magnetoplasma dispersion curve in the presence of disorder for the non-interacting system ($V_{coul} = 0$) and for the interacting system ($V_{coul} \neq 0$). The dashed curves denote the width of the modes. The broadening at long wavelengths is much smaller in the interacting system because there are many fewer modes nearby in energy.

It follows from the above discussion, that electron-electron and electron-impurity scattering must be treated together (and because of the singular nature of the density of states of the pure, non-interacting system, they must be treated self-consistently) in order to see what influence correlation effects have on the cyclotron resonance lineshape. The main part of this chapter describes a formalism which does exactly this and which was first developed to study the case where an integral number of Landau levels are occupied.[13] This theory can be generalized to treat other filling factors if one has a good approximation to the ground state.[14] The formalism is described in Sec. IV, but first we start with some general comments about the impurity potential in these two-dimensional systems, the role of screening (Sec. II) and the dynamical conductivity in the presence of disorder (Sec. III).

The following analysis is expected to be applicable more to GaAs heterojunctions than to silicon inversion layers for several reasons. The valley degeneracy in silicon, which we have not taken into account, will increase the density of excited states near the cyclotron frequency. Also, the effective mass is larger in silicon than in GaAs, and, therefore, Landau level mixing is more important in silicon. Both of these properties are expected to reduce the effect of electron-electron interactions in silicon, and, in particular, the effect of valley degeneracy could be very important since it may introduce new modes near the cyclotron frequency. This means that to see correlation effects, one would need much weaker electron-impurity scattering in silicon inversion layers than in GaAs heterojunctions. However, the scattering due to impurities is typically stronger, not weaker, in the silicon MOSFETs, and, hence, it seems unlikely that one could observe many body effects in these systems. Therefore, we will concentrate on electron-impurity potentials which are reasonable models for scattering in GaAs heterojunctions.

II. IMPURITY POTENTIAL

The model we consider is an interacting, disordered two-dimensional electron gas in a strong perpendicular magnetic field, at zero temperature, described by the Hamiltonian:

$$\mathcal{H} = \frac{1}{2m^*}\sum_j (\mathbf{p}_j - e\mathbf{A}_j/c)^2 + g\mu_B \sum_j \mathbf{B}\cdot\mathbf{S}_j + \sum_{i<j} v(\mathbf{r}_i - \mathbf{r}_j) + \sum_j u(\mathbf{r}_j), \quad (2.1)$$

where $\mathbf{A}_j$ is the vector potential due to the magnetic field $\mathbf{B}$, $v(\mathbf{r}) = e^2/\epsilon r$, and u is the electron-impurity potential. If the finite thickness of the electron layer is taken into account, the potentials u and v are modified,[1] but otherwise the formalism is unchanged.

The dominant scattering mechanism in GaAs/AlGaAs heterojunctions in zero magnetic field is believed to be scattering due to the ionized impurities, which are typically set back from the inversion layer by several hundred angstroms.[15] In this case, the unscreened impurity potential is the Coulomb potential,

$$u(\mathbf{r}) = \sum_k \frac{e^2}{\epsilon(|\mathbf{r} - \mathbf{R}_k|^2 + Z_k^2)^{1/2}}, \quad (2.2)$$

for impurities located at $(\mathbf{R}_k, Z_k)$. Since screening is quite different in the presence of a magnetic field, it does not immediately follow that scattering from ionized impurities is the dominant scattering mechanism in the strong field limit. However, when screening is taken into account, the relative importance of Coulomb scattering from ionized impurities is further enhanced by the presence of a strong magnetic field, at least in the integral and fractional quantum Hall regimes. In zero magnetic field, screening is strongest at long wavelengths, tending to suppress the effect of Coulomb scattering. On the other hand, for the special case of integer filling factors, the potential is not screened in the low temperature, high field limit.[10] At other fill-ings, there is screening and one can calculate the static dielectric function in the sin-gle mode approximation at fillings where the fractional quantum Hall effect is ob-served, $\nu = 1/3$, etc. In this case, there is no screening at long wavelengths, and the maximum screening is at wavelengths comparable to $2\pi\ell_o$.[14] Thus, as pointed out in Ref. 14, screening at these densities will tend to enhance the relative importance of Coulomb scattering from the remote ionized impurities, since it is predominantly long-wavelength, compared to other scattering mechanisms and in particular, com-pared to short-range scatterers. Because of these considerations, in our numerical analysis we concentrate on Coulomb scattering from remote ionized impurities.

In spite of the statements made above, it is also useful for pedagogical reasons to consider the case of short-range scatterers in the inversion layer, where the impurity potential is

$$u(\mathbf{r}) = \frac{a\hbar^2}{2m^*} \sum_k \delta(\mathbf{r} - \mathbf{R}_k)\delta_{Z_k,0}, \qquad (2.3)$$

and a is a dimensionless constant. This potential is often used to model electron-impurity scattering in silicon MOSFETs, where disorder at the silicon-oxide surface is important, and is sometimes used for modelling GaAs heterojunctions. However, as emphasized above, long-range Coulomb scattering due to the ionized impurities is believed to be more important in the GaAs systems, particularly at the fillings which we will consider.

III. DYNAMICAL CONDUCTIVITY

The absorption of infrared radiation, normally incident on the electron layer, due to cyclotron resonance is proportional to the real part of the dynamical conductivity,

$$\sigma_+(\omega) = \frac{1}{\omega} \int_0^\infty dt \, e^{i\omega t} \langle [J_+(t), J_-(t)] \rangle, \qquad (3.1)$$

where $\mathbf{J} = \frac{e}{m^*} \sum_j (\mathbf{p}_j - e\mathbf{A}_j/c)$ and $J_\pm = J_x \pm iJ_y$.

If the impurity potential is weak, one may expand the current-current correlation function in the above expression to lowest order in the impurity potential, to find[7]

$$\sigma_+(\omega) = \frac{in_e e^2/m^*}{\omega - \omega_c} + \frac{in_e e^2/m^*}{\omega(\omega - \omega_c)^2} \cdot n_I \int \frac{d\mathbf{q}}{2\pi} |U(\mathbf{q})|^2 \mathbf{q}^2 \left[\chi(\mathbf{q}, \omega) - \chi(\mathbf{q}, 0) \right], \qquad (3.2)$$

where n_e is the two-dimensional electron density and χ is the density response func-tion of the system in the absence of disorder. Here, U is related to the configuration average of the impurity potential,

$$\langle u(\mathbf{q})u(-\mathbf{q}') \rangle_c = n_I |U(\mathbf{q})|^2 (2\pi)^2 \delta(\mathbf{q} - \mathbf{q}'), \qquad (3.3)$$

and n_I is the average density of ionized impurities.

The magnetoplasma modes of the pure system correspond to poles of the density response function, i.e. $\chi(\mathbf{q}, \omega) \to \infty$. Therefore, from Eq. (3.2), one expects peaks in the conductivity at frequencies which correspond to a large density of magnetoplas-mon states. Fig. 2 shows the absorption to lowest order in the impurity potential for a single filled Landau level and short-range scatterers. The (unphysical) square-root divergences in the absorption correspond to extrema in the magnetoplasma disper-sion curve, which in turn correspond to divergent densities of states for magnetic excitons.

To lowest order in the impurity potential, there is no broadening due to electron-impurity scattering of the main cyclotron peak, as seen in Fig. 2, since this is a higher order effect. Thus, this approximation tells one nothing about how many-body effects alter the cyclotron resonance lifetime, although it does illustrate how correlation effects can enter into the absorption. Sometimes, terms higher order in the impurity potential are introduced by using the memory function approach to lowest order.[16,17] Basically, this involves putting the lowest order correction, as calcu-

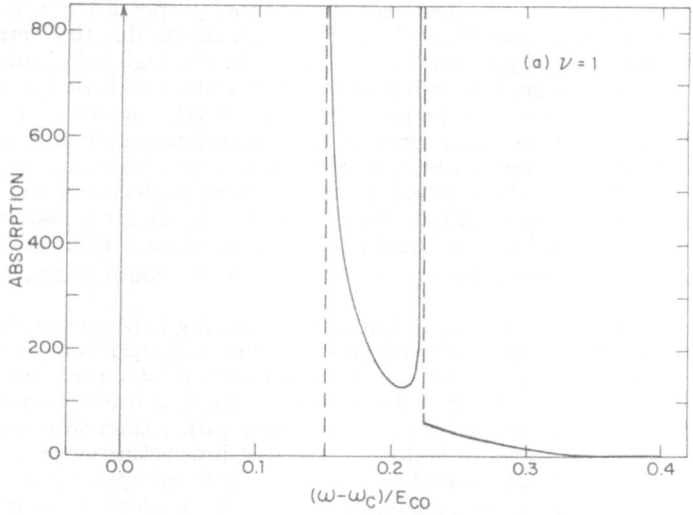

Fig. 2. The absorption near the cyclotron frequency to lowest order in the impurity potential for a single filled Landau level and short-range scatterers. The strength of the impurity potential corresponds to a mobility of 7×10^4 cm^2/Vs.

lated above, into the denominator of the expression for the conductivity. This approach will then include some terms of all orders in the impurity potential and will also reduce to the lowest order expression in the limit of weak impurity potential. However, in general, it will still result in unphysical divergences because it does not include the broadening of the magnetoplasmon peaks in a self-consistent way. It is possible to remove these divergences by arbitrarily introducing a single particle lifetime into the expression for the conductivity.[7] Although such an approach may be reasonable for silicon inversion layers, where the scattering is predominantly short-ranged, it is not a reasonable approach for GaAs heterojunctions, where long-range scatterers are important. In particular, for Coulomb scatterers, the single particle lifetime may be several orders of magnitude smaller than the transport lifetime.[18] In this case, one needs to treat the broadening in a self-consistent way.

IV. SELF-CONSISTENT EXCITON APPROXIMATION

In this section we describe a self-consistent treatment of impurity scattering for the interacting system. In general, it is very difficult to treat both electron-electron interactions and electron-impurity scattering together in a self-consistent way. The approximation described here treats electron-electron interactions exactly in the strong magnetic field limit for integral fillings and in the single mode approximation for non-integral fillings, as described in the previous chapter.[12] Electron-impurity scattering is treated in the simplest self-consistent approximation, similar to the self-consistent Born approximation which is often used for non-interacting electrons.[19] This approximation, which includes vertex corrections as well as self-energy corrections, is exact in the limit of weak electron-impurity scattering and is at least sensible for arbitrary strengths of the impurity potential (in that the unphysical divergences associated with lowest order calculations are removed).

It would certainly be desirable to treat impurity scattering in a more realistic way. For example, impurity scattering is known to localize many states in these systems, which has a dramatic effect on the transport properties and, in particular, gives rise to the quantized Hall effect. Although effects due to electron localization are not expected to play an important role in the high field limit where the cyclotron frequency is large, it is still possible that a more realistic treatment of electron-impurity scattering may be necessary to correctly interpret the experimental results. On the other hand, the present approximation is sufficient to show that electron-electron interactions can have a drastic effect on cyclotron resonance in high mobility samples and that these interactions cannot be neglected if one hopes to interpret the experiments. In addition, the interacting theory gives rise to theoretical predictions as to how the linewidth and position depend on various parameters (such as temperature, filling factor, density, etc.) which can be compared to experiment.[20]

If mixing of Landau levels due to impurity scattering is neglected, then the magnetoplasmon or exciton states with energies near the cyclotron frequency form a complete set of states which can be used to study the cyclotron mode in the presence of impurity scattering. In the interacting system, it is useful to consider the scattering of these magnetic excitons by impurities, rather than working directly with the single electron states as is done in the non-interacting system. The exciton-impurity scattering has been treated self-consistently in an approximation which neglects multiple scattering from a single impurity and neglects most correlations between impurities.[13] This approximation, which we refer to as the self-consistent exciton approximation (SCEA), is shown diagrammatically in Fig. 3.

The exciton Green's function in the presence of impurities, D, is related to the Green's function in the absence of impurities, D_0, by Dyson's equation:

$$D^{-1}(\mathbf{k},\omega) = D_0^{-1}(\mathbf{k},\omega) - \Pi(\mathbf{k},\omega), \tag{4.1}$$

where

$$D_0^{-1}(\mathbf{k},\omega) = \omega - E(\mathbf{k}) - i\epsilon, \tag{4.2}$$

and $\Pi(\mathbf{k},\omega)$ is the exciton self-energy. In general D is a matrix, but for simplicity we are assuming that for each wavevector there is only a single magnetoplasma mode with energy $E(\mathbf{k})$ near $\hbar\omega_c$, in which case D is a scalar. The equations are easily generalized to the case where there are several modes near $\hbar\omega_c$.

The integral equation which the self-energy satisfies in the SCEA is

$$\Pi(\mathbf{k},\omega) = n_I \int \frac{d\mathbf{q}}{2\pi} |U(\mathbf{k}-\mathbf{q})|^2 |M(\mathbf{k},\mathbf{q})|^2 D(\mathbf{q},\omega). \tag{4.3}$$

The matrix element between exciton states is

$$(2\pi)^2 \delta(\mathbf{k}-\mathbf{q}-\mathbf{q}') M(\mathbf{k},\mathbf{k}-\mathbf{q}) = \langle \mathbf{k}| \sum_j e^{i\mathbf{q}\cdot\mathbf{r}_j} |\mathbf{q}'\rangle, \tag{4.4}$$

where $|\mathbf{q}\rangle$ is the many-electron state with a single magnetic exciton of momentum $\mathbf{q}$. One can show that, both for integral filling and for rational filling (i.e. where the quantum Hall effect is observed and the ground state is isotropic), in the limit $q \to 0$, the matrix element satisfies[14]

$$M(\mathbf{k},\mathbf{k}-\mathbf{q}) \to i(\mathbf{q} \times \mathbf{k}) \cdot \hat{\mathbf{z}}. \tag{4.5}$$

This reflects the fact that the magnetic excitons have a well defined dipole moment which is proportional to and perpendicular to their wavevector.[13]

Finally, the absorption is simply related to the imaginary part of the exciton Green's function:

$$\text{Re}[\sigma_+(\omega)] = \frac{n_e e^2}{m^*} \frac{\omega_c}{\omega} \text{Im}[D(0,\omega)].$$

(4.6)

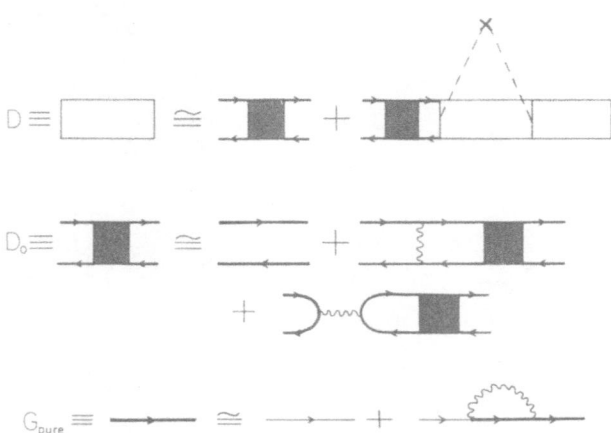

Fig. 3. Self-consistent exciton approximation for the particle-hole Green's function $D(\mathbf{q}, \omega)$. D_0 is the particle-hole Green's function in the absence of impurity scattering. The dashed lines denote the electron-impurity interaction and the wiggly lines denote the unscreened electron-electron interaction. The lines with arrows denote the single-particle Green's function of the noninteracting (thin lines) and interacting (thick lines) systems without impurities.

V. RESULTS

The Green's function in the absence of impurities and the exciton matrix elements M can be calculated analytically for the simple case of integral filling as described in an earlier chapter.[12] For rational fillings, where the single mode approximation is used, the pure Green's function is calculated numerically. In either case, for a particular choice of the impurity potential, the radial part of the integral equation (4.3) must be iterated numerically to achieve self-consistency. (The angular part can be treated analytically for central potentials.)

As discussed earlier, a reasonable model for the impurity potential at integral fillings is the unscreened Coulomb potential. If one assumes that the ionized impurities are uniformly distributed in a layer of width t which is set back a distance α from the electron layer, then the quantity which appears in the integral equation is

$$|U(\mathbf{q})|^2 = \left(\frac{2\pi e^2}{\epsilon}\right)^2 \frac{e^{-2\alpha q}}{q^2} \left(\frac{1 - e^{-2tq}}{2tq}\right),$$

(5.1)

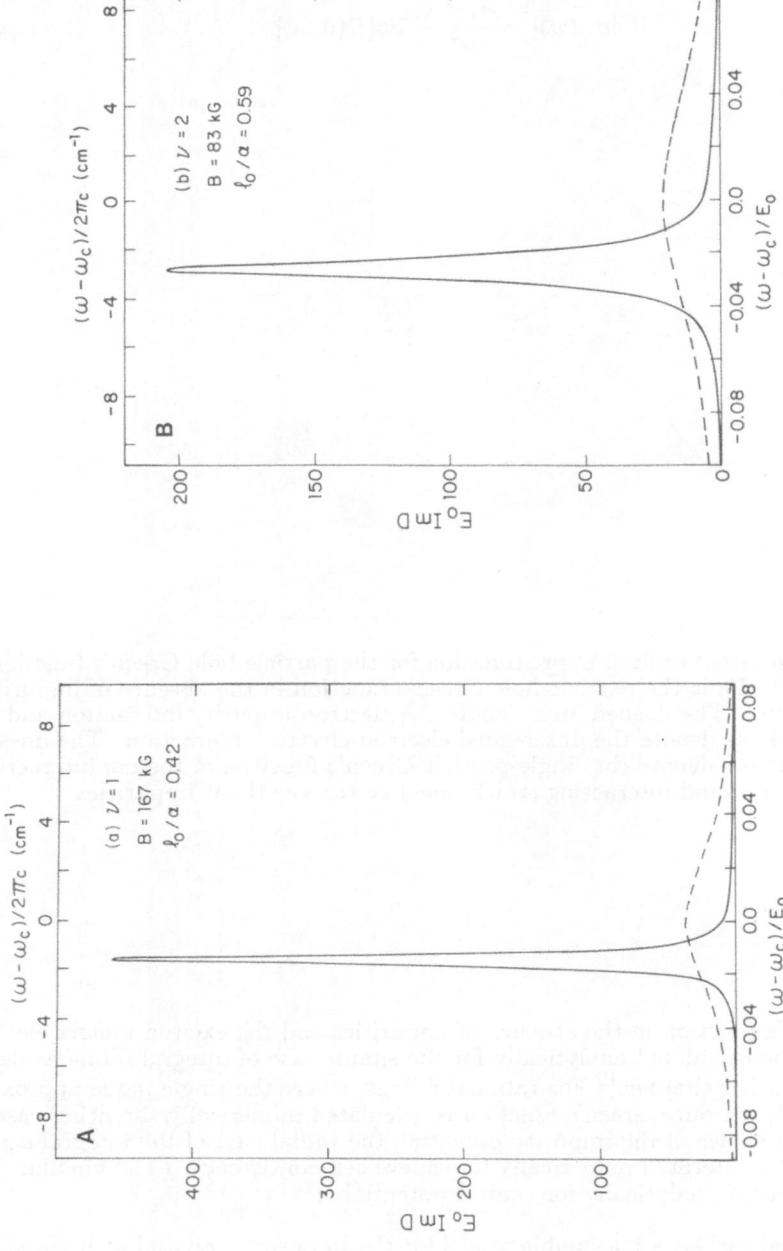

Fig. 4. The calculated infrared absorption near ω_c is shown for long-range scatterers and filling factors (a) $\nu = 1$ and (b) $\nu = 2$. The electron density is 4×10^{11} cm^{-2}, the impurity setback α is 150Å, and the strength of the impurity potential corresponds to a mobility of 3×10^5 cm^2/Vs.

where α and t are in units of the magnetic length. The impurity potential in GaAs is not well enough characterized to determine the value of t and often it is set equal to zero. We use $t = 100\text{\AA}$, which we believe is a typical experimental value. However, the results are fairly insensitive to t and are much more sensitive to the value of α. If one fixes both α and t, there is still a remaining parameter n_I which characterizes the impurity potential. This can be fixed by setting the mobility equal to a value which is typical for the setback distance α. The mobility μ is related to the impurity potential by

$$\frac{1}{\mu} = \frac{m^{*2}}{\pi e \hbar^3} \int d\mathbf{k}' |U_s(\mathbf{k} - \mathbf{k}')|^2 \delta(k^2 - k'^2)(1 - \cos\theta), \qquad (5.2)$$

where U_s is the screened potential (in zero magnetic field).

Fig. 4 shows the calculated absorption curves for filling factors 1 and 2, with electron density of 4×10^{11} cm^2, $\alpha = 140\text{\AA}$ and a mobility of 3×10^5 cm^2/Vs. These curves are compared to the curves calculated in the absence of electron-electron interactions, with the same approximation used for electron-impurity scattering. The main features are that the line is narrowed by an order of magnitude and the effective cyclotron mass is increased if one includes electron-electron interactions. (Since we are concentrating on the differences between the interacting and non-interacting systems, we have not included band nonparabolicity, which will also renormalize the cyclotron mass.) There is a small asymmetry in the lineshape, with somewhat more absorption on the high frequency side. This is due to the fact that the magnetoplasma modes are shifted to higher energies by correlation effects.

For a given mobility, short range scatterers give a much narrower absorption line than long range scatterers in both interacting and non-interacting systems. This is because small angle scattering does not contribute to the mobility. If one considers the case of short-range scatterers for a non-interacting system, one finds linewidths which are comparable to the linewidths for long-range scatterers in the interacting system. If one considers short-range scatterers for an interacting system (in the SCEA) one finds the unusual result that the cyclotron line is not broadened, but it is shifted to lower frequencies. For example, the case of $\nu = 1$ and $\mu = 7 \times 10^4$ cm^2/Vs is shown in Fig. 5. Notice that this is on a very different scale than Fig. 4, and most of

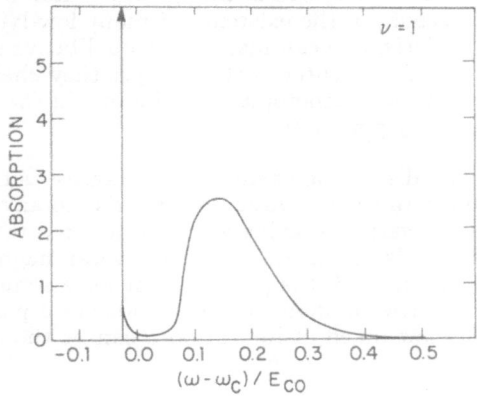

Fig. 5. The calculated absoption near ω_c is shown for $\nu = 1$ and short-range scatterers corresponding to a mobility of 7×10^4 cm^2/Vs. About 85% of the weight is in the delta function peak.

the weight (85%) is in the delta function peak. Clearly this unphysical result is an artifact of the approximation. For example, one can argue that non-perturbative contributions to the electron-impurity scattering will certainly broaden the delta function peak.[13] However, the result suggests that broadening due to short-range scattering will be very small in the interacting system for high mobilities. (Recall that this result cannot be applied directly to silicon inversion layers, where the dominant scattering mechanism may in fact be short-ranged, because we have not included valley degeneracy. Furthermore, this result disappears at lower mobilities where the weight of the delta function becomes negligible.)

VI. COMPARISON TO EXPERIMENT AND DISCUSSION

If one assumes that scattering is due to remote ionized impurities, the calculated cyclotron linewidths for $\nu = 1, 2,$ or 3 and mobilities of the order of a few times 10^5 cm^2/Vs, are about 1 or 2 wavenumbers, in agreement with experiments on GaAs heterojunctions at low temperatures.[11,21] On the other hand, if one neglects electron-electron interactions, for the same impurity potential, one calculates linewidths which are 10 or more wavenumbers at these mobilities. (See Fig. 4.) Since short-range scatterers are much less efficient at broadening the cyclotron line (for a fixed mobility) it is possible to calculate linewidths in agreement with experiments on GaAs by assuming short-range scatterers and neglecting electron-electron interactions. However, since the dominant scattering mechanism is believed to be long-ranged in GaAs and since electrons certainly interact via the Coulomb potential, such agreement would clearly be fortuitous.

The calculations presented here have concentrated on the special case of integral filling because of its simplicity. The analysis is essentially identical for the other special case of rational fillings where the fractional quantum Hall effect is observed, except that the impurity potential is then screened. The results are similar to the integral case, except the cyclotron line is narrowed even more because the effective impurity potential is weaker due to screening. This again is in agreement with experiments, where the linewidth is observed to narrow as the filling is reduced below a filled Landau level.[8,20,22] For example, in Fig. 6, the calculated lineshapes and experimental data are compared for $\nu \leq 1$. One sees that the main experimental features, which are a substantial linewidth narrowing and a downward shift in resonant field as the filling factor is decreased from a filled Landau level, are explained by the SCEA theory.

Calculations of infrared absorption at fillings other than integer or special fractions are complicated because of the existence of many low-lying excitations. However, from the above analysis, one can make some qualitative statements. Electron-electron interactions, in general, enter in three ways: they change both the magnetoplasmon dispersion and the magnetoplasmon lifetime in the pure system and they screen the electron-impurity potential.

The magnetoplasmon dispersion changes continuously as the density is varied and one would not expect this contribution to give rise to any anomalous behaviour, in general, as the filling is varied near integral or fractional values. (However, when more than two Landau levels are filled, there are several magnetoplasmon branches near the cyclotron frequency, and it is possible that for a some impurity potentials, the dispersion would give rise to anomalous behaviour at a particular filling. This is essentially the point of view taken in Ref. 11, and can be thought of as a level crossing.)

The effect of electron-electron scattering on the broadening of the magnetoplasma modes and the effect of screening both change dramatically as one moves away from the special (integral and fractional) fillings since both effects are absent at these densities. They have competing effects on the cyclotron linewidth, since the first will broaden the line and the second will narrow the line. Although screening can be calculated at all densities in either the random phase approximation or the Hartree Fock approximation, neither of these approximations can be used to estimate the lifetime of the magnetoplasma modes due to electron-electron scattering. Therefore, there presently is no theory which can be used to say whether one expects to see linewidth maxima or minima (or, indeed, either) at integral and fractional fillings where the quantum Hall effect is observed, and in fact the answer will

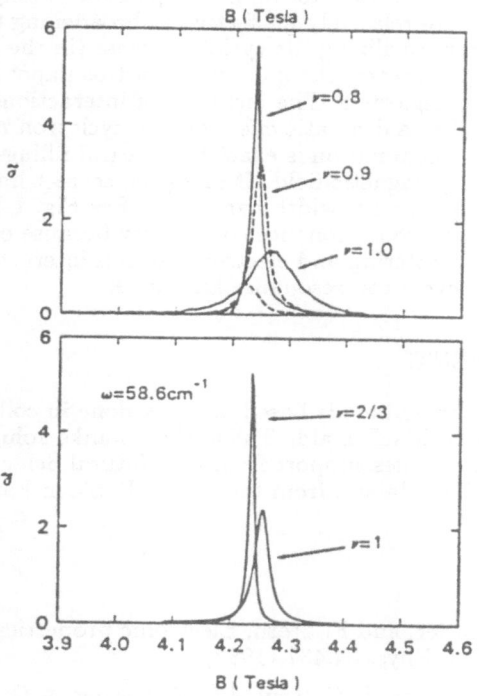

Fig. 6. Comparison of experimental and theoretical cyclotron resonance lineshapes for $\nu \leq 1$. The quantity $\tilde{\sigma}$ is proportional to the real part of the dynamical conductivity. The upper curves show the experimental data taken at $T = 0.4K$ and 58.6 cm^{-1}, and the lower curves are calculated in the SCEA with an impurity setback distance between 250 and 350 Å. The unrenormalized cyclotron resonance occurs at $B = 4.19T$. (This figure is taken from Ref. 20.)

depend on what assumptions are made about the impurity potential. (Broadening due to electron-electron scattering has been estimated at low densities and high temperatures,[23] but this theory is not readily extended to low temperatures.) Experimentally, linewidth maxima at integral fillings have been observed,[8,9,11] and there has also been some evidence of variations in linewidth at fillings corresponding the the fractional quantum Hall effect.[24]

One can also study the cyclotron resonance lineshape as a function of temperature, and the considerations at integral and fractional filling are similar to those for varying the density. As the temperature is increased, both broadening of the magnetoplasma modes and screening come into play. The theory described in this chapter has been used to interpret the temperature dependence observed in both the linewidth and the shift in resonance frequency at densities $\nu \leq 1$.[20] These experiments imply that the effect of screening is dominant at these densities.

Clearly, cyclotron resonance in high mobility heterojunctions is not well understood. In order to make quantitative comparison between theory and experiment, one needs to know what the impurity potential is in the experimental system since different potentials can have quite different effects on the absorption lineshape. For example, as discussed above, in comparison to long range scatterers, short range scatterers are relatively ineffective at broadening the cyclotron line, but they are effective at renormalizing the cyclotron mass (in the presence of electron-electron interactions). However, the main point of this paper is that electron-eletron interactions cannot be neglected. The inclusion of interactions, even in the random phase approximation, has a dramatic effect on the cyclotron resonance lineshape. (The random phase approximation is exact for integral fillings in the limit of long-wavelengths and strong magnetic field. It gives the correct linear dispersion, which is largely responsible for the linewidth narrowing. See Fig. 1.) This fact has largely been missed by the cyclotron resonance community because of the necessity of treating electron-impurity scattering and electron-electron interactions self-consistently in the calculation of the cyclotron resonance lineshape.

ACKNOWLEDGEMENTS

The material presented here is based on work done in collaboration with B.I. Halperin and with A.H. MacDonald. The author thanks John Berlinsky for useful discussions and acknowledges support from the Natural Sciences and Engineering Research Council of Canada and from the Alfred P. Sloan Foundation.

REFERENCES

1. T. Ando, A.B. Fowler, and F. Stern, Electronic properties of two-dimensional systems, Rev. Mod. Phys. 54:437 (1982).

2. H.L. Stormer, A. Chang, D.C. Tsui, J.C.M. Hwang, A.C. Gossard, and W. Wiegmann, Fractional quantization of the Hall effect, Phys. Rev. Lett. 50:1953 (1983).

3. R.B. Laughlin, Anomalous quantum Hall effect: An incompressible quantum fluid with fractionally charged excitations, Phys. Rev. Lett. 50:1395 (1983).

4. W. Kohn, Cyclotron resonance and de Hass-van Alphen oscillations of an interacting electron gas, Phys. Rev. 123:1242 (1961).

5. J.P. Kotthaus, G. Abstreiter, and J.F. Koch, Subharmonic structure of cyclotron resonance in an inversion layer on silicon, Solid State Comm. 15:517 (1974).

6. T. Ando, Mass enhancement and subharmonic structure of cyclotron resonance of cyclotron resonance in an interacting two-dimensional electron gas, Phys. Rev. Lett. 36:1383 (1976).

7. C.S. Ting, S.C. Ying, and J.J. Quinn, Infrared cyclotron resonance in semiconducting surface inversion layers, Phys. Rev. Lett. 37:215 (1976); Theory of cyclotron resonance of interacting electrons in a semiconducting surface inversion layer, Phys. Rev. B 16:5394 (1977).

8. K. Muro, S. Narita, S. Hiyamizu, K. Nanbu, and H. Hashimoto, Far-infrared cyclotron resonance of two-dimensional electrons in an AlGaAs/GaAs Heterojunction, Surf. Sci. 113:321 (1982).

9. Th. Englert, J.C. Maan, Ch. Uihlein, D.C. Tsui, and A.C. Gossard, Observation of oscillatory linewidths in the cyclotron resonance of GaAs-AlGaAs, Solid State Comm. 46:545 (1983).

10. S. Das Sarma, Self-consistent theory of screening in a two-dimensional electron gas under strong magnetic field, Solid State Comm. 36:357 (1980).

11. Z. Schlesinger, S.J. Allen, J.C.M. Hwang, P.M. Platzman, and N. Tzoar, Cyclotron resonance in two dimensions, Phys. Rev. B 30:5655 (1984).

12. C. Kallin, Magnetoplasma modes of the two dimensional electron gas, in:"Interfaces, Quantum Wells and Superlattices,"

13. C. Kallin and B.I. Halperin, Many-body effects on the cyclotron resonance in a two-dimensional electron gas, Phys. Rev. B 31:3635 (1985).

14. A.H. MacDonald, K.L. Liu, S.M. Girvin, and P.M. Platzman, Disorder and the fractional quantum Hall effect: Activation energies and the collapse of the gap, Phys. Rev. B 33:4014 (1986).

15. M.A. Paalanen, D.C. Tsui, A.C. Gossard, and J.C.M. Hwang, Temperature dependence of electron mobility in GaAs-AlGaAs heterostructures from 1 to 10K, Phys. Rev. B 29:6003 (1984).

16. W. Götze and P. Wölfle, Homogeneous dynamical conductivity of simple metals, Phys. Rev. B 6:1226 (1972).

17. Y. Shiwa and A. Isihara, On the memory-function formulation of the dynamic conductivity for two-dimensional electrons in a magnetic field, J. Phys. C 16:4853 (1983).

18. S. Das Sarma and F. Stern, Single-particle relaxation time versus scattering time in an impure electron gas, Phys. Rev. B 32:8442 (1985).

19. T. Ando, Theory of cyclotron resonance lineshape in a two-dimensional electron system, J. Phys. Soc. Jpn. 38:989 (1975).

20. Z. Schlesinger, W.I. Wang, and A.H. MacDonald, Dynamical Conductivity of the GaAs two-dimensional electron gas at low temperature and carrier density, Phys. Rev. Lett. 58:73 (1987).

21. K. Muro, S. Mori, S. Narita, S. Hiyamizu, and K. Nanbu, Cyclotron resonance of two-dimensional electrons in AlGaAs/GaAs heterojunctions, Surf. Sci. 142:394 (1984).

22. B.A. Wilson, S.J. Allen, Jr., and D.C. Tsui, Evidence for a collective ground state in Si inversion layers in the extreme quantum limit, Phys. Rev. Lett. 44:479 (1980); Evidence for a magnetic field induced Wigner glass in the two-dimensional electron system in silicon inversion layers, Phys. Rev. B 24:5887 (1981).

23. H. Fukuyama, Y. Kuramoto, and P.M. Platzman, Many-body effects on level broadening and cyclotron resonance in two-dimensional systems under strong magnetic field, Phys. Rev. B 19:4980 (1979).

24. G.L.J.A. Rikken, H.W. Myron, P. Wyder, G. Weimann, W. Schlapp, R.E. Horstman and J. Wolter, Anomalous cyclotron resonance linewidth in heterojunctions displaying the fractional quantum Hall effect, J. Phys. C 18:L175 (1985).

STARK SHIFTS AND EXCITONIC EFFECTS IN SEMICONDUCTOR QUANTUM WELLS AND SUPERLATTICES

G. Bastard

Groupe de Physique des Solides de l'Ecole Normale
Supérieure
24 rue Lhomond F
75005 Paris
France

I. INTRODUCTION

In this set of lectures I shall give a brief theoretical description of Stark shifts and excitonic effects in semiconductor quantum wells and superlattices. These two topics have stimulated a significant amount of academic research. They are also at the heart of novel opto-electronic devices [1,2] (e.g. fast electro-optical modulators). For the first time the room temperature operation of some semiconductor devices is possible only due to the existence of bound electron-hole pairs: the excitons. The excitons are more stable in quantum wells than in bulk materials, due to their confinement in a narrow slab by large potential barriers. They can also be shifted in energy by an external electric field without being bleached. These two factors explain their technological relevance.

In the following, I shall use the envelope function formalism of heterolayer energy levels as discussed in this Summer School by Altarelli [3]. In most of the discussions, however, I shall neglect important features such as band mixing effects in the valence subbands. These lectures do not aim at a quantitative accuracy but rather at pointing out qualitative features and trends.

The physics of unperturbed semiconductor quantum wells is summarized on fig. (1). Along the growth (z) axis there exist bound motions. The in-plane (x,y) motions are free and their eigenstates are plane waves. To these unbound motions one may associate dispersion relations. In the simplest case, these dispersion relations are isotropic upon the in-plane wavevector $\vec{k}_{\perp} = (k_x, k_y)$. The resulting density of states is the sum of

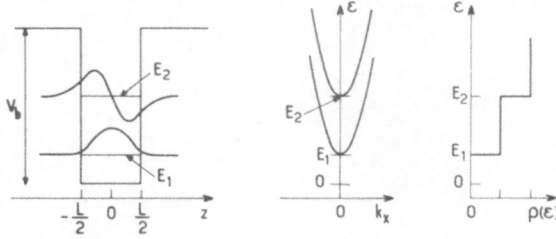

Fig. 1. A recollection of the main properties of isolated quantum well
bound states.

staircases, each being associated with a bound state along the z direction. The envelope functions of the single, rectangular, quantum wells can
be written

$$\Psi(\vec{r}) = \chi_n(z)(1/\sqrt{S})\exp(i\vec{k}_\perp \cdot \vec{r}_\perp) \tag{1}$$

corresponding to the energy

$$E_n(\vec{k}_\perp) = E_n + \hbar^2 k_\perp^2/2m^* . \tag{2}$$

In eqs. (1,2), S is the sample area, E_n the subband edge of the nth subband and m^* the carrier effective mass which I shall assume to be
position-independent.

A superlattice consists of an infinite (in practice large) sequence
of identical wells separated by identical barriers (fig. 2). As a result

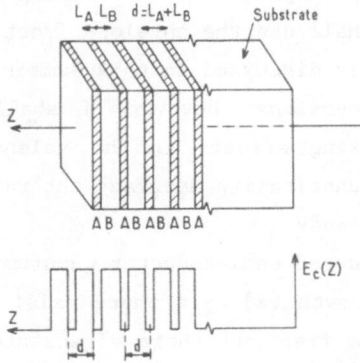

Fig. 2. Semiconductor superlattices: material structure and conduction
band edge profile.

of finite tunnel effects between the wells, the single well levels hybridize to give rise to superlattice minibands. Apart from spin, a superlattice state can be labelled by an in-plane wavevector $\vec{k}_\perp$, a subband index, and a superlattice wavevector q. The latter can be restricted to the first superlattice Brillouin zone. Thus $-\pi/d \leq q < + \pi/d$ where d is the superlattice period. When the coupling between consecutive wells is not too large the lowest superlattice subband has the following approximate dispersion relation:

$$\varepsilon_1(q) = E_1 + 2\lambda\cos qd \tag{3}$$

where $4|\lambda|$ is the band width.

II. ELECTRIC FIELD EFFECTS IN QUANTUM WELLS AND SUPERLATTICES

As pointed out in the introduction there is a rich technological potential in application and switching of a longitudinal ($F||z$) electric field to quantum wells and superlattice structures. Actually, the latter were proposed by Esaki and Tsu [4] as substitutes for the bulk materials for realizing the Bloch oscillator. So far, much of the effort has been directed towards the single or multiple uncoupled quantum wells, possibly due to their easier growth. The usefulness of these structures lies in their capability of withstanding large electric fields ($F \leq 10^5$ V/cm), which produces large Stark shifts while still displaying quasi discrete bound states. We first describe these two effects and then briefly sketch the electric field effects on superlattices (i.e. coupled quantum wells).

II-1. Isolated Quantum Wells We shall assume that the field is uniform over the whole structure, as approximately realized when the quantum well is inserted in the intrinsic part of a reverse-biased p-i-n junction or in the depletion length of a reverse biased Schottky diode. In the case of multiple quantum wells, the barrier separating two consecutive wells will be assumed thick enough to prevent any sizeable coupling between their eigenstates. Thus, the Schrödinger equation we have to investigate is

$$\left[-\frac{\hbar^2}{2m^*}\frac{d^2}{dz^2} + eFz + V_b(z)\right]\chi = \varepsilon\chi . \tag{4}$$

In eq. (4) the in-plane motion has been dropped and $V_b(z)$ is the potential energy profile of a single rectangular quantum well

$$V_b(z) = V_b Y[z^2 - L^2/4] \ . \tag{5}$$

In eq. (5) V_b is the barrier height, $Y(x)$ the unit step function and L the quantum well thickness. In eq. (4) the electrostatic potential eFz has been set equal to zero at the center of the quantum well. It may sometimes be more convenient to take its origin on the left hand side (lhs) corner of the well. A constant electric field leads to pathological behavior of the eigenstates of eq. (4) versus F. At zero field the Schrödinger equation admits at least one bound state ($\varepsilon < V_b$) while an arbitrarily small F is sufficient to transform the allowed energy spectrum into a continuum. This is because the potential energy is arbitrarily large and negative at large and negative z. Despite the lack of true bound states we expect, if F is not too large, that there will exist in the continuous spectrum some particular energies where the carrier wave-function piles up in the quantum well. Moreover, these particular energies will smoothly extrapolate to the true quantum well bound states when $F \to 0$. In fact, we do know from experiments that, over a significant field range (typically $0 \leq F \leq 10^5 V/cm$), the quantum well structures support states which behave as if truly bound. Thus, it is worth trying to convince ourselves that some stationary eigenstates of eq. (4) are indeed peculiar in that they display an accumulation of their wavefunctions in the well. An alternative description, which is often more revealing, is to consider them as metastable (i.e. non stationary) solutions of the time dependent tilted quantum well problem. This kind of description is relevant when the decay time of the quasi bound states is long. Hereafter, we shall also denote the peculiar stationary solutions of eq. (4) as metastable states, since it can be shown [5] that the real part of the complex eigenenergies of the time-dependent problem coincide with the (real) energies of the peculiar solutions of eq. (4).

Apart from the exact, Airy-like solutions, there are various ways to find the metastable states of eq. (4). Firstly, one may cut the electric field at some large distance from the investigated quantum well (fig. 3) and impose the existence of an infinite barrier. The problem becomes that of finding true bound states in a complicated band edge profile. The outcome of such calculations is that there exists a large number of states which are essentially localized in the large triangular well and show a very small probability of being in the well. A small number of states are found to display an enhanced probability of being in the well. These are the metastable states whose energy positions extrapolate smoothly to those of the zero field bound states of the well.

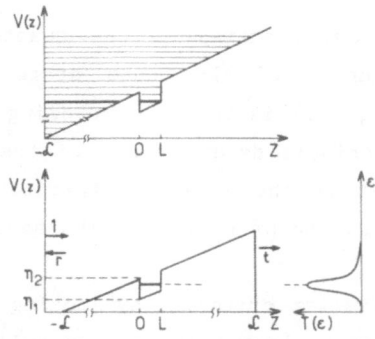

Fig. 3. Two approximate ways to calculate the metastable states of a
quantum well tilted by an electric field (see text).

One may also cut the electric field at some large distances on both
sides of the well and investigate the transmission coefficient $T(\varepsilon)$ =
$|t(\varepsilon)|^2$ of a plane wave with unity amplitude impinging at $z = -L$ on the
barrier and being finally transmitted at $z = +L$. $T(\varepsilon)$ is very low
(because L is large) except in the vicinity of some energies ε_n which
belong to the segment $[\eta_1, \eta_2]$ where it displays sharp peaks ($T \leq 1$). As
usual, these transmission resonances correspond to the trapping of the
particle inside the quantum well (see Mendez's lectures [17] on resonant
tunnelling). If the resonances are narrow this means that the corre-
sponding trapping times τ_n are long: $\tau_n \Delta E_n \sim \hbar$ where ΔE_n is the width
of the resonance.

Another way to depict these resonances is to consider them as
virtual (or metastable) bound states, i.e. as bound states of the quantum
well with a complex energy $\varepsilon_n - i\hbar/2\tau_n$. The reason why the energies of
these states have to be complex and not purely real is that the solutions
of the Schrödinger equation (4) with the boundary conditions
corresponding to a piling up of the wavefunction at $t = 0$ in the quantum
well and to an out-going wave at $z \rightarrow -\infty$ do not fulfill the requirement
of the probability current conservation, i.e. are not stationary. On the
other hand, it can be shown [5] that such solutions are approximate
eigenstates of eq. (4) provided their energies are complex.

In order to set up a criterion allowing us to neglect, in a first
approximation, the escape of the particle outside the quantum well, we
proposed [6] some time ago the condition

$$eF\kappa_b^{-1} \ll V_b - E_1 \qquad (6)$$

where E_1 is the confinement energy in the quantum well and κ_b^{-1} the characteristic decay length of the bound state wavefunction at zero field. The meaning of eq. (6) is that the lowering of the barrier by the field over the characteristic decay length of the wavefunction remains negligible with respect to the effective barrier height $V_b - E_1$ which would enter in the evaluation of κ_b if the lhs barrier were flat instead of being tilted.

One can derive a similar criterion by looking at the problem from a different angle. Suppose we know that we have built up at t = 0 a quasi bound state E_1, which, if the lhs barrier were flat, would be truly bound. Correspondingly, the particle would (classically) oscillate back and forth in the well forever. Since, however, the lhs barrier is tilted the electron, which oscillates in the well since t = 0, progressively escapes to infinity. Semi-classically, this means that while hitting the lhs wall the electron has a finite probability to leave the well by the tunnel effect instead of going backward to execute another bound oscillation. If T is the period of the classical motion, the number of impacts per unit time on the lhs wall is T^{-1}. Thus, the inverse of the E_1 level lifetime $1/\tau(E_1)$ is the product of T^{-1} by the transparency coefficient of the barrier $D(E_1)$

$$\tau^{-1}(E_1) = T^{-1}(E_1)D(E_1) \ . \tag{7}$$

For a triangular barrier in the semiclassical approximation

$$D(E_1) = D_0 \exp\left[-\frac{4}{3eF} \left(\frac{2m^*}{\hbar^2}\right)^{1/2} \left(V_b - E_1\right)^{3/2}\right] \tag{8}$$

with $D_0 \approx 1$. As for the evaluation of $T^{-1}(E_1)$ one gets:

$$T^{-1}(E_1) = (1/2)eF(2mE_1)^{-\frac{1}{2}} \qquad \text{if } E_1 < eFL \ , \tag{9}$$

$$T^{-1}(E_1) = (1/2)\ (1 + \sqrt{1-eFL/E_1})(1/2L)(2E_1/m^*)^{\frac{1}{2}} \qquad \text{if } E_1 \geq eFL \ .$$

The lifetime $\tau^{-1}(E_1)$ will be long if the argument of the exponential is large and negative, which means:

$$V_b - E_1 \gg (3/4)eF\kappa_b^{-1} \ , \tag{10}$$

which (apart from the factor 3/4) is just the criterion we derived previously on the smallness of the barrier lowering by the field over the distance κ_b^{-1} where the wavefunction is important in the barrier.

Assuming that the lower bound for $\gg$ is 10 and taking $V_b - E_1 = 0.125$ eV, $m^* = 0.07m_0$ we find that eq. (10) is satisfied if $F < 7.93 \times 10^4$ V/cm. If, in addition, L = 100 Å, $D_0 \sim 1$ and $E_1 = 70$ meV, we get

$\hbar/2\tau(E_1) \sim 2.5 \ 10^{-4}$ meV and $\tau(E_1) \sim 1.3$ ns. For these typical sample parameters the level broadening is still small with respect to the confinement energy. The escape lifetime is relatively long but is not much longer than a recombination lifetime (~ 1 ns). The capture of the carrier by some shallow impurity may eventually be more efficient than the field-induced tunnelling. Notice finally the strong dependence (F exp($-F_0/F$)) of τ^{-1} with the field. It arises from the transparency coefficient eq. (8). If instead of applying $\sim 8 \times 10^4$ V/cm we only apply 50 kV/cm, $\tau(E_1)$ increases by several decades and the field-induced tunnelling becomes a completely negligible effect with respect to recombination or recapture phenomena. Most of the quantum well structures which have been so far investigated (e.g. in electro-absorption) [2] have shown excellent performances (i.e. pronounced exciton peaks) in fields up to $\sim 10^5$ V/cm. The previous considerations have aimed to point out why it could be so in spite of the field induced tunnelling. From now on we shall neglect this effect and investigate the second useful feature of the electric field effects on quantum well states which is the existence of large Stark shifts.

The zero-field eigenstates of eq. (4) may be classified according to the parity operator. Since eFz is odd in z it only couples the zero-field eigenstates of opposite parities. Since we forget about field-induced tunnelling (which can be neatly done by imposing two infinite barriers at $z = \pm \mathcal{L}$ where $\mathcal{L} \gg L$) we may have recourse to perturbation or variational approaches to calculate the field dependence of the eigen-energies. The basic physics is that the quantum well bound states become polarized by the electric field which in turn leads to a shift of their eigenenergies by a quantity $(1/2)D_n F$ where D_n is the average value of the dipole operator in the nth state. In the low field limit $D_n = X_n F$ where X_n is a c-number. This results in a quadratic (Stark) shift upon F. For larger fields X_n becomes F-dependent displaying some saturation: $D \sim eL/2$. This is the carrier accumulation regime where the wavefunction piles up near the lhs of the well for electrons and the right-hand-side (rhs) for holes (see fig. 4). This carrier accumulation is the useful feature for electro-optical devices: by inhibiting the carrier escape towards $\pm \infty$ (which is unavoidable in bulk materials) the quantum well walls give access to large (~ 30 meV) energy shifts while preserving the carrier localization and thus the beneficial action of enhanced excitonic binding. This effect, also called Quantum Confined Stark Effect by the ATT Bell Laboratories group has led to a number of electro-optic devices, e.g. fast optical modulators [2].

In the small field limit we can use a second order perturbation approach to obtain

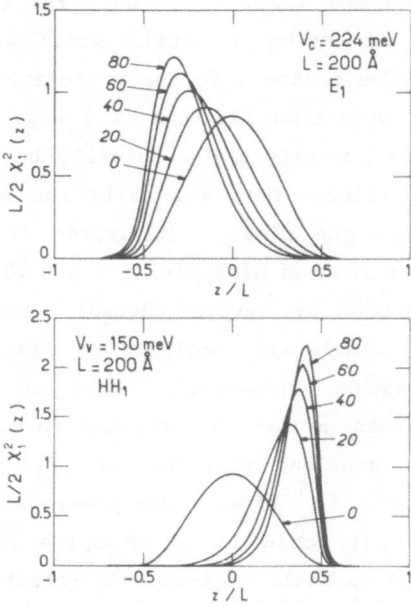

Fig. 4 Calculated envelope functions for the electron and hole ground (metastable) state in a quantum well tilted by an electric field (L = 200 Å).

$$\varepsilon_n(F) = \varepsilon_n(0) + e^2 F^2 \sum_{m \neq n} |\langle m|z|n \rangle|^2 / (\varepsilon_n - \varepsilon_m) \qquad (11)$$

where $|n \rangle$ is the nth zero field bound eigenstate with energy $\varepsilon_n(0)$ and where the summation over m runs over the zero field bound and unbound states. The latter give a small contribution and are usually neglected. From eq. (11) we see that the ground state experiences a red shift, as always. Since $\varepsilon_1 - \varepsilon_m \sim L^{-2}$ and $\langle n|z|1 \rangle \sim L$, $\varepsilon_1(F) - \varepsilon_1(0)$ scales like $L^4 F^2 m^*$. But the domain of validity of eq. (11), which is that the field induced shift remains small with respect to the unperturbed energy splittings, narrows like $m^* F L^3 = $ constant. Once the field is too large to use eq. (11), one may use variational approaches. A linear variational treatment [7] consists of expanding $\psi(z)$ on the uncomplete basis spanned by the zero-field bound eigenstates of eq. (4). In this way, one obtains the field dependences of all the states. In fig. (5) we show the outcome of such a calculation. If one is interested in the ground state only (as often in device applications), a non linear variational wavefunction like [6]

$$\psi_1(z) = \psi_1^{(0)}(z) \exp(-\beta z) \qquad (12)$$

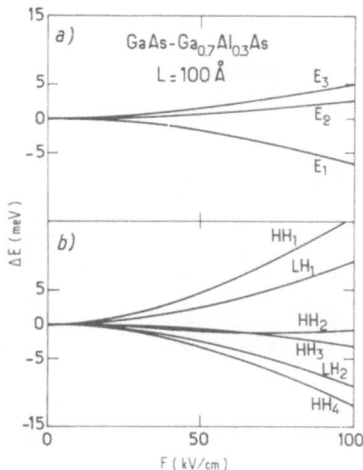

Fig. 5. Calculated Stark shifts of conduction and valence metastable levels in a GaAs-Ga$_{0.7}$Al$_{0.3}$As quantum well (L = 100 Å).

is quite accurate, as it describes the tendency towards accumulation ($\beta > 0$ for electrons, $\beta < 0$ for holes), and simple to use. In eq. (12) $\psi_1^{(0)}(z)$ is the ground bound solution of (4) at zero field. The wavefunction given in eq. (12) also contains the signature of significant field-induced tunnelling: as $|\beta|$ increases with F it happens that it becomes larger than $\kappa_b^{(0)}$ the zero-field wavevector characterizing the evanescent wing of the ground bound state. The minimization procedure becomes impossible and one may rightfully consider that the very notion of quasi discrete bound state fades away. To exemplify the tunability range of energy level shifts upon the electric field we show in fig. (6) the calculated field dependence of the E_1-HH$_1$ inter subband energy for various well thicknesses L. It is seen that considerable shifts can be produced for L ~ 100Å.

Since at the same time the electron and hole remain localized in the well, one may expect the excitonic binding to be little affected by the field. Let us recall that in bulk materials the exciton is dissociated if eFa* ~ R* where a* and R* are the effective Bohr radius and Rydberg respectively. For bulk GaAs this means that ~ 10 kV/cm are sufficient to ionize the exciton. A simple trial wavefunction [2,7] for excitons in quantum wells in an electric field is:

$$\psi_{exc}(z_e, z_h, \rho, \vec{R}_\perp) = \frac{N}{\sqrt{S}} \exp(i\vec{K}_\perp \cdot \vec{R}_\perp) \psi_q^{(0)}(z_e) \psi_q^{(0)}(z_h)$$

$$\exp(-\beta_e z_e - \beta_h z_h) \exp(\rho/\lambda) \tag{13}$$

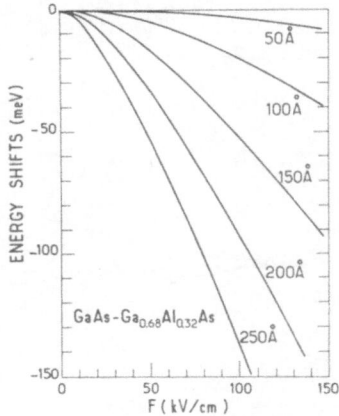

Fig. 6. The calculated exciton resonance energy shift $\hbar\omega(F) - \hbar\omega(0)$ is
plotted versus the electric field intensity F for
$GaAs-Ga_{0.68}Al_{0.32}As$ quantum wells of different thicknesses.

where the plane wave takes care of the in-plane free motion of the
exciton center of mass. ρ is the in-plane relative distance of the
electron-hole pair and the z_e, z_h dependences of the exciton wavefunction
are assumed to be those of the uncorrelated electron and hole. This
means that eq. (13) describes a situation where the z_e, z_h quantizations
arise from the tilted quantum well effect, while the electron-hole
coulombic interaction binds the relative motion of the pair in the layer
plane. Thus eq. (13) should not work either in the very thin or in the
very thick quantum well cases where the subband separation is not much
larger than the exciton binding energy. Eq. (13) implicitly describes
the weakened exciton stability due to the spatial separation of the electron and the hole towards the opposite interfaces of the quantum well.
In fact, the minimization procedure over λ involves an effective
bi-dimensional coulombic potential which is the average over the electron
and hole charge distributions along the z axis of the three dimensional
coulombic potential. Thus, the excitonic binding weakens because, on the
average, the electron and the hole orbit on two separated planes at
finite F, while at zero field they were, on the average, orbiting on the
same plane [7]. The exciton dissociation is inhibited however, since the
quantum well walls prevent the electron and hole charge distributions to
be separated by more than ~ L. If $L < a^*$, the bulk effective Bohr
radius, there is still enough binding to give rise to pronounced excitonic peaks, even at room temperature. An example of calculated field
dependences of the E_1-HH_1 exciton binding energy is shown on fig. (7)

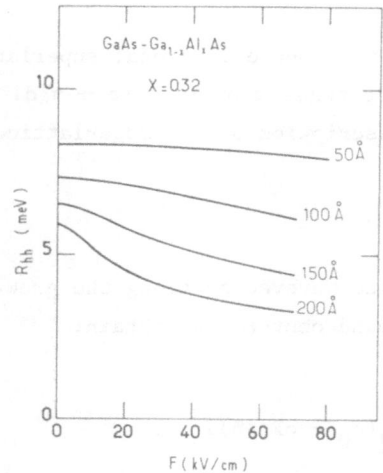

Fig. 7. Field dependence of the E_1 - HH_1 exciton binding energy for
GaAs-$Ga_{0.68}Al_{0.32}As$ quantum wells of different thicknesses.

where it is seen that the overall variations of the excitonic binding is
indeed small.

One may improve eq. (13) in many ways, especially by taking into
account the complications of the valence subband dispersion relations.
With the improvement of GaAs quantum well samples, a rich spectroscopy of
the field dependence of the ground and excited states of the excitons is
becoming available [8]. The experimental data are satisfactorily
described by the calculations.

II-2. Superlattices

A long time ago, Bloch pointed out that a constant electric field
superimposed on a periodic potential would lead to a very fast
oscillatory motion. A simple derivation of the Bloch oscillations relies
on a semi-classical treatment [9] of the electric field effects. In the
semi-classical approximation one assumes that:

i)   the band index n is constant: no field-induced interband transi-
     tions (or Zener tunnelling) are allowed
ii) the carrier wavevector varies with time according to the Newton law

$$\hbar \, d\vec{k}/dt = -e\vec{F} \tag{14}$$

iii) the carrier velocity is related to the dispersion relation $\varepsilon_n(\vec{k})$ by:

$$\vec{v} = (1/\hbar)\delta\varepsilon/\delta\vec{k} \ . \tag{15}$$

Let us apply these rules to a one dimensional superlattice with period d. The first Brillouin zone extends from $-\pi/d$ to $+\pi/d$. For simplicity let us use a tight binding description of the superlattice dispersion relations:

$$\varepsilon_1(q) = E_1 + 2\lambda\cos qd \tag{16}$$

where q is the superlattice wavevector along the growth axis, $4|\lambda|$ is the bandwidth and E_1 is the band center. We obtain:

$$q(t) = q_0 - eFt/\hbar \tag{17}$$

$$v_z(t) = -(2\lambda d/\hbar)\sin\left[(q_0 - eFt/\hbar)d\right] \ . \tag{18}$$

Thus, the carrier motion is periodic with a period $2\pi\hbar/eFd$. For $F = 10^4$ V/cm and d = 200 Å we obtain a period of $T \approx 2 \times 10^{-13}$ s. The carrier will execute periodic oscillations with an amplitude of 250 Å if the bandwidth 4λ is equal to 50 meV. To execute its Bloch oscillations the carrier should not be scattered during a period and thus, at least, its mobility relaxation time τ_r should be > T. For the previous example this means a d.c. mobility of ~ 5000 cm^2/Vs for the z motion, if one assumes an effective superlattice mass of 0.07 m_0. This is one or two decades too large for actual GaAs-Ga(Al)As samples [10]. In addition to the technological limitations (fluctuations of the well thicknesses, field non-uniformities over a thick superlattice etc.) one should also wonder about the validity of the semi-classical description. It is easily proved [9] that the Newton law ii) is exactly valid in the case of a constant electric field. On the other hand, there is no warrant that the assumption i) will hold.

A superlattice in an electric field may be viewed as a succession of identical tilted quantum wells which are shifted by eFd from one cell to the next. This means that the first order effect of an electric field is to kill the resonant tunnelling which exists at zero field (and thereby gives rise to minibands) by misaligning the single well energy levels.

To illustrate these considerations we treat the simplest possible case which is that of a symmetrical double quantum well (well thickness L, barrier thickness h). Again, we neglect the field-induced tunnelling in the lhs barrier. We denote by E_1, t, s the ground quantum well bound state when the wells are isolated, the transfer integral $t = \langle \varphi_1|V_1|\varphi_2 \rangle$ and the shift integral $s = \langle \varphi_1|V_2|\varphi_1 \rangle$ respectively. Furthermore, we assume that the field is small enough to admit only the two lowest levels of the double quantum well at zero field. The two isolated well wavefunctions φ_1 and φ_2 will be taken as orthogonal for the sake of convenience. Thus by writing

$$X = a\varphi_1 + b\varphi_2 \qquad (19)$$

and by projecting onto $\langle\varphi_1|$ and $\langle\varphi_2|$ we get

$$\begin{bmatrix} \varepsilon_1 + s - eF\frac{d}{2} - \varepsilon & t \\ t & \varepsilon_1 + s + \frac{eFd}{2} - \varepsilon \end{bmatrix} \begin{bmatrix} a \\ b \end{bmatrix} = 0 \qquad (20)$$

and therefore

$$\varepsilon - E_1 = \pm |t| \sqrt{1 + x^2} \qquad (21)$$

with

$$E_1 = \varepsilon_1 + s; \quad x = eFd/2|t|; \quad d = L + h \quad . \qquad (22)$$

In the vanishing field limit, the two degenerate levels of the isolated wells split into a symmetric S ($\varepsilon_S = E_1 - |t|$) and an antisymmetric A($\varepsilon A = E_1 + |t|$) combination. The integrated probability of finding the particle in either well is 1/2 in both the A and S levels. When the field is turned on the doublet splitting increases with F (fig. 8). Correlatively, the lower and upper level wavefunctions become increasingly localized in the lhs and rhs wells respectively: the field suppresses the tunnelling effects between the wells and localizes the carrier wavefunction in a given well. For very large fields, it is no longer possible to neglect the admixture with the E_2 levels of both wells. This would lead to a Stark shift of E_1, similar to the one discussed previously, and also to the recovery of resonant tunnelling phenomenon from one well to the next when eFd matches the energy difference between E_2 and E_1. In addition to these purely electronic effects, one can think about resonant or non-resonant phonon-assisted tunnelling where the energy difference $E_2 - E_1$ is partly accommodated by the eFd term and partly by either an acoustical phonon or an optical one [11].

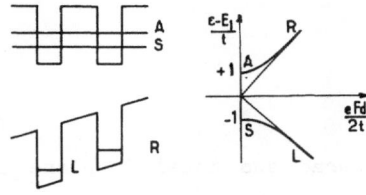

Fig. 8.   Double quantum well tilted by an electric field.

III. EXCITON STATES IN TYPE 1 QUANTUM WELLS AND SUPERLATTICES

The lowest lying excited states of undoped semiconductor quantum wells are the excitons, i.e. the shallow bound states formed between a conduction electron and a valence hole (Wannier excitons). If the valence subbands were as simple as the conduction ones, the exciton problem would resemble very much that of the coulombic donor problem [12]. Actually, despite the fact that the hole kinematics considerably obscure the exciton algebra, it remains true that the exciton states in quantum wells share many features with the coulombic impurity levels. Noticeably, the trend of the binding energy versus the quantum well thickness and the occurrence of sharp excitonic resonances below the edge of excited subband transitions (i.e. $HH_n \rightarrow E_m$ or $LH_n \rightarrow E_m$; $n + m$ even, n or $m > 1$) in addition to the $HH_1 \rightarrow E_1$ exciton bound state are identical to what is found in the impurity problem. The enhanced binding energies (in type 1 quantum wells where the electron and hole are mostly confined within the same layer) allow the observation of excitonic absorption peaks at room temperature in $GaAs-Ga_{1-x}Al_xAs$ and other type 1 quantum well structures.

The excitons are <u>delocalized</u> entities: the electron-hole reduced motion can be bound or unbound but the center of mass (or its equivalent) is free to move in the layer plane of perfect heterostructures.

We shall only consider idealized quasi bi-dimensional excitons formed between an electron and a hole whose subbands display a quadratic dispersion upon the in-plane wavevector. For a single quantum well the exciton Hamiltonian is then

$$H_{exc} = \frac{p_{z_e}^2}{2m_{e||}} + \frac{p_{z_h}^2}{2m_{h||}} + \frac{p_\perp^2}{2\mu_\perp} + \frac{p_\perp^2}{2M_\perp} + V_s Y[z_e^2 - \frac{L^2}{4}] +$$

$$|V_p| Y[z_h^2 - \frac{L^2}{4}] + \varepsilon_A - \frac{e^2}{\kappa} [r_\perp^2 + (z_e - z_h)^2]^{-\frac{1}{2}} \quad (23)$$

with:

$$\mu_\perp = \frac{m_{e\perp} m_{h\perp}}{m_{e\perp} + m_{h\perp}} \quad (24)$$

$$M_\perp = m_{e\perp} + m_{h\perp} \quad (25)$$

$\mu_\perp$ and $M_\perp$ are the reduced and total in-plane effective masses of the electron and the hole respectively. ε_A is the bandgap of the well-acting material (thickness L) and κ the relative dielectric constant of the heterostructure.

The two-dimensional vectors:

$$\vec{r}_\perp = \vec{r}_{e\perp} - \vec{r}_{h\perp} \tag{26}$$

$$\vec{R}_\perp = (m_{e\perp}\vec{r}_{e\perp} + m_{h\perp}\vec{r}_{h\perp})/M_\perp \tag{27}$$

are the in-plane projections of the reduced electron-hole position vector and of the center of mass position vector respectively. It is clear that the in-plane wavevector $(\vec{K}_\perp)$ of the exciton center of mass is a good quantum number, its associated wavefunction being the plane wave $\exp(i\vec{K}_\perp \cdot \vec{R}_\perp)$. Since the center of mass motion decouples from the internal degrees of freedom, we shall drop the former from now on. The in-plane reduced motion and the electron and hole longitudinal motions do not separate. Thus we shall again resort to the variational method to obtain the exciton binding energies. Notice that several sets of excitons may be formed with the electron and the hole belonging to the various conduction and valence quantum well subbands. A single set is truly bound: the excitons which are formed between electrons and holes both belonging to the ground subbands. The other excitons are resonances superimposed on the two-dimensional continua of the lower lying electron-hole subbands. The broadenings which are associated with the auto-ionization of the excited exciton states are in fact small since it is possible to observe excitonic spectra associated with optical transitions between highly excited electron and hole subbands (fig. 9). Here, we shall neglect these broadenings. Consequently, if ψ_{nm} denotes the trial

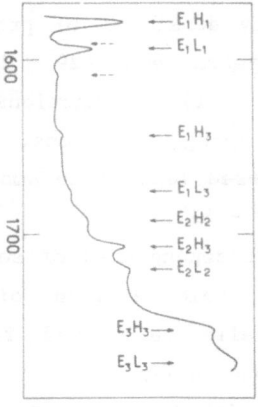

PHOTOLUMINESCENCE INTENSITY (a.u.)

Fig. 9. An example of prominent excitonic structures showing up in the low temperature photoluminescence excitation spectrum of a Separate Confinement Heterostructure (after reference [16]).

exciton wavefunction with the electron in the conduction subband (E_n) and the hole in the mth valence subband (H_m) and R_{nm} the ($H_m - E_n$) exciton binding energy then:

$$R_{nm} = \varepsilon_A + E_n + H_m - \min \langle \psi_{nm} | H_{exc} | \psi_{nm} \rangle \tag{28}$$

where the minimization is performed over all the variational parameters of the ψ_{nm} trial wavefunction.

Several ψ_{nm} have been used. A very convenient one works rather well over the range of GaAs quantum well thicknesses which is the most invest-igated, i.e. $0.5a^*_\perp < L < 2a^*_\perp$, where $a^*_\perp$ denotes the in-plane effective bulk Bohr radius (i.e. calculated with the in-plane reduced mass $\mu_\perp$). This simple wave function is:

$$\psi_{nm}(\vec{r}_\perp, z_e, z_h) = \chi_n(z_e)\chi_m(z_h)g_{nm}(\vec{r}_\perp) \tag{29}$$

$$g_{nm}(\vec{r}_\perp) = f_{nm}(r_\perp)\exp(ij\varphi); \quad j = 0, \pm1, \pm2, \ldots \quad ; 0 < \varphi < 2\pi \tag{30}$$

where the χ's are the E_n and H_n envelope functions and where we have made use of isotropicity of the coulombic interaction in the layer plane. For the ground (quasi 1S) exciton state $j = 0$ and:

$$f_{nm}(r_\perp) = \left(\frac{2}{\pi\lambda^2_{nm}}\right)^{1/2} \exp\left(-\frac{r_\perp}{\lambda_{nm}}\right) \tag{31}$$

where λ_{nm} is a variational parameter. To write eqs. (29 - 31) amounts to assuming that the longitudinal electron and hole motions are forced by the quantum well potential while the bound states for the in-place reduced motion are provided by the coulombic electron-hole interaction averaged over the probability densities of presence $\chi_n^2(z_e) \chi_m^2(z_h)$. Besides leading to simple algebra, eqs. (29 - 31) have the advantage of providing excited state (n, m > 1) wavefunctions which are automatically orthogonal to the lower lying solutions. More elaborate trial wavefunctions have been proposed leading to small improvements over the results derived from eqs. (29 - 31).

We show in fig. 10 the L dependences of several ground (i.e. quasi 1S) light hole and heavy hole exciton binding energies R_{nm} in GaAs-Ga$_{0.7}$Al$_{0.3}$As quantum wells calculated from eqs. (29-31) [13]. Several features are noticeable in fig. 10:

i) R_{nm} (L) first increases with decreasing L and then decreases. This behavior is the same as found for coulombic impurity states. It reflects the increasing quasi bi-dimensionality of the coulombic problem when L decreases until one of the two particles (or both) loses this quasi bi-dimensionality by having a confinement energy

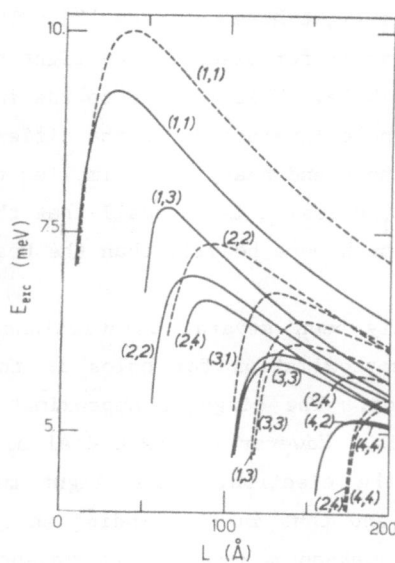

Fig. 10. Calculated thickness dependence of the binding energies of the (n, m) exciton formed between the nth electron subband and the mth hole subband $(n, m = 1, 2, \ldots)$ of a GaAs-Ga$_{0.7}$Al$_{0.3}$As quantum well. The full (broken) curves correspond to the heavy (light) hole exciton respectively.

which approaches the top of the confining well. For this reason all the excitions $E_n - HH_m$, $E_n - LH_m$ disappear at small enouhg L, unless $n = m - 1$. Notice however that it is in principle possible to form an exciton between a true bound state (say a valence state) and virtual bound state (say a conduction state) of the quantum well if the resonant state is narrow enough. The magnitude of the binding energy for such an exciton and its stability against dissociation have however never been calculated, to our knowledge.

ii) The binding energies R_{nm} decrease with increasing n or m, but relatively slowly. In a first approximation, the exciton binding energies R_{nm} are nearly the same for all n (granted that both the E_n, HH_n, or E_n, LH_n pairs of levels are tightly bound in the well). This arises from the weak dependence of the averaged electron-hole interaction upon the electron and hole quantum well bound state.

iii) The ground light hole exciton $LH_1 - E_1$ is found more bound than the heavy hole exciton $HH_1 - E_1$, unless L is very small. This is a consequence of the mass reversal effect for the hole subbands in the diagonal approximation: the heavy hole subbands have a lighter in-plane mass than the light hole subbands [3]. Thus, the in-plane reduced mass is heavier for electron and light hole than for electron and heavy hole. This, joined to the insensitivity of the averaged electron-hole interaction to the difference between the χ_1's of the light hole and heavy hole, implies that $LH_1 - E_1$ is more bound than $HH_1 - E_1$ unless L is so small that the light hole wavefunction χ_1 leaks much more heavily than the heavy hole χ_1 in the barrier.

There have recently been several calculations [14] attempting to improve the diagonal approximation for holes in the evaluation of the exciton binding energies. The diagonal approximation poorly describes the actual hole subbands. However, a great deal of the exciton characteristics arise from the electron, whose light mass prevails in the exciton reduced masses and thus in the binding energy. In fact, calculations including the subband mixing in the valence band improve very little ($\approx$ 1meV) over the evaluation of the $HH_1 - E_1$ binding energy performed in the diagonal approximation for the holes. On the other hand, it appears that the actual camel back shape of LH_1 may significantly enhance (2 - 3 meV) the binding energy of the $E_1 - LH_1$ excitons over the value obtained in the diagonal approximation. In no case, however, have these refined exciton calculations obtained $E_1 - HH_1$ excitons which were more bound than the $E_1 - LH_1$ excitons. This is in agreement with optical data but disagrees with magneto-optical data.

In a superlattice the quantum well bound states hybridize to give rise to minibands. Correlatively, the electron (or hole) wavefunctions become increasingly delocalized when the superlattice period d decreases. Since only the excitons with a zero center of mass momentum are optically active we write the trial exciton wavefunction for a $E_1 - HH_1$ exciton as [15]

$$\psi(\vec{\rho}, z_e, z_h) = N \exp\left(-\frac{\rho}{\lambda}\right) \sum_{q_e} \exp\left(\frac{-q_e^2}{2\beta^2}\right) \chi_{q_e}(z_e) \chi_{-q_e}(z_h) \qquad (32)$$

where N is a normalization constant, λ and β are variational parameters, q_e is the z component of the electron wavevector ($-\pi/d < q_e < \pi/d$) and $\chi_{q_e}(z_e)$ and $\chi_{-q_e}(z_h)$ are the E_1 and HH_1 electron and hole envelope functions in the superlattice. It can be checked that eq. (32) reduces to the separable exciton envelope function of isolated wells in the multiple quantum well limit (i.e. thick barriers) and gives near vanishing d a bulk exciton binding energy which is about 85% of the bulk Rydberg.

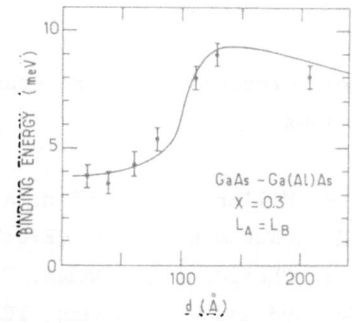

Fig. 11. Exciton binding energy versus period in GaAs-Ga$_{0.7}$Al$_{0.3}$As
superlattices with equal layer thicknesses. Solid line:
theory. Symbols: experiments.

Fig. 11 shows the calculated variations of the "1S" heavy-hole
exciton binding energy B_{1S} as a function of the superlattice period d
when the well and barrier widths are equal. When the period tends to
zero B_{1S} tends to the GaAs bulk value. B_{1S} reaches a maximum near d $\approx$
140 Å and then decreases. The latter behavior is the same as that of
single quantum wells. The experimental values are also reported in Fig.
12 and their variations upon d are correctly reproduced by the
calculations. Two domains are apparent on Fig. 11: (i) d < 94 Å (for
the chosen set of parameters), where the exciton spreads over several
layers and displays a 3-D character and ii) d > 94 Å, where the exciton
is found to be confined in the wells ($\beta \to \infty$) and has a 2-D character as in
multiple quantum wells structures. The localized-delocalized transition
is found to be abrupt in our calculations. In fact the curves $E(\lambda,\beta)$
have always two minima which correspond either to a finite β or a very
large one. The transition at d $\approx$ 94 Å results from the interchange of
the relative positions of these two minima. A more elaborate trial
wavefunction may eventually smooth the transition.

ACKNOWLEDGEMENTS

It is a pleasure to thank J.A. Brum, C. Delalande, M.H. Meynadier
and J. Orgonasi for their active participation in the work reported here.
This work has been supported in part by the GRECO "Expérimentations
Numériques".

REFERENCES

The following list of references is by no means exhaustive. Most of the quoted papers are reviews.

1. D. S. Chemla, D. A. B. Miller, P. W. Smith, A. C. Gossard and W. Wiegmann, IEEE J. Quantum Electron. QE20:265 (1984).

2. D. A. B. Miller, D. S. Chemla, T. C. Damen, T. H. Wood, C. A. Burrus, A. C. Gossard and W. Wiegmann, IEEE J. Quantum Electron. QE21:1462 (1985). See also D. A. B. Miller, D. S. Chemla, T. C. Damen, A. C. Gossard, W. Wiegmann, T. H. Wood and C. A. Burrus. Phys. Rev. B32:1043 (1985). See also D. A. B. Miller, J.S. Weiner and D. S. Chemla, IEEE J. Quantum Electron. QE22:1816 (1986).

3. M. Altarelli, this volume.

4. L. Esaki and R. Tsu, IBM J. Res. Develop. 14:61 (1970).

5. D. Bohm "Quantum Theory" (Prentice Hall, New York, 1951).

6. G. Bastard, E. E. Mendez. L. L. Chang and L. Esaki, Phys. Rev. B28:3241 (1983).

7. J. A. Brum, Thèse de Doctorat d'Etat, Paris 1987 unpublished. See also G. Bastard and J. A. Brum, IEEE J. Quantum Electron. QE22:1625 (1986). See also J. A. Brum and G. Bastard, Phys. Rev. B31:3893 (1985).

8. L. Vina, R. T. Collins, E. E. Mendez and W. I. Wang, Phys. Rev. Lett. 58:832 (1987).

9. N. W. Ashcroft and N. D. Mermin, "Solid State Physics" (Holt, Rinehart and Winston, New York, 1976).

10. J. F. Palmier in "Heterojunctions and Semiconductors Superlattices" edited by G. Allan, G. Bastard, N. Boccara, M. Lannoo and M. Voos (Springer Verlag, Berlin, 1986).

11. F. Capasso, K. Mohammed and A. Y. Cho, J. Quantum Electron, QE22:1853 (1986).

12. G. Bastard, Phys. Rev. B. 24:4717 (1981). See also W. T. Masselink, Y. C. Chang, H. Morkoc, D. C. Reynolds, C. W. Litton, K. K. Bajaj and P. W. Yu, Solid State Electron. 29:205 (1986).

13. G. Bastard, E. E. Mendez, L. L. Chang and L. Esaki, Phys. Rev. B26:1974 (1982). See also J. A. Brum and G. Bastard, J. Phys. C 18:L789 (1985).

14. G. E. W. Bauer and T. Ando in "Proceedings of the 18th International Conference on the Physics of Semiconductors" (World Scientific, Singapore, 1987). See also M. Altarelli in "Excitons in Confined

Systems" edited by R. Del Sole (Springer Berlin, to be published).

15. A. Chomette, B. Lambert, B. Devaud, A. Regreny and G. Bastard, Europhysics Lett. (1987) in press.

16. M. H. Meynadier, C. Delalande, G. Bastard, M. Voos, F. Alexandre and J. L. Liévin, Phys. Rev. B31:5539 (1985).

17. E. E. Mendez, this volume.

18. G. E. W. Bauer and T. Ando, Proceedings of the MSS3 Conference (Montpellier, France, 1987) to be published by the Journal de Physique.

ULTRAFAST NONLINEAR OPTICAL PHENOMENA IN SEMICONDUCTOR QUANTUM WELLS

Stefan Schmitt-Rink

AT&T Bell Laboratories
Murray Hill, New Jersey 07974

INTRODUCTION

Following the discovery of optical bistability in bulk semiconductors in 1979, two recent developments have opened new areas in nonlinear semiconductor optics, namely the advances in layered semiconductor growth and ultrashort optical pulse generation. The combination of both techniques proved to be a fertile ground for novel physics, helping to increase our understanding of basic light-matter interactions and many-body effects in open two-component Coulomb systems in semiconductors. It is this rapidly evolving field that is the subject of the present paper.

After introducing in Section II some of the basic concepts in nonlinear semiconductor optics,[1] we discuss in Sections III and IV the ultrafast excitonic optical nonlinearities that arise under generation of i) real but nonthermal and ii) virtual electron-hole populations in semiconductor quantum wells (QWs). We describe the experimental studies,[2-7] their interpretation,[8,9] and, in Section V, some of the potential applications.[10] The emphasis is on the physics of these processes, and not on rigorous treatments, which can be found elsewhere.[1,8,9]

We have chosen to discuss only the most basic phenomena and those (original) experiments and theories which are particularly demonstrative. As a consequence, the list of references is rather incomplete. An exhaustive list of references to other work in the field can be found in Ref. 1.

EXCITONIC OPTICAL NONLINEARITIES IN SEMICONDUCTORS

In semiconductors, the absorption of intense laser light promotes electrons (e) from bonding into antibonding states. The excited electrons and the holes (h) left behind are charged and mobile quasiparticles that interact with the light field (transverse photons) and through the Coulomb forces (longitudinal photons). They can form individual bound (excitons, trions, biexcitons) or collective unbound (e-h plasma) states that modify in turn the electronic and optical properties, each species in its own characteristic way, which is the basic working principle of nonlinear optical semiconductor devices (in combination with a feedback).

The details of these renormalizations can be drastically different i) in different dimensions, ii) depending on whether the excited species are in quasiequilibrium or not, and iii) depending on whether real or virtual excited state populations are generated. However, they always rely on a few simple mechanisms that result from very general principles. Before discussing the effects of i)-iii) and the recent respective investigations in QWs, it is therefore instructive to reconsider these basic mechanisms.[1] For that purpose, we shall limit ourselves to exciton and e-h plasma induced effects, assuming local equilibrium.

Most obviously, when e-h pairs are created in a semiconductor, they fill states in phase space and prevent further absorption into these states by Pauli exclusion, in much the same way as in atomic (two-level) systems. However, the nature of the e-h pair states themselves is also modified, due to the Coulomb interaction. This makes the determination of the nonlinear optical response of a semiconductor a genuine many-body problem, unlike conventional quantum optics.

In most semiconductors, in particular in QWs, the optical spectra in the vicinity of the absorption edge are profoundly influenced by excitonic effects. If (and only if) the excitons are well resolved, it is the vanishing of these with increasing excitation intensity, i.e., e-h pair density, which is the dominant nonlinearity.

The inset of Fig. 1 shows the typical behavior of the absorption spectrum of a bulk semiconductor as a function of excitation intensity. For low excitation intensities, i.e., for a small number of photoexcited e-h pairs, one usually sees the lowest e-h bound state, the 1s exciton, resolved. With increasing excitation intensity, i.e., with increasing number of photoexcited e-h pairs, this bound state becomes less and less stable, the exciton unbinds, and consequently the excitonic optical absorption diminishes. Simultaneously, exchange and correlation effects among the photoexcited pairs lead to a lowering of the

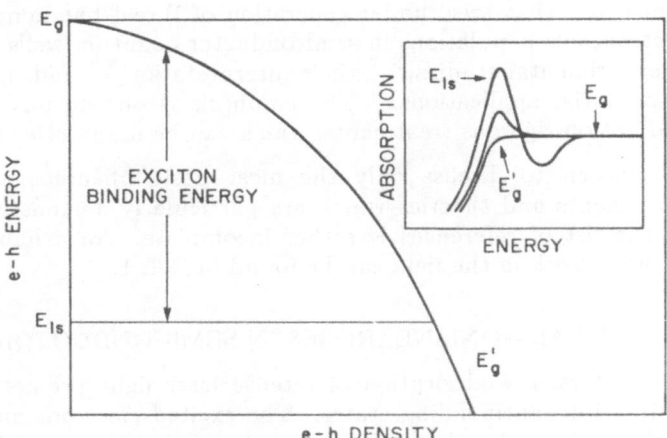

Fig. 1.   Renormalized exciton energy E_{1s} and band gap E_g' in 3D as a function of e-h pair density. The inset shows the respective changes of the absorption spectrum.

individual e and h energies and thus to a shrinkage of the gap, as shown in Fig. 1. In bulk semiconductors, this decrease of the band gap (due to self-energy corrections) and the decrease of the exciton binding energy (due to vertex corrections) cancel each other in such a fashion that the absolute exciton energy stays almost constant up to the point where the exciton ceases to exist. Hence, with increasing excitation intensity, there is a strong reduction of the exciton oscillator strength but no significant shift of the exciton resonance. The renormalized e-h scattering continuum eventually remains the only spectral feature, it shall however not concern us here.

If the photoexcited e-h pairs consist in form of free e and h, characterized by local Fermi distributions, two effects can be identified as the major source of this bleaching of the exciton resonance, i) the effects of the exclusion principle, "phase space filling" (PSF) and fermion "exchange" (E), and ii) the effects of "screening" (S). Because of the Pauli principle, the states in phase space that are already occupied by free e and h are no longer available for the formation of bound states or accessible in optical transitions. Put in simple terms, one cannot optically create e-h pairs on top of each other, if they require the same momentum states. In the fermion picture, this gives rise to a reduction of the attractive e-h interaction (and corresponding e and h exchange self-energies) and thus contributes to the bleaching of the excitonic optical absorption. Simultaneously, it causes a blocking of optical transitions, similar to that in two-level systems, which also contributes to the absorption bleaching. (In the absence of Coulomb interaction, this is the only bleaching mechanism, corresponding to the well-known Burstein-Moss shift of the absorption edge in doped semiconductors.)

A further reduction of the attractive e-h interaction (and a corresponding band gap renormalization) is due to S. In 3D and at high temperatures, this mechanism alone accounts already for the bleaching of the exciton resonance. This is because in 3D one needs a critical coupling strength in order to form a bound state. If one models the screened Coulomb interaction by a Yukawa potential $[e^2\exp(-\kappa r)]/(\epsilon_0 r)$, the unbinding of the exciton takes place for $\kappa a_0 \sim 1$, where a_0 is the exciton Bohr radius, ϵ_0 the static dielectric constant and κ^{-1} the screening length. This is Mott's original criterion. For noninteracting e and h, the screening wavenumber κ reduces to its Debye-Hückel or Thomas-Fermi value, depending on the plasma degeneracy, and in the nondegenerate limit the corresponding critical plasma density as derived from the simple Mott criterion is in reasonable agreement with experimental observations.[1] PSF and E play no role, because in 3D and at high temperatures, the screening wavenumber reaches the inverse Bohr radius long before overlap effects become important. In this context, it should be noted that in 3D and in the nondegenerate limit the simple Debye mean field theory is exact in leading order in the plasma parameter, so that the band gap renormalization is simply given by the Debye shift $-(e^2\kappa_{DH})/\epsilon_0$, where κ_{DH} is the Debye-Hückel screening wavenumber. Obviously, the leading change in the exciton binding energy and this shift cancel each other exactly, which accounts for the behavior shown in Fig. 1. This cancellation reflects the charge neutrality of a bound state, i.e., inside an exciton the Debye clouds of the e and h compensate each other.

Even if the photoexcited e-h pairs consist in form of neutral excitons, one can still use the same language in order to describe the bleaching of the exciton resonance. If only the exciton ground state population needs to be considered, which is the case at low temperatures, the picture given above can be easily extended. It is only necessary to note that an exciton is built up from a linear combination of fermion states, distributed according to the exciton orbital wavefunction $\phi_{1s}(k)$, so that in the dilute limit the creation of N excitons corresponds to an occupation probability $(N|\phi_{1s}(k)|^2)/2$ in the fermion space. At low temperatures, the resulting bleaching of the exciton resonance due to excitonic PSF and E is of course smaller than that due to free e and h PSF and E, because the latter have a larger overlap with a given exciton than other excitons, due to their large thermal wavelength. At high temperatures, when the thermal wavelength is small compared to the exciton Bohr radius, the opposite holds.

Most remarkably, the nature of S changes completely, if only excitons are present. This is because excitons are neutral bound states. They can be only polarized, corresponding to transitions from the exciton ground state to excited states, the polarizability increasing with increasing spatial extent of the exciton or with decreasing binding energy. This should be contrasted with the perfect metallic screening discussed above, which is characterized by an infinite polarizability. Correspondingly, the Mott transition in an exciton gas may be viewed as a dielectric polarization catastrophe.

Clearly, only a self-consistent treatment can describe the unbinding of an exciton embedded in an exciton gas. PSF, E and S depend on the exciton binding energy, but the exciton binding energy depends itself on these effects. Since the influence of other excitons on a given exciton is weaker than that of free e and h of the *same* density and temperature, one expects that at low temperatures the unbinding of the exciton and the saturation of the excitonic optical absorption occur at a higher pair density than predicted by the simple Mott criterion $\kappa_{TF} a_0 \sim 1$, where κ_{TF} is the Thomas-Fermi screening wavenumber. This has been confirmed experimentally.[12] It corresponds to a stabilization of the exciton phase by the charge fluctuation feedback, similar to the stabilization of ^{3}He—A by the spin fluctuation feedback.

A more rigorous description of the absorption saturation involves of course the exact solution of the corresponding few-body problems, the exciton-exciton, exciton-electron and exciton-hole scattering. If one treats excitons as structureless point bosons, the corresponding ab initio *effective* interactions (including virtual transitions to excited states) are closely related to the many-body effects discussed above. The exclusion principle effects show up as a short-range hard-core repulsion, while the screening (mutual polarization) effects show up as a long-range Van der Waals attraction. The observed remarkable constancy of the exciton energy shows that in 3D the respective blue and red shifts cancel each other almost exactly.

In summary, for fixed e-h pair density and temperature, the effects of free e and h on a given exciton are always much larger than those of other excitons. In 3D and at high temperatures, the unbinding of the exciton and the saturation of the excitonic optical absorption are mainly due to S, while at low temperatures they are both due to S and PSF and E.[1,11-13]

ULTRAFAST NONLINEARITIES DUE TO EXCITATION OF REAL BUT NONTHERMAL E-H DISTRIBUTIONS

Excitons can be generated selectively by resonant excitation. At low temperatures, i.e., temperatures smaller than the exciton binding energy, they are thermally stable and only disappear by recombination, while at high temperatures, i.e., temperatures larger than the exciton binding energy, they are unstable against thermal ionization. In the dilute limit, the latter process is due to collisions with thermal LO phonons and the corresponding ionization time τ can be estimated from the homogeneous absorption linewidth. In 100 Å GaAs QWs one finds at room temperature $\tau \sim 0.4$ ps. Femtosecond spectroscopic techniques are thus necessary to resolve exciton-induced nonlinear optical effects at high temperatures.[2,4] In these experiments, a weak broad-band fs continuum probe pulse is used to measure over a large spectral range the transmission of a sample at tunable delay from a strong narrow-band fs pump pulse.

According to the discussion given above, in 3D and at high temperatures, S is the dominant exciton saturation mechanism. Therefore, resonant exciton creation at room temperature should produce a weak bleaching of the excitonic optical absorption that would increase substantially as the dielectric excitons transform into a metallic e-h plasma. Experiments designed to observe this behavior in 2D QWs lead to different and surprising results that we discuss now.[4,8]

Figure 2 shows room temperature differential transmission spectra of 100 Å GaAs QWs measured with a broad-band 50 fs probe pulse in 50 fs intervals before and after resonant n=1 hh exciton excitation by a 100 fs pump pulse.[4] The pump intensity is moderate and creates approximately $N \sim 2 \cdot 10^{10}$ cm^{-2} excitons. Initially, a strong bleaching of the n=1 hh exciton absorption lasting less than 1 ps occurs, then the absorption recovers partly and settles at the

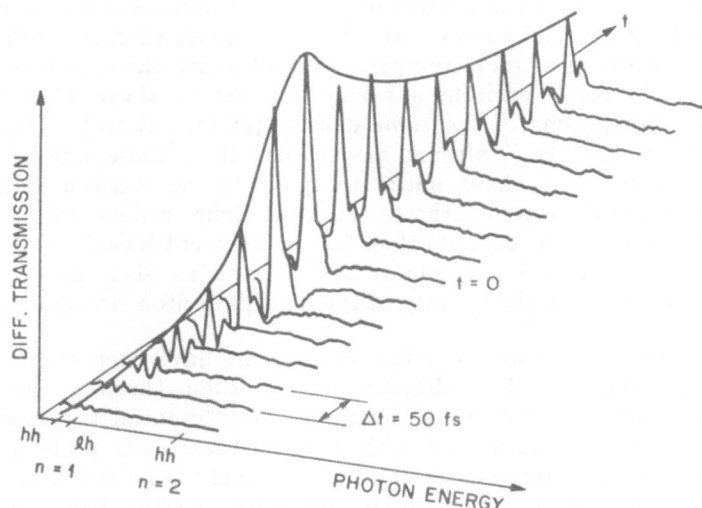

Fig. 2.   Differential transmission spectra as measured with a broad-band 50 fs probe pulse in 50 fs intervals before and after resonant n=1 hh exciton excitation with a 100 fs pump pulse. Initially, the absorption is reduced efficiently as excitons are created in the sample. It then recovers partly in ~ 300 fs as the excitons are ionized by collisions with LO phonons. The absorption then settles at the same value as it would under cw excitation.

same value as it would under cw excitation. The dynamics of the n=1 hh exciton bleaching can be described assuming the instantaneous generation of n=1 hh excitons (the number of excitons following the integral of the pump pulse), which transform with a ~300 fs time constant into free n=1 e and h, which then live for long times. The excitons ionize because of the large temperature, and, as discussed above, this thermal ionization is due to collisions with LO phonons. The measured ionization time of ~300 fs is in excellent agreement with the value obtained from the homogeneous absorption linewidth.

The excitons produce, before they ionize, a bleaching of the n=1 hh exciton absorption about *twice as large* as that due to the free e and h at long times. Also, the changes of the absorption spectrum in the vicinity of the n=2 hh exciton resonance are an order of magnitude smaller than those in the vicinity of the n=1 hh exciton resonance. Both experimental findings are at first surprising and the latter one does not fit at all into the picture given above, because both the n=1 and the n=2 hh exciton are subject to S. Since the n=1 hh exciton is in addition subject to PSF and E (contrary to the n=2 hh exciton), one can already conclude at this point that the effects of S on the excitonic resonances are weak and that the observed strong bleaching of the n=1 hh exciton is mainly due to PSF and E. This explains then also the fact that the bleaching due to the resonantly excited excitons is larger than that due to the free e and h released via thermal exciton ionization.[8] Between the absorption event and the first collision with a LO phonon the resonantly excited excitons have not yet had time to reach thermal equilibrium and therefore, for this very short time of ~300 fs, they are essentially at zero temperature. Then, because the LO phonon energy is much larger than the exciton binding energy, the first collision with a LO phonon ionizes the excitons and produces free e and h of rather large thermal energy. At long times, these thermally activated free e and h are essentially at room temperature and hence the experiment compares the effects of "cold" excitons (at short times) to those of a "hot" room temperature e-h plasma of the *same* density (at long times). The PSF and E effects due to the "hot" plasma are smaller than those due to the "cold" excitons, because the phase space sampled by an exciton decreases with decreasing thermal wavelength to exciton Bohr radius ratio, as already mentioned above. This accounts then for the different bleaching of the n=1 hh absorption at short and long times and it explains also the changes of the absorption spectrum in the vicinity of the n=1 lh exciton resonance.[8]

The conclusions drawn from this experiment have been confirmed recently by creating free e and h directly, rather than through thermal exciton ionization. Figure 3 shows room temperature differential transmission spectra of 100 Å GaAs QWs measured with a broad-band 50 fs probe pulse in 50 fs intervals before and after excitation of nonthermal n=1 e and h distributions by a 100 fs pump pulse.[5] The pump intensity creates again approximately $N \sim 2 \cdot 10^{10}$ cm^{-2} e-h pairs. Initially, the pump beam burns a spectral hole in the absorption spectrum, which then thermalizes within ~200 fs. Because the carriers are generated with an excess energy less than one LO phonon energy, this thermalization is due to carrier-carrier collisions and there is essentially no exchange of energy with the lattice.

The temporal evolution of the bleaching of the exciton resonances gives further evidence of the relative unimportance of S as compared to PSF and E under quantum confinement. Because the mean energy of the plasma does not change during the experiment, S is essentially the same at all times. The constant small changes of the absorption spectrum in the vicinity of the n = 2 hh exciton resonance serve again as an internal probe of the effects of S and show that they are small and independent of the carrier distribution. The n=1 resonances exhibit again a completely different behavior. Initially, only S is effective, giving rise to a small reduction of the absorption only, but as the carriers thermalize and occupy the band states out of which the n=1 excitons are constructed PSF and E become effective and the bleaching increases by almost an order of magnitude.

The experiments discussed above (and similar experiments on modulation-doped QWs[14]) show unambiguously that in 2D the effects of S on excitonic resonances are much less significant than in 3D and that the unbinding of excitons and the bleaching of the excitonic optical absorption are

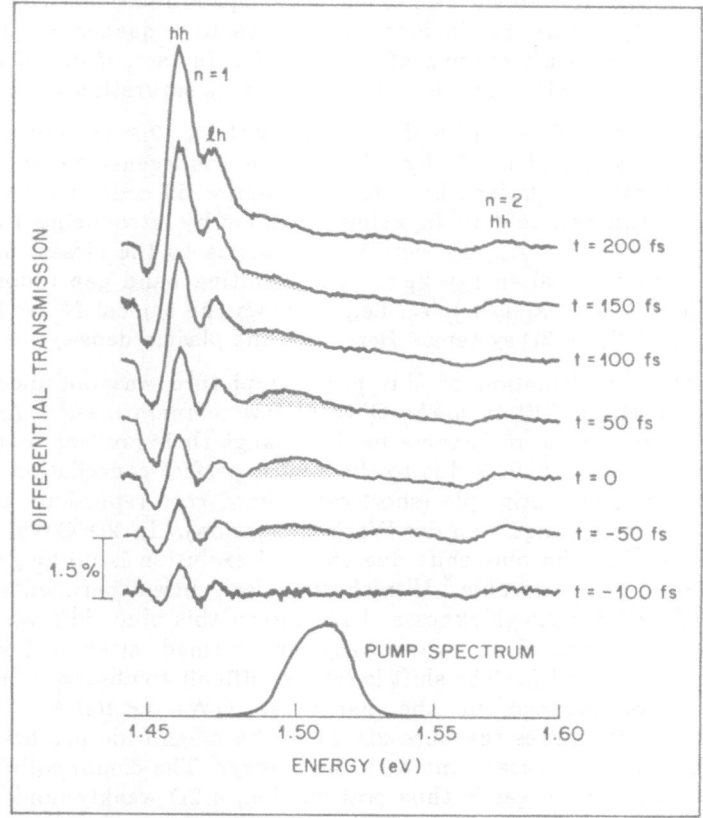

Fig. 3. Differential transmission spectra as measured with a broad-band 50 fs probe pulse in 50 fs intervals before and after excitation of nonthermal n=1 e and h distributions with a 100 fs pump pulse. Initially, the pump burns a spectral hole in the absorption spectrum, which then thermalizes in ∼ 200 fs. During thermalization, the effects of the Pauli principle on the n=1 excitons are turned on.

mainly due to PSF and E, i.e., overlap effects. Various facts must be considered to understand this behavior. First of all, in 1D and 2D an arbitrarily small attractive potential always supports at least one bound state, no matter how weak it is (similarly, all states are localized in 1D and 2D disordered systems). In addition, in 2D S is small and saturates at high densities due to the constant density of states, so that even an ultimately statically screened Coulomb interaction would always have a bound state with a significant (i.e. *not* exponentially small) binding energy, leading to an extension of the 3D Mott criterion to 2D ad absurdum. Unlike in 3D, an unbinding of excitons due to S is thus not possible. Moreover, in 2D the $1/r$ interaction is less long-ranged than in 3D, which has further consequences, the best known of which is the vanishing of the plasma frequency in the long-wavelength limit. More important in the present context, S is reduced far beyond its mean field value, "cut off" at distances corresponding to the size of the exchange-correlation hole surrounding each particle. Quantum confinement leads thus to a quenching of S and its effects on excitons, while overlap effects persist. In fact, if one decreases the dimension further (to 0D), PSF is the only surviving saturation mechanism.[15]

The breakdown of mean field theories of S in 2D is reflected by the logarithmic divergence of the Debye shift in the nondegenerate, weak coupling limit. Short-range correlations have to be employed in order to remove it. As implied above, this can be done in a simple fashion by introducing a momentum transfer cutoff $q_c^{-1} = e^2/(k_B T)$, which corresponds to the closest approach of two e or h with thermal energy $k_B T$. The resulting band gap renormalization varies like $[e^2 \kappa_{DH} \log (\kappa_{DH}/q_c)]/\epsilon_o$, i.e., it shows the typical N log N behavior characteristic of dilute 2D systems. Here, N is the plasma density.

Yet another confirmation of this physical picture was obtained from the study of the exciton shift in GaAs QWs at low temperatures.[2,3] As discussed above, in 3D the exciton resonances hardly change their position as the density of excitons increases, which is due to the almost perfect cancellation of the blue shift due to the Pauli principle (short-range hard core repulsion) and the red shift due to S (long-range Van der Waals attraction). In 2D QWs, because of the quenching of S, the blue shift due to Pauli exclusion is no longer balanced and thus should be measurable.[8] Ultrashort optical pulse experiments performed on GaAs QWs of various thicknesses have shown this blue shift when excitons are selectively generated or when they are formed after a few ps from photoexcited free e and h. The shift is rather difficult to observe in thick layers but is very pronounced in the narrowest QWs (<100 Å), where the dimensionality approaches the pure 2D limit. Its magnitude has been found to be in surprisingly good agreement with the theory.[3] The compressibility of a 2D weakly nonideal exciton gas is thus positive, i.e., a 2D weakly nonideal exciton gas is thermodynamically stable.

ULTRAFAST NONLINEARITIES DUE TO EXCITATION OF VIRTUAL E-H DISTRIBUTIONS

The nonresonant application of intense monochromatic laser pulses in the transparency region of a semiconductor, i.e., *below* the absorption edge, results in optical nonlinearities that are drastically different from the resonant renormalizations discussed above. One of these is the "lamp shift" or optical Stark effect, the high-intensity analogue of the Lamb shift. It has been extensively studied in atomic vapours, but it was not until recently that such light-induced renormalizations have been observed in a semiconductor.[6,7]

In time-resolved studies of the exponential absorption tail and phonon sidebands in 100 Å GaAs QWs at 70 K, using 6 ps pump and probe pulses, strong and transient changes of the probe beam transmission, as shown in Fig. 4, have been observed (by accident).[7] These changes can be positive or negative depending on the spectral position of the probe beam with respect to the n=1 hh exciton resonance. *They last only as long as the pump pulse.* In addition, if the pump detuning from the absorption edge is not too large, as in Fig. 4, much weaker effects due to an e-h plasma generated through phonon-assisted absorption are also seen. This process is easily distinguished from the fast effects. It persists for the lifetime and completely disappears for large detunings of the pump beam. The fast component of the change in probe

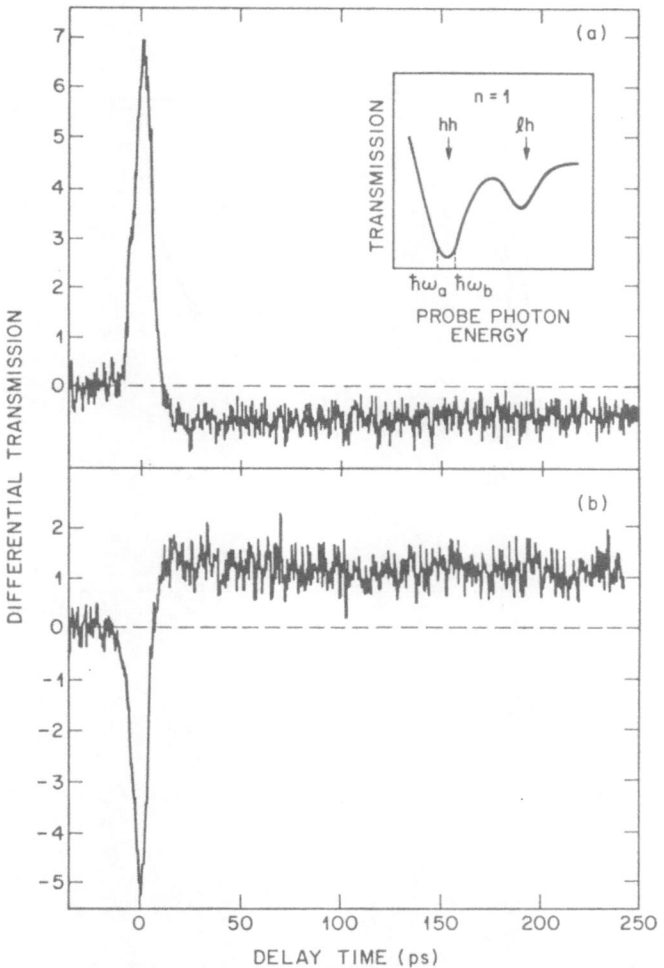

Fig. 4.   Differential transmission as a function of the pump-probe delay. The pump is tuned ∼ 25 meV below the n=1 hh exciton resonance. The probe is tuned ∼ 1 meV below the n=1 hh exciton resonance in (a) and ∼ 1 meV above the n=1 hh exciton resonance in (b).

beam absorption was studied as a function of the pump parameters. It was found that it is linear in the inverse pump detuning from the n=1 hh exciton resonance $(E_{ls} - \omega_p)^{-1}$ for a fixed pump intensity (Fig. 5a), and linear in the pump intensity $|E_p|^2$ for a fixed pump detuning (Fig. 5b). Moreover, a lineshape analysis of the fast component shows that the differential transmission corresponds to a transient blue shift of the exciton resonances. For a pump detuning ~ 30 meV and a pump intensity ~ 8 MW/cm^2, the magnitude of the shift is ~ 0.2 meV for the n=1 hh exciton peak and ~ 0.05 meV for the n=1 lh exciton peak. The same effects have also been observed using fs pump and continuum probe pulses, although at much higher pump intensity.[6] Again, they were found accidentally in the study of the exciton blue shift due to exciton-exciton interactions in narrow QWs discussed above. In addition, in the

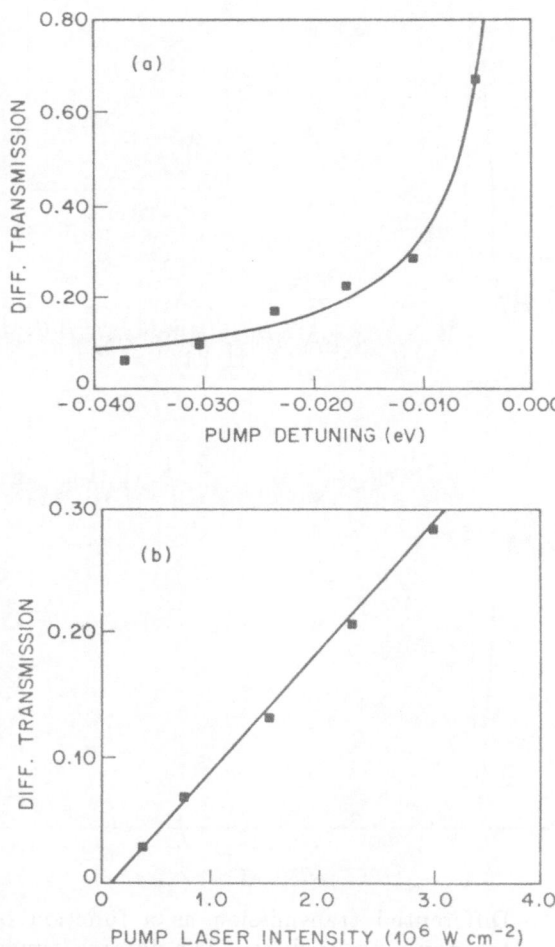

Fig. 5. Differential transmission as a function of the pump detuning from the n=1 hh exciton resonance (a) and pump intensity for a fixed pump detuning ~ 18 meV (b).

latter experiments, there is clear evidence of a loss of excitonic oscillator strength. In conclusion, the main effects of the strong below-gap excitation on the transmission of a probe beam are a shift and bleaching of the exciton resonances, i.e., an optical (AC) Stark effect.

Physicswise, this excitonic optical Stark effect is closely related to certain phenomena encountered in superfluid ^{4}He, ^{3}He and superconductors. We shall exploit these analogies in order to explain it.[9] Admittedly, this is not the simplest explanation, but it is the only correct one, covering *all* aspects of the problem. (As we shall see, the properties of noninteracting and localized atomic systems (Frenkel excitons) follow trivially.)

It is appropriate to visualize below-gap excitation as coherent generation of *virtual* e-h pairs, with no net energy transfer to the medium. The virtual transitions occur in a time $(E_{1s}-\omega_p)^{-1}$, much faster than the exponentially long relaxation time in the spectral region of the Urbach tail. They correspond to (an)harmonic undamped displacements of the valence charges, coherently driven by the laser field.

If we neglect all anharmonic interactions for the moment, all virtual e and h occupy the same quantum state, namely the one with energy ω_p and zero center of mass momentum. They are generated in this state by the pump and, in the absence of relaxation, they simply stay in it as long as the pump is applied. Hence, we are dealing with a macroscopic occupation of the q=0 state, just like in He II, for example.

The fact that this "Bose condensate" is virtual and externally imposed (rather than real and spontaneous) reflects itself in the chemical potential and in the phase of the virtual e-h pairs. The chemical potential μ, i.e., the energy required to create an additional virtual e-h pair, is identical to the pump frequency ω_p and thus below the bottom of the lowest exciton band, E_{1s}; the phase of the virtual e-h pairs is pinned, it is the same as that of the pump field. This leads inevitably to a gap in the q=0 collective excitation spectrum, as measured e.g. by a probe beam, unlike in He II. This gap is nothing but the minimum energy required to create *real* e-h pairs out of the condensate, i.e., the difference between the 1s exciton energy and the pump frequency, $E_{1s}-\omega_p$.

For moderate pump intensities and not too large detunings from the lowest exciton resonance, the virtual e-h pairs will be of the bosonic 1s exciton variety. If we now include anharmonic interactions, the problem will be simply that of a weakly nonideal virtual 1s exciton gas. Due to the anharmonic interactions, the condensate will be depleted and the gap $E_{1s}-\omega_p$ renormalized, which shows up as the AC Stark shift of *real* 1s excitons excited out of the condensate.

The anharmonic interactions can be split into two parts, both of which have already been discussed above. First of all, there is the effective exciton-exciton interaction, symbolically written as (ψ_{1s} is the 1s exciton field operator)

$$H_{int} = \frac{1}{2}V_{eff}\psi_{1s}^+\psi_{1s}^+\psi_{1s}\psi_{1s} \tag{1}$$

It follows simply from the elimination of the fermionic degrees of freedom (E) and the allowance for virtual exciton ionization (S). Secondly, there is the effective exciton-photon interaction

$$H_{dipole} \propto \psi_{1s}^+ E_p e^{-i\omega_p t} + c.c. \longrightarrow \psi_{1s}^+ E_p e^{-i\omega_p t}(1 - \lambda \psi_{1s}^+ \psi_{1s}) + c.c. \qquad (2)$$

Here, we have to make sure that, when we destroy a photon and create an exciton, we do not occupy the same states in phase space twice (PSF). This is done by projecting out the states that are already occupied and leads to an anharmonic exciton - photon coupling.

Given these two effective interactions, the problem is readily solved using the Bogolubov-Beliaev theory of a weakly nonideal Bose gas. We consider the two basic processes shown in Fig. 6, which give rise to exciton self-energies Σ_{11} and Σ_{12}, respectively. Σ_{11} describes the renormalization of 1s excitons excited out of the condensate due to anharmonic interactions with pump photons or virtual 1s excitons in the condensate, while $\Sigma_{12}(\Sigma_{12}^*)$ describes pair creation (annihilation); two pump photons and thus two virtual 1s excitons out of the condensate are destroyed and two renormalized excited 1s excitons are created, and vice versa. In terms of Σ_{11} and Σ_{12}, the spectrum of collective excitations has the well-known form

$$\omega_{1s} = \sqrt{(E_{1s} + \Sigma_{11} - \mu)^2 - |\Sigma_{12}|^2} \qquad (3)$$

The renormalized 1s exciton energy is simply given by $\mu + \omega_{1s}$. Note again that $\mu = \omega_p$.

Figure 7 shows a sketch of the resulting theoretical absorption spectrum, as measured by a weak test beam. The dashed line gives the unperturbed 1s exciton absorption at $\omega = E_{1s}$, the full lines the absorption in the presence of a pump beam at frequency $\mu < E_{1s}$. Optical absorption occurs at $\omega = \mu + \omega_{1s}$ and corresponds to the creation of a renormalized, i.e., Stark shifted and bleached, real exciton. Optical gain occurs symmetric about μ at $\omega = \mu - \omega_{1s}$ and corresponds to the simultaneous emission of a test photon and a collective excitation, or, with $\mu - \omega_{1s} = 2\mu - (\mu + \omega_{1s})$, to the depletion of the condensate; two pump photons or two virtual 1s excitons out of the condensate are destroyed and two renormalized excited 1s excitons are created, one of which transforms into a test photon and leaves the crystal. Note that this optical gain is i) due to photon exchange between pump and probe beam through the medium, and *not* due to a population inversion, and (ii) at least quadratic in the pump intensity, because two pump photons are required.

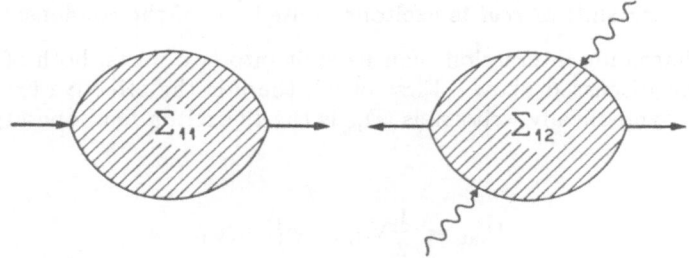

Fig. 6.   Basic interaction processes in a weakly nonideal virtual exciton gas. Σ_{11} describes exciton renormalization, Σ_{12} (Σ_{12}^*) exciton pair creation (annihilation).

The exciton self-energy due to anharmonic exciton-photon interactions (PSF) is[9]

$$\Sigma_{11}^{PSF} = |\Sigma_{12}^{PSF}| = \frac{1}{2}\frac{|er_{vc}E_p|^2}{E_{1s}-\omega_p}\frac{|\phi_{1s}(0)|^2}{N_S^{PSF}} \tag{4}$$

where er_{vc} is the (atomic) interband dipole matrix element, $\phi_{1s}(0)$ the 1s exciton orbital wavefunction at the origin and N_S^{PSF} the 1s exciton saturation density due to PSF.[8] Substituting this expression into Eq. (3), we recover exactly the standard expression for the Rabi frequency of a two-level system off resonance, with suitable (Wannier) exciton modifications

$$\omega_{1s} = \sqrt{(E_{1s}-\omega_p)^2 + |er_{vc}E_p|^2|\phi_{1s}(0)|^2/N_S^{PSF}} \tag{5}$$

As for the Stark shift, an exciton behaves thus in the same fashion as N_S^{PSF} independent two-level atoms, if exciton-exciton interactions can be neglected.

In leading order in the pump intensity, $|E_p|^2$, the Stark shift, $\omega_{1s} - E_{1s} + \omega_p$, is given by Eq. (4), which explains the experiments discussed above, without any adjustable parameter. The theoretical hh exciton shift is 0.15 meV and the ratio of the hh exciton and lh exciton shifts 4, in excellent agreement with the experimental data. The first factor in Eq. (4) expresses the

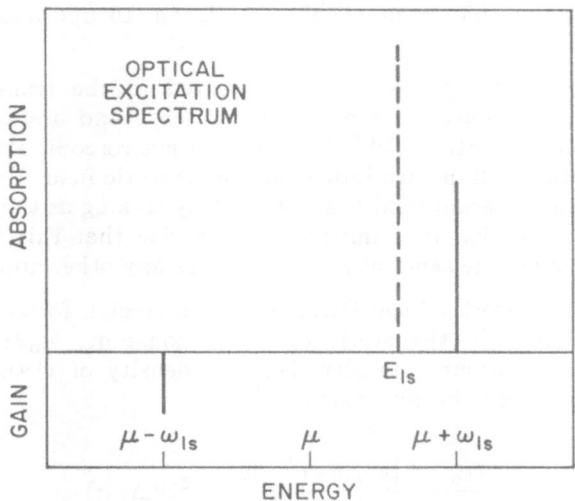

Fig. 7.   Optical excitation spectrum as a function of frequency ω. The dashed line gives the unperturbed exciton absorption (at $\omega=E_{1s}$), the full lines the absorption (at $\omega=\mu+\omega_{1s}$) and gain (at $\omega=\mu-\omega_{1s}$) in the presence of a pump beam at frequency μ. ω_{1s} is the frequency of collective excitations.

Stark shift of the atomic s and p states that form the conduction and valence bands. The second factor describes the renormalization of this atomic shift due to excitonic effects. Its numerator reflects the fact that an exciton is build up from a linear combination of Bloch states that originate themselves from the atomic states. It expresses the enhancement $|\phi_{1s}(0)|^2$ of the oscillator strength due to the correlation in the excitonic state. (The same factor appears in Elliot's formula for excitonic linear absorption.) The denominator contains the saturation density N_S^{PSF}, above which the concept of excitons becomes invalid. In the case of Frenkel excitons, we recover atomic behavior.

GENERATION OF ULTRASHORT ELECTRICAL PULSES THROUGH SCREENING BY VIRTUAL POPULATIONS IN BIASED QWs

One can imagine many potential applications of these ultrafast optical nonlinearities in high-speed optoelectronics. A very intriguing one is the combination of the optical (AC) Stark effect with the Quantum Confined (DC) Stark[16] or Franz-Keldysh[17] Effect (QCSE or QCFKE) to generate ultrashort voltage pulses in electrically biased QWs.[10]

In QWs biased by an electric field F normal to the plane of the layer the e and h wavefunctions are strongly distorted, positive and negative charges are pushed against opposite walls of the QW, as illustrated in the inset of Fig. 8. This gives rise to large red shifts of the absorption edge, an effect known as the QCSE[16] or QCFKE.[17] Because of the potential barriers, substantial charge separation can be obtained while keeping the tunneling time out of the QW significantly long, so that even at high fields an abrupt absorption edge is retained.

When a biased QW is strongly photoexcited in the transparency region below the absorption edge, the spatially separated e and h states are virtually populated so that the optical field E_p induces a macroscopic static polarization P. This polarization will in turn induce an electrostatic field $\delta F = 4\pi P/\epsilon_0$ which tends to screen the external field F and lasts only as long as the optical field E_p is applied to the sample. It is important to realize that this field is *real* and therefore will produce the same physical effects as any other applied field.

Within the framework of the QCFKE,[17] it is straightforward to derive the corresponding change in the static dielectric constant, $\delta\epsilon_0/\epsilon_0 = -\delta F/F$. In leading order in the pump intensity, i.e., the density of virtually excited e-h pairs, one finds assuming infinite barriers[10]

$$\frac{\delta\epsilon_0}{\epsilon_0} = \frac{|er_{vc}E_p|^2}{4\,E_0^2}\left(\frac{E_0}{W}\right)^{5/2}U(\Delta,\,\sigma) \tag{6}$$

where $W = \pi^2/(2mL^2)$ is the confinement energy (L is the QW thickness) and E_0 the excitonic Rydberg, m being the reduced e-h mass. The universal function $U(\Delta,\,\sigma)$ is plotted in Fig. 8 as a function of the dimensionless pump detuning from the lowest QW transition $\Delta = (E_g + W - \omega_p)/W$ for two different values of the electron to hole mass ratio $\sigma = m_e/m_h$.

For example, for a 100 Å GaAs QW and a pump detuning of 42 meV from the absorption edge, Eq. (6) yields a third order nonlinear susceptibility $\chi^{(3)} \sim 10^{-9}$ e.s.u. The importance of excitonic effects, finite barriers, etc. can

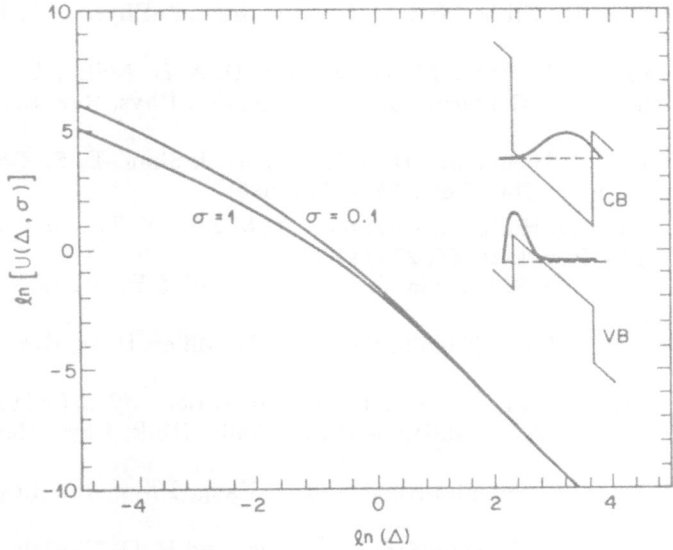

Fig. 8. Plot of the log of the universal function $U(\Delta, \sigma)$, which determines the change in the static dielectric constant, versus the log of the normalized pump detuning $\Delta = (E_g + W - \omega_p)/W$. Two curves are shown for electron to hole mass ratio $\sigma = m_e/m_h = 0.1$ and 1. The inset shows the $n=1$ electron and hole wavefunctions schematically, in the direction normal to the layer for a biased quantum well. (CB-conduction band, VB-valence band.)

be independently estimated from semiempirical electrorefraction calculations, using real QW spectra and the permutation relations for nonlinear optical susceptibilities. One finds an enhancement of $\chi^{(3)}$ by a factor of ~ 15.[10] In a diode structure containing fifty wells with 100 Å barriers reverse-biased to 10 V and with an incident intensity of 10 GW /cm^2, the voltage change generated would be -2.2 V. Such an optical intensity could be generated by an unamplified short pulse dye laser producing 100 pJ, 100 fs optical pulses focussed to a $10 \mu m^2$ area. Most importantly, this voltage would be generated directly inside a micron-thick semiconductor structure, offering new opportunities in high-speed electrical measurements and optoelectronic devices.

REFERENCES

1. H. Haug and S. Schmitt-Rink, J. Opt. Soc. Am. B *2*, 1135 (1985); and the recent collection of articles in: "Optical Nonlinearities and Instabilities in Semiconductors", H. Haug, ed. (Academic, New York, 1987).
2. N. Peyghambarian, H. M. Gibbs, J. L. Jewell, A. Antonetti, A. Migus, D. Hulin, and A. Mysyrowicz, Phys. Rev. Lett. *53*, 2433 (1984).
3. D. Hulin, A. Mysyrowicz, A. Antonetti, A. Migus, W. T. Masselink,

H. Morkoc, H. M. Gibbs, and N. Peyghambarian, Phys. Rev. B *33*, 4389 (1986).

4. W. H. Knox, R. L. Fork, M. C. Downer, D. A. B. Miller, D. S. Chemla, C. V. Shank, A. C. Gossard, and W. Wiegmann, Phys. Rev. Lett. *54*, 1306 (1985).

5. W. H. Knox, C. Hirlimann, D. A. B. Miller, J. Shah, D. S. Chemla, and C. V. Shank, Phys. Rev. Lett. *56*, 1191 (1986).

6. A. Mysyrowicz, D. Hulin, A. Antonetti, A. Migus, W. T. Masselink, and H. Morkoc, Phys. Rev. Lett. *56*, 2748 (1986).

7. A. Von Lehmen, D. S. Chemla, J. E. Zucker, and J. P. Heritage, Opt. Lett. *11*, 609 (1986).

8. S. Schmitt-Rink, D. S. Chemla, and D. A. B. Miller, Phys. Rev. B *32*, 6601 (1985).

9. S. Schmitt-Rink and D. S. Chemla, Phys. Rev. Lett. *57*, 2752 (1986).

10. D. S. Chemla, D. A. B. Miller, and S. Schmitt-Rink, Phys. Rev. Lett., in press.

11. J. P. Loewenau, S. Schmitt-Rink, and H. Haug, Phys. Rev. Lett. *49*, 1511 (1982).

12. G. W. Fehrenbach, W. Schaefer, J. Treusch, and R. G. Ulbrich, Phys. Rev. Lett. *49*, 1281 (1982).

13. L. Schultheis, J. Kuhl, A. Honold, and C. W. Tu, Phys. Rev. Lett. *57*, 1635 (1986).

14. A. E. Ruckenstein, S. Schmitt-Rink, and R. C. Miller, Phys. Rev. Lett. *56*, 504 (1986).

15. S. Schmitt-Rink, D. A. B. Miller, and D. S. Chemla, Phys. Rev. B *35*, 8113 (1987).

16. D. A. B. Miller, D. S. Chemla, T. C. Damen, A. C. Gossard, W. Wiegmann, T. H. Wood, and C. A. Burrus, Phys. Rev. Lett. *53*, 2173 (1984).

17. D. A. B. Miller, D. S. Chemla, and S. Schmitt-Rink, Phys. Rev. B *33*, 6976 (1986).

APPLICATIONS OF RESONANT TUNNELING IN SEMICONDUCTOR HETEROSTRUCTURES

E. E. Mendez

IBM Thomas J. Watson Research Center
Yorktown Heights, New York 10598, U.S.A.

INTRODUCTION

These notes are the third part of a set of three lectures presenting the basic concepts related to resonant tunneling in semiconductor heterostructures, and their application to electronic devices.
The first two lectures, published elsewhere ', were devoted to the
physics of resonant tunneling and to fundamental topics of current
interest. This part deals primarily with the implementation of those
ideas into devices, with a short introductory review of relevant
physics concepts.

Many structures that take advantage of the idea of resonant
tunneling have been proposed in the literature, and some have already
been realized experimentally. Here I will consider only some representative devices demonstrated in the laboratory, discussing their
advantages and limitations. The purpose has been to illustrate the
basic ideas rather than to give a comprehensive review of the field,
a task beyond the scope of these lectures.

PHYSICS OF RESONANT TUNNELING IN SEMICONDUCTORS: A RESUME

A particle of energy E can pass unattenuated through a symmetric
double-barrier potential of height V_0 for certain values of E. When
$E<V_0$, we talk of <u>resonant tunneling</u> at those particular energies, that
correspond to the eigenstates of a quantum well of depth V_0. The
concept was demonstrated in 1974 in semiconductor heterostructures[2],
in which the potential barriers were provided by two thin $Ga_{1-x}Al_xAs$

layers with a GaAs region in between, as electrons impinged on the barriers from a heavily-doped GaAs electrode. (See Fig. 1(a).) Since the maximum electronic energy was determined by the electrode Fermi energy, an electric field was applied between the source electrode and a similar counterelectrode to bring the incident electron in resonance with the eigenstates of the well. For certain resonant voltages a significant current flowed between the electrodes, followed by regions of negative conductance.

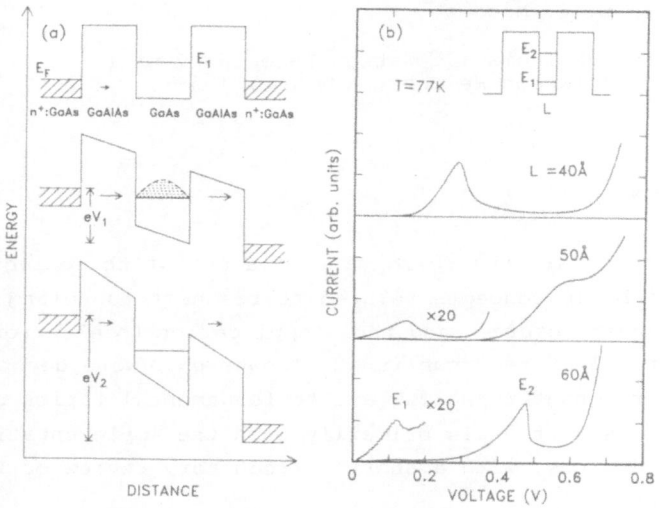

Fig. 1. (a) Double-barrier GaAlAs-GaAs-GaAlAs heterostructure clad
between heavily-doped GaAs electrodes, under different bias
conditions. When the voltage is such that the energy of the
quantum state E_1 is within the Fermi energy of one of the
electrodes, resonant tunneling occurs and the tunneling
current increases by several orders of magnitude. When E_1
falls below the conduction-band edge of the electrode the
current drops to zero.
(b) Current-voltage characteristics, at 77K, for three
double-barrier $Ga_{0.60}Al_{0.40}As$-GaAs-$Ga_{0.60}Al_{0.40}As$
heterostructures that differ only on the width, L, of the
quantum well, ranging from 40Å to 60Å. The width of the
barriers was in all cases 100Å.

Improvements in materials quality have made possible the observation of negative-conductance structures with large peak-to-valley ratios, like the ones shown on Fig. 1(b) for three devices with dif-

ferent well widths, ranging from 40Å to 60Å. The 40Å-well
heterostructure shows only one negative-resistance region, whereas
the 60Å well exhibits two, corresponding to the number of quantum
states confined in the well. The ratio of the peak-to-valley currents
exceeds 10 in some of the structures of Fig. 1(b), and increases even
further at 4.2K. Ratios over 20 have been reported recently[3]. This
large value reflects the high quality of the materials that it is
possible to prepare nowadays with techniques like molecular beam
epitaxy.

When the two barriers are identical, the voltages V_i at which
negative conductance occurs are related to the eigenenergies of the
quantum well by

$$V_i \simeq \frac{2E_i}{e} .$$ (1)

Experimentally, the measured voltages are usually larger than V_i because of additional voltage drops in the heterostructure outside the
barriers and the well, e.g., in the counterelectrode depletion layer.

Although the main effort has been on the GaAs-GaAlAs system,
other materials, e.g. InGaAs-InP, have also revealed negative resistance associated with resonant tunneling[4]. Recently, features in
the current-voltage characteristics of double-barrier
heterostructures made out of amorphous $Si-Si_3N_4$, have also been attributed to resonant tunneling[5]. This is a surprising observation
since resonant tunneling is seen as a process in which the momentum
of the electron parallel to the interfaces is conserved. It is unlikely that this condition holds in amorphous materials, in which the
mean free path is smaller than typical heterostructure dimensions.

An alternative mechanism that can give rise to negative resistance has been proposed by Luryi[6], who has considered the possibility
of a sequential process, by which the electron tunnels from the
electrode into the well, losing phase coherence through inelastic
scattering events. Eventually the electron leaks out of the well,
and a current is collected at the counterelectrode. The criterion
that determines which of the two extreme mechanisms -resonant or sequential tunneling- is responsible for negative resistance in a given
heterostructure, is the transit time of the electron through the
barriers and well, relative to the inelastic scattering time. If the
former is much shorter than the latter then the transport is coherent
and we can talk of resonant tunneling. In the opposite case, sequential tunneling is dominant[7].

The total transit time is mostly determined by the time it takes for the electron wavefunction to build up in the well and to leak out of it, since the tunneling time through the barriers is short compared with the build-up time. Semiclassically,

$$\frac{1}{\tau} = \frac{\sqrt{\frac{2E_i}{m}}}{2w} \, T(E_i),$$

[2]

where E_i is an eigenenergy of the well and $T(E_i)$ is the transmission probability for an individual barrier. It can be shown that τ as determined by Eq.2 is in reasonable agreement with a better estimate given by the uncertainty principle

$$\tau = \frac{\hbar}{\Delta E_i} \, ,$$

[3]

where ΔE_i is the energy width of the resonant state E_i , that can be determined from the transmission probability as a function of energy. Since this width can be quite narrow, e.g. for large electronic effective mass or wide barriers, resonant tunneling can be a slow process.

The energy width also determines the current densities involved in tunneling and in particular the peak value, J_{peak}. Using a very simple model it is shown[1] that

$$J_{peak} \sim \frac{em\Delta E}{2\pi\hbar^3} \, T_0 E_F.$$

[4]

Consequently, the peak current is small when either high or wide potential barriers are involved. The current also depends on the tunneling probability at resonance, T_0, which in a coherent process is of the order of unity. Since T_0 is much smaller in sequential tunneling, from Eq. 4 it would be expected that the peak current should be smaller in a non-coherent process than in a coherent one. However, recent calculations by Weil and Vinter[8], and by Jonson and Grincwajg[9], show that both mechanisms should give rise to the same current densities. In contrast, Büttiker, using a formalism that includes inelastic scattering, has concluded that the current density associated with resonant tunneling should be much larger than for sequential tunneling[10]. Although preliminary experimental results seem to support the latter view[1], systematic measurements are still lacking.

A consequence of the wavefunction build-up in the well is a corresponding charge accumulation. This charge can be estimated from the current density, since

$$Q = \tau J.$$

[5]

Then,

$$Q_{max} \sim \frac{em}{2\pi\hbar^2} E_F T_C,$$ [6]

which is independent of any structural parameters. In this simple estimation, for typical values of GaAs-GaAlAs structures it is $\frac{Q_{max}}{e} \sim 5 \times 10^{11} cm^{-2}$. Much more elaborate calculations have been performed by Frensley, who finds some dependence on the dimensions of the barriers[11].

The observation of negative resistance in double-barrier structures is a macroscopic manifestation of a quantum-mechanical effect that provides insight to fundamental questions regarding electronic transport in heterostructures. But, in addition, negative-resistance effects -also observed in the tunnel diode and the metal-insulator-semiconductor diode- are interesting for applications as high-frequency oscillators and amplifiers. Some of these possibilities are being actively explored, and new devices are being proposed based on the concept of resonant tunneling. In the following, I will review some of them, which, for convenience, I have divided into two categories: two-terminal, and three-terminal devices.

TWO-TERMINAL DEVICES

Two fundamental parameters that characterize the performance of a negative-resistance diode are its maximum frequency of oscillation, and the amount of power it can deliver. The device can be described by a capacitance, C, and a resistance, -R, in parallel. The inductance L_s and resistance R_s of the wires are elements in series with the diode[12]. It is easily shown that the equivalent circuit can sustain oscillations up to a maximum frequency, f_{max},

$$f_{max} = \frac{1}{2\pi RC} \sqrt{\frac{R}{R_s} - 1} \, .$$ [7]

As the negative resistance depends on the peak current density and on the peak-to-valley ratio, it is common to introduce a figure of merit called the speed index, defined as the ratio of the peak current to the capacitance at the valley voltage. Thus, for fast switching a large speed index (or small RC product) is required. Since, in addition, the maximum power delivered by the device[13] is $3/16 \, I_{max} V_{max}$ (V_{max} is the voltage at which the peak current I_{max} occurs), it is important to maximize the peak current, the peak voltage and the peak-to-valley ratio, and to minimize its capacitance.

Let us look now at the material parameters that determine those electrical quantities. The peak current increases with decreasing barrier height and width. On the other hand, the peak-to-valley ratio depends on the thermionic current and therefore the larger the barrier the larger that ratio. The optimum thickness to accomodate the two conflicting requirements has not been investigated systematically. Table I summarizes the relation between the quantities to be optimized with the parameters that can be varied in the design.

TABLE 1. Some variables relevant to the design of resonant-tunneling devices.

Optimization	Parameter	Problem
Current	Narrow barriers	Increases capacitance
	Low barriers	Decreases P.-V. ratio
P.-V. Ratio	High barriers	Decreases current
Voltage	Low doping	Increases resistance
	Diodes in series	
Capacitance	Low doping	Increases resistance

Regarding barrier height, it is worth a short digression about $Ga_{1-x}Al_xAs$-GaAs-$Ga_{1-x}Al_xAs$ double-barrier heterostructures, when $Ga_{1-x}Al_xAs$ becomes an indirect-gap semiconductor (x>0.45), with the bottom of the conduction band at the X point. Recent experiments by Tsuchiya and Sakaki[14] show that the resonant current decreases monotonically with x, confirming that, indeed, the Γ-Γ barrier determines resonant tunneling, regardless of the Γ-X barrier[15]. Then, for device applications, the only reason to use AlAs as the barrier material would be to reduce the thermionic component, so that the peak-to-valley ratio is increased. However, experiments by Solomon, Lanza and Wright[16] indicate that the thermionic current is controlled by the lowest barrier, the Γ-X barrier in GaAs-AlAs. This last result is puzzling, in view of substantial peak-to-valley ratios (>3:1) obtained in AlAs-GaAs-AlAs at room temperature[17].

The answer to this apparent contradiction may lie in the fact that Solomon et al.'s measurements[16] were done in relatively thick barriers (>300Å) whereas the resonant tunneling experiments were performed in heterostructures with very thin (~15Å) barriers[17]. Let us assume that two channels exist for tunneling, through the X and Γ points, so that for a given energy the total current is the sum of two terms, each proportional to the tunnel probability of either the Γ-Γ or Γ-X barrier. Since the Γ-X barrier is five times smaller than the Γ-Γ barrier, and the transverse mass is only three times heavier, the tunneling probability through the former is larger than through the latter. Consequently, for thick barriers, Γ-X tunneling should dominate.

On the other hand, Γ-X tunneling requires inelastic scattering and, therefore, it should be less likely than Γ-Γ, that conserves momentum. In mathematical terms it can be stated that the prefactor of the tunneling probability through X should be smaller than the one for tunneling through Γ. It is easy to show that in this case the relative contribution of Γ-tunneling to the total current is larger the thinner the barrier, and there could be situations where Γ-tunneling might even dominate over X-tunneling. This would explain the large peak-to-valley ratios for 15Å AlAs barriers.

The intrinsic speed of a tunnel diode can be larger than that of a resonant-tunnel diode, since in the former case we have to consider only the transit time through a potential barrier, whereas in the latter we should also take into account the time it takes for the electronic wavefunction to be built in the quantum well, which can be long. However, the large capacitance of a very heavily doped p^+-n^+ junction reduces the effective speed of the tunnel diode. In contrast, the resonant-tunneling diode does not suffer from that limitation. Although the thin barriers required to have a large current density tend to increase the device capacitance, it is possible to overcome this problem partially by moderate doping of the electrode regions next to the barriers.

So far, the largest effort to use resonant-tunneling diodes as oscillatory devices has come from Lincoln Labs, where fundamental frequencies up to 56 GHz and powers up to 60μW have been achieved[17]. As a comparison, the best tunnel diodes reported had oscillation frequencies as high as 103 GHz, but with an output power of less than 1μW (Ref.13). At the moment, resonant-tunneling diodes, or arrays of them in series, are quite promising as practical oscillators able to deliver moderate powers. To our knowledge, the possible use of resonant-tunneling diodes as amplifiers remains unexplored. Recently, Capasso et al.[18] have used two resonant-tunneling diodes in parallel, monolithically integrated, to build a three-state memory cell that can be used in multiple-valued logic applications.

Negative-resistance effects have been also observed in semiconductor heterostructures that differ from the simple double barriers described so far, such as the CHIRP[19,20] and the superlattice tunnel diode[21]. A CHIRP (coherent heterointerfaces for reflection and penetration) superlattice is made of alternating layers of two semiconductors with periodicity that changes gradually, as seen in Fig.2(a). Its band structure is such that, in equilibrium, the energies of the bottom and top of the miniband gap increase along the direction of the superlattice and current transport is possible for small applied electric fields. However, for a critical field the bandgap position is constant throughout the heterostructure and the current cannot be sustained, leading to negative resistance. Because of the nature of the effect the I-V characteristics are highly asymmetric, the negative resistance being observed for one of the voltage polarities but not for the other (Fig.2(b)).

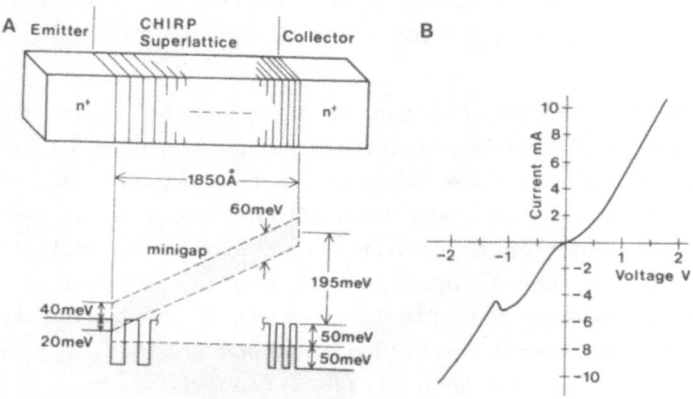

Fig. 2. (a) Schematic diagram of a CHIRP superlattice (top) and its calculated band profile (bottom). For electric fields such that the minigap becomes "parallel" to the emitter Fermi level, conduction is not possible, giving rise to negative resistance.
(b) Experimental current-voltage characteristics, at 85K, for a CHIRP superlattice. (After Ref. 20.)

A superlattice tunnel diode is analogous to a metal-insulator-metal diode, but the metal electrodes are replaced by semiconductor superlattices. Under a certain bias between the superlattices, current can tunnel through the barrier as long as there is overlap between the minibands of the superlattices. Beyond a certain voltage the overlap ceases, and tunneling is not possible because of the absence of any available states at the drain side. Further increase of the voltage restores the band overlap with the next miniband and makes tunneling current possible again, as shown in the energy diagrams of Fig.3.

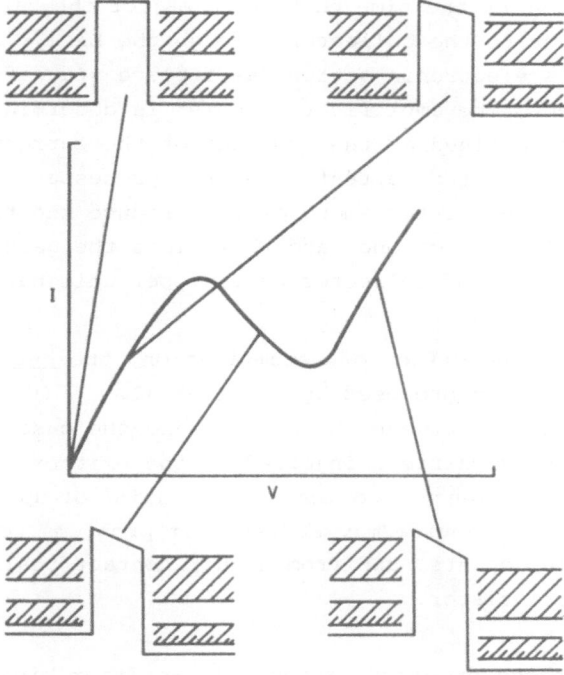

Fig. 3. Schematic current-voltage characteristic of the superlattice
        tunnel diode, showing its principle of operation in the
        various voltage regions. (After Ref. 21.)

One of the driving forces behind the pioneering work on
superlattices was the feasibility of producing Bloch oscillations at
much lower electric fields than in bulk semiconductors, as a result
of a reduced Brillouin zone[22]. However, this goal of ultra-high frequency oscillations seems experimentally elusive because the electronic mobilities in the superlattice direction obtained so far are
orders of magnitude below the critical mobility for their observation. This topic is treated in some detail in Bastard's notes in this
volume and it will not be discussed here any further.

THREE-TERMINAL DEVICES

Since the invention of the transistor, in which a large modulation of the collector current can be controlled by a small change
of the base current -thus producing amplification-, numerous devices
have been proposed that improve certain aspects of the original
junction device. I will review here, briefly, some of the recent attempts based on the concepts of tunneling and resonant tunneling.

Two of the most important device parameters of a transistor are
its switching speed and its amplification power. The former depends
to a large extent on the time that it takes for the carriers to travel
from the emitter to the collector through the base. In a bipolar
transistor this electronic motion takes place via diffusion along a
density gradient. The amplification power is determined by the
transfer ratio, defined as the quotient of the current at the collector over the emitter current. Several processes, recombination
of carriers in the emitter among others, reduce the transfer ratio
below its ideal value of one, and thus limit the gain of the device,
which is the change of collector current per unit base current.

To diminish the effect of recombination, the heterojunction bipolar transistor was proposed by Dumke et al.[23]. In this device the
barrier that exists between the emitter and the base reduces the
number of minority carriers injected at the emitter, thus improving
the collector efficiency. Common-emitter gains of up to 350 have been
achieved[24], but the device may suffer from problems related to traps
associated with dopants, and from higher contact resistance than the
homojunction transistor.

To solve these problems, a tunneling-emitter bipolar device has
been proposed by Xu and Shur[25], that takes advantage of the large
difference in the tunneling probabilities of electrons and holes.
In the original proposal, a graded GaAlAs layer was inserted between
a n-GaAs emitter and a p-GaAs base. This layer would present an additional barrier for holes injected from the base, tunneling through
which is very improbable because of the heavy hole mass. This device
has been reduced to practice recently by Najjar et al.[26], who have
even combined the features of a tunnel bipolar with those of a
heterojunction bipolar. Current gains exceeding 400 were achieved
with a $n-Ga_{0.76}Al_{0.24}As$ emitter followed by a thin AlAs layer that acted
as an effective barrier for holes from the base.

The concept of tunneling is also at the core of the many hot-
electron transistors that have been proposed and operated. The earliest one consisted of three metal layers separated from each other
by insulator films acting as barriers for electrons injected from one
of the metals[27]. Since then the trend has been to replace the metals
by semiconductors and to form the potential barriers either by doping,
like in the camel transistor[28] and the planar-doped barrier
transistor[29] (PDBT), or by compositional changes, like in the
tunneling hot electron transfer amplifier[30] (THETA). A hybrid transistor, mixture of the PDBT and the THETA devices, has been implemented by Long et al.[31].

236

In the THETA device, sketched in Fig.4, the emitter is a n⁻ semiconductor, which for simplicity we will assume to be GaAs. The electrons are injected into the base, a thin n-GaAs layer, via tunneling through a GaAlAs barrier. Their average energy is well above kT, and most of them can reach the collector (a n-GaAs film separated from the base by another GaAlAs barrier) while they are still hot. If the base is shorter than the electron mean free path, then many electrons reach the collector ballistically, that is without suffering any collision.

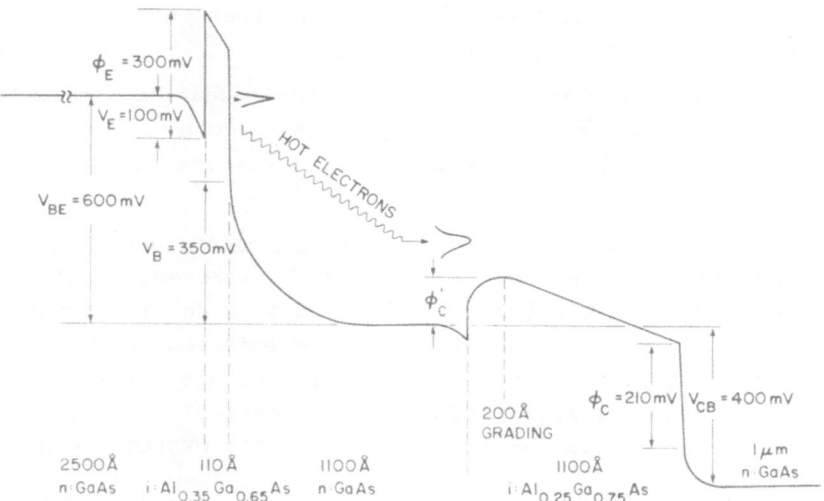

Fig. 4. Band profile of a THETA device. Electrons from the GaAs region on the left (emitter) are launched via tunneling into a thin GaAs base, by applying a voltage between them. Some electrons can traverse the base ballistically and be collected on the GaAs region on the right (collector). The GaAlAs layer between the base and the collector can act as a selective barrier for spectroscopic studies of the incident electron beam. (After Ref. 33.)

The first experimental demonstration of transistor action in a THETA structure was due to Yokoyama et al.[32], who achieved only a modest transfer ratio, $\alpha = 0.28$, in a device with a base 1000Å thick. By reducing this dimension to 300Å, Heiblum et al. have reached a value of α as high as 0.90, and have shown that a substantial amount of electrons traverse the base ballistically[33]. Quasi-ballistic motion in doped planar devices has been demonstrated also by Levi et al[34]. In spite of the potential of the THETA device for high speed, it has several limitations:

1. Its fabrication is complicated by the difficulty of making separated ohmic contacts to the emitter and the base. A very shallow contact at the emitter is required to avoid shorting the base.

2. The narrow base implies that the total number of carriers in it is small, which increases the base resistance. Selective doping of the GaAlAs barrier next to the collector has been proposed as a partial solution to that problem.

3. The gain decreases with increasing emitter current because as the electrons become very hot they can suffer scattering to other Brillouin-zone minima (at the L and X points). By using other semiconductor systems, e.g. InGaAs, in which those minima are much higher in energy relative to the Γ point than in GaAs, transfer ratios up to 0.94 have been achieved[35].

4. The transfer ratio is reduced by quantum-mechanical reflections at the base-collector barrier. By making a compositionally-graded barrier, the problem can be practically eliminated.

A variation of the THETA device, the resonant-tunneling hot-electron transistor, has been implemented by Yokoyama et al.[36], in which the emitter-base barrier is replaced by a double-barrier heterostructure, so that the electrons are injected into the base by resonant tunneling. As seen on Fig.5, this feature allows the device to act as a frequency multiplier and as an exclusive-NOR gate. An implementation of a similar logic function with MESFETs requires eight transistors.

A limitation common to the various hot-electron transistors just described is that their ultimate gain is not high because a significant fraction of the hot electrons cannot surmount the collector barrier, due to electron-phonon scattering in the base. To overcome this drawback, Bonnefoi et al. have proposed structures similar to the THETA devices in which the roles of the base and the collector are inverted[37], and Futatsugi et al. have proposed and demonstrated a resonant-tunneling bipolar transistor[38]. This device is a tunnel heterostructure bipolar transistor similar to the one of Najjar el al.[26], except that a double-barrier structure replaced the tunnel barrier at the emitter-base interface. Although the electrons also lose energy at the base they can reach the collector more easily because of the absence of the potential barrier. At 77K current gains of 20 have been reported. Capasso et al. have shown transistor action with gains up to 7 at room temperature, in a similar device in which the double-barrier structure was placed directly in the base[39].

I hope that the prototype devices illustrated here have given the flavor of a very active research field. Before they become a reality in commercial systems, work remains to be done for optimum designs. Efforts have to be devoted also to the elimination of marked differences frequently found in the characteristics of devices prepared from the same wafer. However, in spite of these difficulties we have reasons to be optimistic, as our understanding of the basic physics underlying resonant tunneling in semiconductors is advancing, and new structures are being proposed that show a large ingenuity in implementing this concept into a new breed of devices.

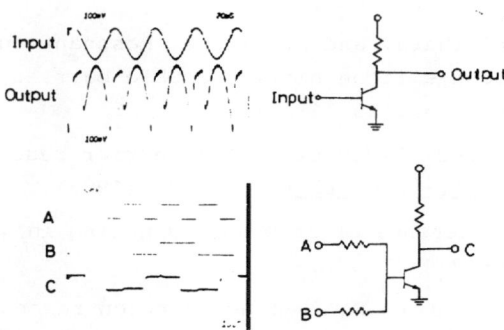

Fig. 5. Some applications of the resonant-tunneling hot electron transistor. In (a) the frequency of an input waveform is multiplied by two as a result of the transfer characteristic of the device. In (b) the circuit acts as an exclusive NOR gate when two digital signals are fed into inputs A and B. (After Ref. 36.)

I am thankful to T. Nakagawa, R. A. Davies, T. M. Kerr, M. J. Kelly, M. Heiblum and N. Yokoyama, for providing most of the figures used in these notes, and to A. C. Williams for a critical reading of the manuscript. This work has been sponsored in part by the Army Research Office.

REFERENCES

1. E. E. Mendez, Physics of resonant tunneling in semiconductors, in: "Physics and applications of quantum wells and superlattices," E. E. Mendez and K. von Klitzing, ed., Plenum, New York (1988).

2. L. L. Chang, L. Esaki, and R. Tsu, Resonant tunneling in semiconductor double barriers, Appl. Phys. Lett. 24:593 (1974).

3. C. I. Huang, M. J. Paulus, C. A. Bozada, S. C. Dudley, K. R. Evans, C. E. Stutz, R. L. Jones, and M. E. Cheney, AlGaAs/GaAs double barrier diodes with high peak-to-valley current ratio, Appl. Phys. Lett. 51:121 (1987).

4. T. H. H. Vuong, D. C. Tsui, and W. T. Tsang, Tunneling in In$_{0.53}$Ga$_{0.47}$As-InP double-barrier structures, Appl. Phys. Lett. 50:212 (1987).

5. S. Miyazaki, Y. Ihara, and M. Hirose, Resonant tunneling through amorphous silicon-silicon nitride double-barrier structures, Phys. Rev. Lett. 59:125 (1987).

6. S. Luryi, Frequency limit of double-barrier resonant-tunneling oscillators, Appl. Phys. Lett. 47:490 (1985).

7. P. J. Price, Coherence of resonant tunneling in heterostructures, Phys. Rev.B 36:1314 (1987).

8. T. Weil and B. Vinter, Equivalence between resonant tunneling and sequential tunneling in double barrier diodes, Appl. Phys. Lett. 50:1281 (1987).

9. M. Jonson and A. Grincwajg, The effect of inelastic scattering on resonant and sequential tunneling in double barrier heterostructures, Proceed. 3rd. Int. Conf. on Superlattices and Microstructures, Chicago, Aug.17-20, 1987,to be published.

10. M. Büttiker, Coherent and sequential tunneling in series barriers, IBM J. Res. and Develop. Jan. 1988.

11. W. R. Frensley, Quantum transport simulation of the resonant-tunneling diode, IEEE IEDM Technical Digest 571 (1986).

12. S. M. Sze, "Physics of Semiconductor Devices," John Wiley&Sons, New York, 2nd. ed., 1981.

13. E. R. Brown, T. C. L. G. Sollner, W. D. Goodhue, and C. D. Parker, Millimeter-band oscillations based on resonant tunneling in a double-barrier at room temperature, Appl. Phys. Lett. 50:83 (1987).

14. M. Tsuchiya and H. Sakaki, Dependence of resonant tunneling current on Al mole fractions in Al$_x$Ga$_{1-x}$As-GaAs-Al$_x$Ga$_{1-x}$As double barrier structures, Appl. Phys. Lett. 50:1503 (1987).

15. E. E. Mendez, E. Calleja, and W. I. Wang, Tunneling through indirect-gap semiconductor barriers, Phys. Rev.B 34:6026 (1986).

16. P. M. Solomon, S. L. Wright, and C. Lanza, Perpendicular transport across (Al,Ga)As and the Γ to X transition, Superlattices and Microstructures 2:521 (1986). See also I. Hase, H. Kawai, K. Kaneko, and N. Watanabe, Current-voltage characteristics through GaAs/AlGaAs/GaAs heterobarriers grown by metalorganic chemical vapor deposition, J. Appl. Phys. 59:3792 (1986).

17. W. D. Goodhue, T. C. L. G. Sollner, H. Q. Le, E. R. Brown, and B. A. Vojak, Large room temperature effects from resonant tunneling through AlAs barriers, Appl. Phys. Lett. 49:1086 (1986).

18. F. Capasso, S. Sen, A. Y. Cho, and D. Sivco, Resonant tunneling devices with multiple negative differential resistance and demonstration of a three-state memory cell for multiple-valued logic applications, IEEE Electron Dev. Lett. EDL-8:297 (1987).

19. T. Nakagawa, N. J. Kawai, K. Ohta, and M. Kawashima, New negative-resistance device by a chirp superlattice, Electron. Lett. 19:822 (1983).

20. T. Nakagawa, H. Imamoto, T. Sakamoto, T. Kojima, K. Ohta, and N. J. Kawai, Observation of negative differential resistance in chirp superlattices, Electron. Lett. 21:882 (1985).

21. R. A. Davies, M. J. Kelly, and T. M. Kerr, Tunneling between two strongly coupled superlattices, Phys. Rev. Lett. 55:1114 (1985). See also M. J. Kelly, R. A. Davies, N. R. Couch, B. Movaghar, and T. M. Kerr, Novel tunneling structures: physics and device implications, in: "Physics and applications of quantum wells and superlattices," E. E. Mendez and K. von Klitzing, ed., Plenum, New York (1988).

22. L. Esaki and R. Tsu, Superlattice and negative differential conductivity in semiconductors, IBM J. Res. Develop. 14:61 (1970).

23. W. P. Dumke, J. M. Woodall, and V. L. Rideout, GaAs-GaAlAs heterojunction transistors for high frequency operation, Solid State Electron. 15:1339 (1972).

24. M. Konagai and K. Takahashi, (GaAl)As-GaAs heterojunction transistors with high injection efficiency, J. Appl. Phys. 46:2120 (1975).

25. J. Xu and M. Shur, A tunneling emitter bipolar transistor, IEEE Electron Dev. Lett. EDL-7:416 (1986).

26. F. E. Najjar, D. C. Radulescu, Y.-K. Chen, G. W. Wicks, P. J. Tasker, and L. F. Eastman, DC characterization of the AlGaAs/GaAs

tunneling emitter bipolar transistor, <u>Appl. Phys. Lett.</u> 50:1915 (1987).

27. C. A. Mead, The tunnel-emission amplifier, <u>Proc. IRE</u> 48:359 (1960).

28. J. M. Shannon, Hot-electron camel transistor, <u>Solid-State and Electron Devices</u> 3:142 (1979).

29. M. A. Hollis, S. C. Palmateer, L. F. Eastman, N. V. Dandekar, and P. M. Smith, Importance of electron scattering with coupled plasmon-optical phonon modes in GaAs planar-doped barrier transistors, <u>IEEE Electron Dev. Lett.</u> EDL-4:440 (1983).

30. M. Heiblum, Tunneling hot electron transfer amplifiers (THETA): amplifiers operating up to the infrared, <u>Solid State Electron.</u> 24:343 (1981).

31. A. P. Long, P. H. Beton, and M. J. Kelly, Hot electron transport in heavily doped GaAs, <u>Semicond. Sci. and Technol.</u> 1:63 (1986).

32. N. Yokoyama, K. Imamura, T. Ohsima, H. Nishi, S. Muto, K. Kondo, and S. Hiyamizu, Tunneling hot electron transistor using GaAs/AlGaAs heterojunctions, <u>Jap. J. Appl. Phys.</u> 23:L311 (1984).

33. M. Heiblum, M. I. Nathan, D. C. Thomas, and C. M. Knoedler, Direct observation of ballistic transport in GaAs, <u>Phys. Rev. Lett.</u> 55:2200 (1985).

34. A. F. J. Levi, J. R. Hayes, P. M. Platzman, and W. Wiegmann, Injected-hot-electron transport in GaAs, <u>Phys. Rev. Lett.</u> 55:2071 (1985).

35. K. Imamura, S. Muto, T. Fujii, N. Yokoyama, S. Hiyamizu, and A. Shibatomi, InGaAs/InAlGaAs hot-electron transistors with current gain of 15, <u>Electron. Lett.</u> 22:1148 (1986).

36. N. Yokoyama, K. Imamura, S. Muto, S. Hiyamizu, and H. Nishi, A new functional, resonant-tunneling hot electron transistor (RHET) <u>Jap. J. Appl. Phys.</u> 24:L853 (1985).

37. A. R. Bonnefoi, D. H. Chow, and T. C. McGill, Inverted base-collector tunnel transistors, <u>Appl. Phys. Lett.</u> 47:888 (1985).

38. T. Futatsugi, Y. Yamaguchi, K. Ishii, K. Imamura, S. Muto, N. Yokoyama, and A. Shibatomi, A resonant-tunneling bipolar transistor (RBT): a proposal and demonstration for new functional devices with high current gains, <u>IEEE IEDM Technical Digest</u> 286 (1986).

39. F. Capasso, S. Sen, A. C. Gossard, A. L. Hutchinson, and J. H. English, Quantum-well resonant tunneling bipolar transistor operating at room temperature, <u>IEEE Electron Dev. Lett.</u> EDL-7:573 (1986).

POLARON EFFECTS AND OPTIC PHONON SCATTERING IN HETEROSTRUCTURES

R.J. Nicholas, M.A. Hopkins, M.A. Brummell and D.R. Leadley

Clarendon Laboratory
Parks Road
Oxford, OX1 3PU, U.K.

1. INTRODUCTION

The interaction of electrons with the optic phonons in semiconductors is one of the most fundamental problems in solid state physics. In ionic crystals we expect a strong interaction with the longitudinal optic (LO) phonon, through the electric field of the polarization wave. The interaction with transverse optic (TO) phonons will be less strong because of their smaller electric field, except at very low wavevector, q, where the electromagnetic coupling may be strong. This coupling was one of the first examples of the use of quantum field theory in solid state physics (1), where the motion of an electron through an ionic solid was described as a composite particle, the polaron, consisting of an electron dressed by a virtual phonon cloud. Work on semiconductor heterostructures has generated new interest in this field, through the possible changes brought about by the electron confinement, the presence of interfaces and the influence of screening. Theoretical work (2-5) predicts that polaron effects in two-dimensional (2-D) systems should be stronger than in the corresponding bulk materials, but the finite wave functions in the third dimension (5-7) and screening (6,8,9) should reduce the coupling. Experiments on various 2-D systems have suggested both enhanced (10) and reduced (9,11-15) effects.

Polaron coupling influences the electron properties through coupling to both real and virtual phonons. At high temperatures LO phonon absorption plays a dominant role in limiting the momentum relaxation time and determining the carrier mobility. Studies of this have been made by looking at the width of the cyclotron resonance absorption as a function of temperature, and the magnetoresistance at high magnetic fields. Resonant optic phonon absorption leads to oscillatory structure in the magnetoresistance known as the magnetophonon effect (16). Optic phonon emission also leads to magnetophonon oscillations under hot electron conditions, where resonant cooling occurs (16). Virtual phonons make up the lattice distortion which forms the renormalised particle known as a polaron. As a result the dispersion relation of the electron is altered, leading to changes in the effective mass. This is studied by the frequency dependence of the effective mass, which is particularly strongly altered when the electron energy becomes resonant with that of the phonons.

Recent results (14,17-20) have suggested that polaron coupling in heterostructures may be modified to give coupling to new modes close to the TO frequency, or be associated with the interface modes (21). Magnetophonon resonance experiments have demonstrated that cross coupling can occur between electrons in one material and phonons in another (22,23). The dominant phonon frequency may also be carrier concentration dependent (24). The experimental picture is summarised below., with a brief introduction to the theory.

2. POLARON COUPLING

The influence of polaron coupling on the energy levels of the conduction electrons can be calculated using the theory introduced by Fröhlich et al. (25). In this the system is described by a Hamiltonian (H_p) having three terms: that of the electrons in the undisturbed lattice (H_O), of the lattice in the absence of the electrons, and the Fröhlich Hamiltonian (H_F) representing the electron-phonon interaction. H_p is then given by

$$H_p = H_O + \hbar\omega_{LO} \sum_{\underline{q}} b_{\underline{q}}^{+} b_{\underline{q}} + H_F \tag{1}$$

where

$$H_F = \sum_{\underline{q}} v_{\underline{q}} e^{i\underline{q}\cdot\underline{r}} (b_{\underline{q}}^{+} + b_{\underline{q}}) \tag{2}$$

$$v_{q}^{2} = \frac{4\pi\alpha\hbar(\hbar\omega_{LO})^{3/2}}{(2m_{o}^{*})^{1/2} V q^{2}} \tag{3}$$

and b_q^{+} and b_q are creation and annihilation operators for LO phonons, V is the crystal volume and α is the dimensionless Fröhlich coupling constant. This is given by (26)

$$\alpha = \frac{e^2}{4\pi\epsilon_o\hbar} \left(\frac{m^*}{2\hbar\omega_{LO}}\right)^{1/2} \left[\frac{1}{\epsilon(\infty)} - \frac{1}{\epsilon(0)}\right] \tag{4}$$

where $\epsilon(0)$ and $\epsilon(\infty)$ are the static and high frequency dielectric constants. In the weakly polar III-V semiconductors α is typically $\sim 0\cdot1$, GaAs having a value of $\sim 0\cdot07$ (27).

Several methods have been used to solve the polaron Hamiltonian. In materials where $\alpha \ll 1$, the electron-phonon interaction is weak and H_F may be reasonably treated as a perturbation on the undisturbed system. For strong coupling perturbation methods fail and it is necessary to employ other methods for calculating the polaron energy, such as variational techniques (28) or the Feynman polaron model (29). However these will not be discussed further as the III-V semiconductors are weakly polar and thus perturbation methods prove quite successful.

In the weak coupling limit at low energies simple first order perturbation theory leads to the results that the polaron coupling reduces the energy of a free electron by an amount $\alpha\hbar\omega_{LO}$ and increases its effective mass by the factor

$$m_{p}^{*} = m_{o}^{*} / (1 - \alpha/6) \tag{5}$$

In practice this result is only valid at energies well below that of the LO phonon, and as the energy increases towards this value the coupling becomes stronger and causes a greater increase in mass. This region is usually called the resonant polaron. The most accurate way of studying the effective mass involves applying a magnetic field and studying transitions between the quantized Landau levels. In this case the resonant coupling occurs at the magnetic field where the cyclotron frequency, ω_c, is equal to the LO phonon frequency. This is illustrated in figure 1. The unperturbed Landau levels are shown as dashed lines, and the polaron states as solid lines. For the lowest Landau level, $n_L = 0$, there is a shift downwards by the polaron self energy, and a change in the slope as a function of magnetic field. Also shown is the energy of the lowest Landau level plus one phonon. The point where this crosses the energy of the second Landau level, $n_L = 1$, defines the resonant polaron coupling, since these two states of the system would become degenerate without the coupling. By studying transitions between the first two Landau levels we thus have an excellent way of testing the predictions of the polaron theory.

Two standard types of perturbation theory, Rayleigh-Schrödinger (RSPT) (30) and Wigner-Brillouin (WBPT) (28) have been applied to calculate the energy dependent polaron coupling. RSPT is most suitable for use in the low field range, $(\omega_c/\omega_{LO} < 0.5)$, and for magnetic fields above this a degenerate perturbation theory such as WBPT should be used. The best results are obtained using an improved version of WBPT (IWBPT) first introduced by Lindemann et al. (31). Here the polaron Hamiltonian is decomposed to include the ground state RSPT correction to the energy. Using an IWBPT calculation Peeters and Devreese (4) showed that for transitions from $n_L = 0$ to $n_L = 1$ in three dimensional systems

$$1/m^*_{CR} = 1/m^*_0 \{1 - \alpha/6 - \frac{3\alpha}{20} (\frac{\omega_c}{\omega_{LO}}) - \frac{47\alpha}{504} (\frac{\omega_c}{\omega_{LO}})^2$$

$$+ \text{ higher terms in } \omega_c/\omega_{LO}\} \tag{6}$$

Using the same calculation techniques for a perfect 2-D system Peeters and Devreese (4) found that

$$1/m^*_{CR} = 1/m^*_0 \{1 - \pi\alpha/8 - \frac{9\pi\alpha}{64} (\frac{\omega_c}{\omega_{LO}}) - \frac{145\pi\alpha}{1024} (\frac{\omega_c}{\omega_{LO}})^2$$

$$+ \text{ higher terms in } \omega_c/\omega_{LO}\} \tag{7}$$

The same result is obtained from RSPT (6) which differs only in higher orders in α and ω_c than those given above. The reduction in dimensionality results in a nearly three fold increase in the polaron mass enhancement. This has come through the interaction coefficient (equation 3). In practice, as will be seen below, there is no experimental evidence for any large enhancement in the polaron non-parabolicity (the energy dependent terms in equation (6) and (7)) for 2-D systems compared with 3-D.

It has been suggested that the 2-D polaron enhancement may be quenched by screening in the highly degenerate 2-D electron gas (6,8,9,32,33). The finite extent of the wavefunction in the third dimension has also been shown to reduce the coupling (5,6,9). This is because the physical size of the polaron is typically only of order 40 Å (8), which is usually significantly less than that of the z-component of the wavefunction. As a result the coupling is closer to that in the bulk material. Zero magnetic field calculations (8,9,32) suggest that screening and wavefunction effects together may reduce polaron interaction effects in 2-D structures by as much as a factor of 8, and so to values well below the corresponding

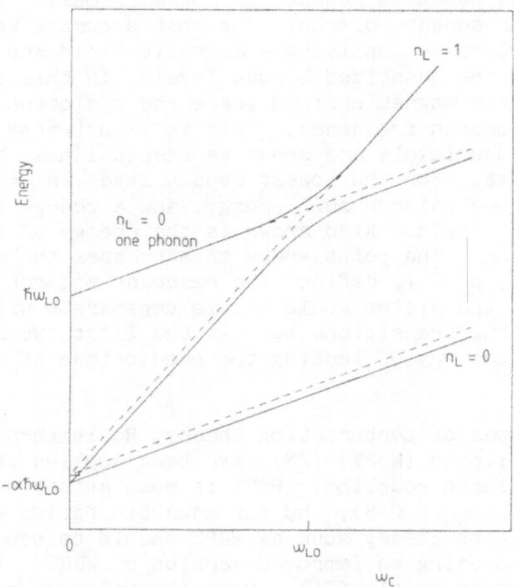

Fig. 1.   A schematic view of the polaron coupling between the first two
          Landau levels. The dashed lines show the unperturbed levels.

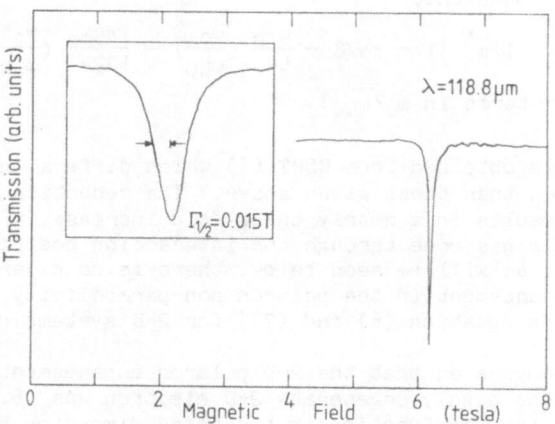

Fig. 2.   An experimental recording of the cyclotron resonance absorption
          in a GaAs/GaAlAs heterojunction at 4·2 K. The inset shows the
          resonance half-width for 50% absorption.

3-D value. The importance of screening can be judged by the length scale of the screening parameter q_S. This is given by

$$q_S = e^2 / 2\epsilon\epsilon_0 \frac{dN}{dE} = e^2 / 2\epsilon\epsilon_0 N/E \tag{8}$$

in 2-D. The static screening length is $1/q_S$, which is of order 50 Å in GaAs.

The screening effects should be even stronger in high magnetic fields, since the effect of the field is to condense states from a range of energies into well defined Landau levels. As a result the density of states can be greatly increased if there is a small amount of level broadening, by approximately a factor $\hbar\omega_c/\Gamma$, where Γ is the broadening parameter. Although calculations based on the use of static screening will overestimate the strength of screening effects, recent calculations by Xiaoguang et al. (34) suggest that there is not a large difference between the results calculated using static and dynamic screening.

3. CYCLOTRON RESONANCE

Cyclotron resonance is a classic tool for studying effective masses in solids. To study polaron coupling it is necessary to work in the Far Infra Red (FIR) region of the spectrum (500 — 30 μm) where the majority of semiconductors are transparent until the Restrahl reflection band is reached in the region of the optic phonons. This makes it intrinsically difficult to study resonant polaron effects. In this respect heterostructures can have some very considerable advantages, since the layer confining the electrons can be extremely thin which means that TO phonon absorption can be much reduced. Figure 2 shows a typical experimental recording of the cyclotron resonance which can be observed in a high quality GaAs/GaAlAs heterojunction, grown at the Philips Research Laboratory, Redhill (35). At low temperatures the linewidth of the resonance can be as small as 0·005 T, allowing extremely accurate determinations of the effective masses to be made (15). The results of this are shown in figure 3, where data has been taken on one sample before and after illumination with light from a red LED. This results in a change of the 2-D carrier concentration. At high carrier concentrations the resultingly high Fermi energy means that at lower magnetic fields a number of different Landau levels may be populated, thus complicating the analysis of the data. At low surface carrier concentrations, $n_s < 2eB/h$, only one level is occupied and we work in what is known as the quantum limit. This condition is satisfied for all of the points shown in figure 3 at the lower carrier concentration. The data show that there is initially an almost linear increase in the effective mass with frequency, followed by a more rapid rise as the optic phonon frequency approaches. In fact the majority of this mass increase is not the result of polaron effects, but is due to band non-parabolicity. This phenomenon has been studied by a number of authors, and has been shown to result in an approximately linear increase in mass with energy. This can be written in an equation of the form for a transition originating on the level n_L,

$$\frac{1}{m^*_{CR}} = \frac{1}{m^*_0} \left[1 + \frac{2K_2}{E_g} (n_L \hbar\omega_c + \langle T_z \rangle) \right] \tag{9}$$

where $\langle T_z \rangle$ is the kinetic energy for motion in the third dimension, which is a result of the quantum confinement in heterostructures. K_2 is a factor characterising the strength of the non-parabolicity and E_g is the band gap. There have been a number of recent calculations of the magnitude of this non-parabolicity using mulitiple band $\underline{k}.\underline{p}$ theory (36-38) and it has been shown that experiment and theory are in good agreement for the cases of bulk GaAs and InP when combined with IWBPT calculations of the polaron effect. The magnitude of the band contribution to the non-parabolicity corresponds to a value of $K_2 = -1.4$ for GaAs. This is shown as a solid line in figure 3, and appears to describe very well the behaviour of the 2-D resonances at low energy, suggesting that any polaron effects are relatively weak. This is illustrated more clearly in figure 4, where experimental results are plotted for both 2-D and 3-D systems (the bulk mass values have been increased by $0.0012\ m_c$ to account for the non-parablicity from the wavefunction confinement in the 2-D system). There is considerably more nonparabolicity in the 3-D results, which are described very well by the theory, whereas the 2-D system has apparently no polaron contribution to its mass until the resonant region. Once the optic phonon frequencies approach however, the coupling appears to be increasing in strength more rapidly for the 2-D system.

The conclusion thus appears to be that there is almost no polaron effect for a 2-D system at low temperatures and low energies, in agreement with the suggestions of the importance of finite wavefunction effects and screening. Brummell et al. have shown however that this changes considerably as the temperature increases (14), while Singleton et al. (39) have found that in quantum wells with very low carrier concentrations polaron effects can be stronger. At high temperatures more than one Landau level will be populated even at magnetic fields beyond the quantum limit. This will influence the value deduced for the effective mass from cyclotron resonance experiments. This can be taken into account by the use of equation (9), and the assumption that the transitions are weighted by the populations of the initial levels. It is convenient to correct the observed mass values back to the effective mass corresponding to the transition between the $n_L = 0$ and $n_L = 1$ Landau levels, m_{01}^*, as seen at $T = 0$ (this correction is $\sim 0.7\%$ at 100 K for GaAs measured at a wavelength of 119 μm).

The values of m_{01}^* deduced by Brummell et al. (14) as a function of temperature are shown in figure 5. Data were taken for three GaAs/$Ga_{0.68}Al_{0.32}As$ heterojunctions with different electron concentrations. There is a very rapid increase in the mass, of order 2%, which has saturated by approximately 100 K. Similar measurements on bulk GaAs are also shown, and have no such change in m_{01}^*. The increase has been attributed to temperature dependent screening of the polaron coupling. In an ideal 2-D system, equation (7) predicts a mass enhancement of 3.9% at this frequency (3,4) (a wavelength of 118.83 μm). However the finite wavefunction effects reduce the coupling (5-7), and the calculations of Das Sarma (6) give enhancements of 1.4%, 1.5% and 1.6% for the three samples studied. The comparison of the non-parabolicities shown above for 2-D and 3-D systems suggest that there is almost no polaron coupling in the 2-D system at low temperatures, largely due to screening. The rise in mass with temperature is attributed to the reduction in importance of screening, and the consequent re-appearance of polaron coupling. The screening should depend upon the density of states in the region of $\sim kT$ around the Fermi energy (40), and at low temperatures the Landau levels are extremely sharp, as reflected by the extremely narrow cyclotron resonance linewidth. This leads to a considerable increase in the importance of screening, as discussed in section 2, as well as to oscillatory cyclotron resonance linewidths (41) which have been attributed to filling factor dependent screening (41,42). At higher temperatures the

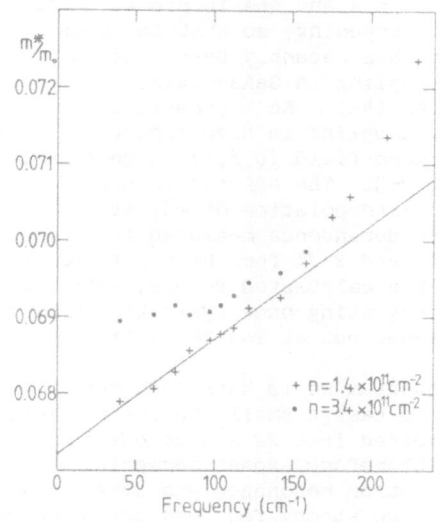

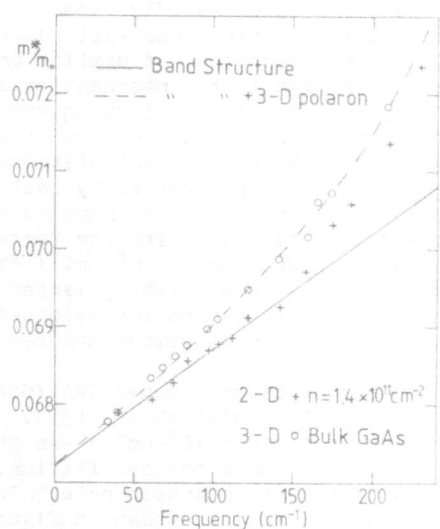

Fig. 3. Plots of the frequency
dependence of the effective
mass for a GaAs/GaAlAs
heterojunction with two
different carrier
concentrations.

Fig. 4. Plots of experimental
results for the frequency
dependence of the
effective mass for 2-D and
3-D GaAs. (The 3-D results
are shifted by 0·0012 m_e).
The solid and dashed curves
show the theoretical values
with and without polaron
coupling.

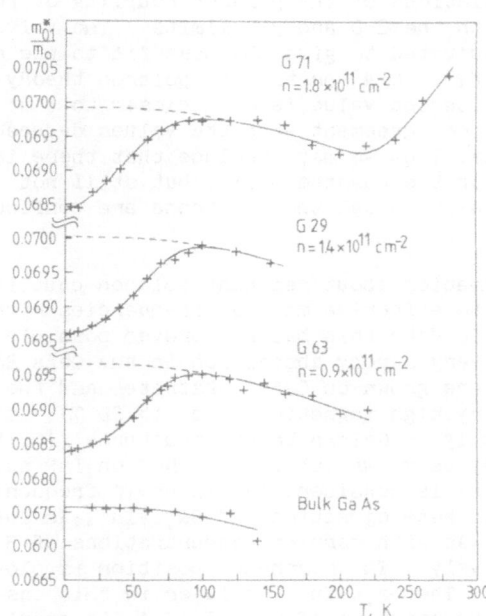

Fig. 5. Experimental plots of the temperature dependence of three
heterojunctions, compared with results for bulk GaAs.

resonances broaden rapidly (see below) and this and the increased thermal distribution of electrons will reduce the screening, so that the mass enhancement reappears. A similar argument has recently been used to explain changes in the resonant polaron coupling in GaAs-GaAlAs heterojunctions in hot-electron experiments (43). No theoretical studies of screening of the electron-optic phonon coupling in high magnetic fields have been reported, but calculations for zero field (6,8,9) suggest that the coupling may be reduced by factors of 2-3. The effects in high fields may thus be expected to be more extreme. Extrapolation of m^*_{o1} at high temperatures to T = 0 with the temperature dependence measured in bulk GaAs gives reductions in m^*_{o1} of 1·8%, 2·1% and 2·2% for the three samples studied. These are slightly larger than the calculated values, with the same electron concentration dependence, suggesting once again that most or all of the mass enhancement has been screened out at low temperatures.

Measurements on undoped GaAs/GaAlAs quantum wells have been reported recently by Singleton et al. (39). In this case a small concentration of electrons (of order 10^7 cm^{-2}) was photoexcited into 22 Å wide quantum wells by continuous photoexcitation. We therefore expect screening to play a much less important role. The cyclotron resonance was detected by photoconductivity, as shown in figure 6. Two strong features are seen in the spectra: the lower sharp peak in the photoresponse can be identified as the bulk cyclotron resonance, presumably due to electrons in the buffer layer on which the multiple quantum well sample was grown; the broader feature at higher fields is identified as being due to cyclotron resonance of the subband electrons in the quantum wells. The effective masses deduced from this data are shown in figure 7. In this case the confinement in the quantum well has introduced a very large kinetic energy (~ 200 meV) which leads to a very considerable increase in the effective mass at the band edge. Knowledge of the band non-parabolicity this high up in the conduction band is rather less precise, so inter-band magneto-optic transitions were also studied simultaneously in the structure. The energy dependence of the effective mass shown in figure 7 was then fitted to the numerical calculations of the polaron coupling of Peeters and Devreese (4), using both the 2-D and 3-D limits. The value of the band non-parabolicity was adjusted to give the best fit to the effective mass data, using values of $K_2 = -0·4$ for the 2-D polaron theory and $K_2 = -1·35$ for the 3-D case. The second value is much closer to that found at lower energies, and was in good agreement with the values deduced from the inter-band measurements. Thus we may conclude that there is evidence for some polaron coupling in the quantum wells, but still not very close to the perfect 2-D limit even though the electrons are confined in a well of only 22 Å width.

More precise information about resonant polaron coupling could be obtained by studying the effective mass at frequencies right through the optic phonon region. To date this has not proved possible in GaAs based structures due to the very strong absorption in the GaAs Restrahl band, since most structures are grown on GaAs substrates and the coupling also requires the use of very high magnetic fields (> 20 T). This region has been studied successfully in GaInAs based structures (17,18) however, since this material may be grown lattice matched on InP substrates, and the Restrahl band in InP is considerably higher in frequency. Figure 8 shows data taken in two heterojunctions of Ga$_{.47}$In$_{.53}$As/InP and Ga$_{.47}$In$_{.53}$As/Al$_{.48}$In$_{.52}$As with carrier concentrations of 3 and 8×10^{11} cm^{-2} respectively. The resonance position is plotted as a function of frequency. The Ga$_{.47}$In$_{.53}$As layer in this case is only ~ 1 μm thick. The optic phonon spectrum of Ga$_{.47}$In$_{.53}$As is complicated by virtue of its being an alloy, and in the middle range of alloy compositions it displays the well known two mode behaviour. Measurements of reflectivity (44) and Raman scattering (45,46) give the frequencies of the 'GaAs-like'

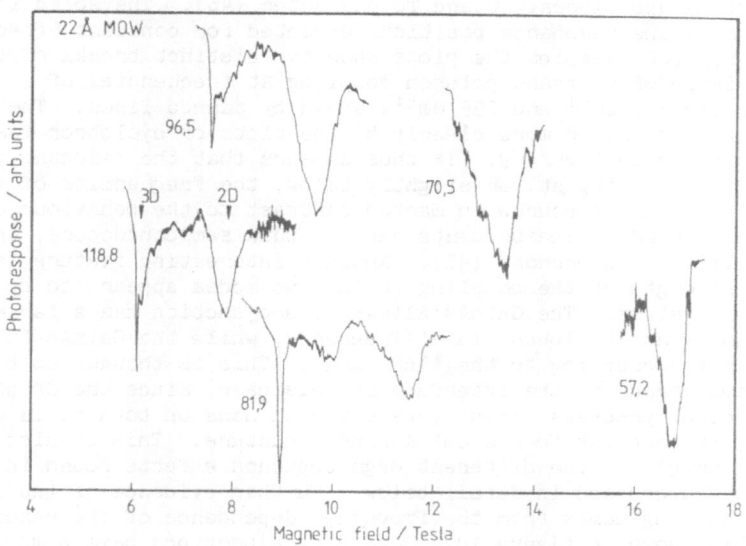

Fig. 6. Experimental plots of photoresponse in a 22 Å GaAs/GaAlAs multi-
quantum well sample at 4·2 K. The carriers are photoexcited by a
red L.E.D..

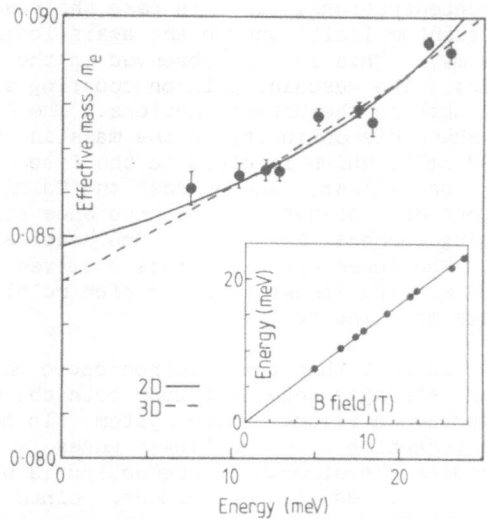

Fig. 7. Plots of the effective mass against energy from the data in
figure 6. The fits use either 2-D or 3-D polaron theory, but
adjust the band non-parabolicity to fit the data.

LO and TO modes to be 272 cm^1 and 255 cm^{-1} and the 'InAs-like' LO and TO modes to be 233 cm^{-1} and 226 cm^{-1}, although it is not clear that these modes display unequivocal LO and TO character (46). The solid lines in figure 8 show the resonance positions expected for constant effective masses. For both samples the plots show two distinct breaks of the form characteristic of resonant polaron coupling at frequencies of approximately 222 cm^{-1} and 255 cm^{-1}, shown by dashed lines. The coupling is demonstrated rather more clearly by the plots of cyclotron mass against frequency shown in figure 9. It thus appears that the resonant polaron coupling is occurring at, or slightly below, the frequencies of the TO phonons. This is of course in marked contrast to the behaviour expected, and to that found in measurements made on bulk semiconductors, which show only coupling to LO phonons (47). Another interesting feature is that the relative strength of the coupling to the two modes appears to be different in the two systems. The GaInAs/AlInAs heterojunction has a larger discontinuity at the lower 'InAs' frequency, while the GaInAs/InP sample shows stronger coupling to the 'InP' mode. This is thought to be evidence for the importance of the interface in this case, since the GaInAs/AlInAs heterojunction posesses 'InAs' type optic phonons on both sides of the junction, in contrast to the GaInAs/InP structure. This is also thought to be the origin of the different magnetophonon effects found in the two systems and discussed in detail below. Further evidence of the resonant polaron coupling comes from the frequency dependence of the resonance linewidths, shown in figure 10. Both heterojunctions have a maximum linewidth in the region between the two coupling frequencies, and once the second coupling is passed the linewidth falls rapidly, decreasing by a factor of ~ 3 in 30 cm^{-1}.

Measurements were also made on quantum wells of the two systems (18), with the electrons confined in 150 Å wells of GaInAs by either InP or AlInAs, and electron concentrations of ~ 1·5 × 10^{11} cm^{-2}. Figures 11 and 12 show the frequency dependence of the resonance positions and the effective masses respectively in the two multiple quantum wells. These show very different behaviour from the heterojunctions, due partly to the much lower electron concentrations. In this case the resonances are almost always in the quantum limit, and we are again looking only at the transitions giving us m_{01}^{*}. This is only observed in the heterojunction at the highest frequencies. The resonant polaron coupling also seems to be rather different from that in the heterojunctions. The GaInAs/AlInAs quantum well shows a sharp discontinuity in the mass and resonance position at around 238 cm^{-1}, which is close to the frequency of the 'InAs' LO mode in both GaInAs and AlInAs. In contrast the GaInAs/InP quantum well shows only a strong displacement of the resonance position to higher fields, with a resulting increase in mass, at frequencies up to that of the 'GaAs' LO phonon. The lower effective mass observed in this sample is due to the GaInAs layers being somewhat indium rich relative to the composition for lattice matching to InP.

These measurements suggest that the electron-optic phonon interactions in heterostructures are strongly dependent upon both the nature of the interfaces and the electronic states of the system. In both GaInAs-AlInAs structures a strong interaction with the 'InAs' modes is seen, whereas in both GaInAs/InP structures the strongest interaction is with the 'GaAs' modes. In both heterostructures all the evidence points to an interaction with optic phonons at the TO frequency, while the superlattices show evidence of coupling at the LO frequency. Some evidence for coupling to optic modes closer to the TO than the LO phonons in GaAs/GaAlAs heterojunctions will be given below from the magnetophonon studies (19,20). The superlattice periodicities are too large to allow the involvement of superlattice optic phonons, and the different behaviour of the superlattices and heterojunctions is likely to be due to size effects,

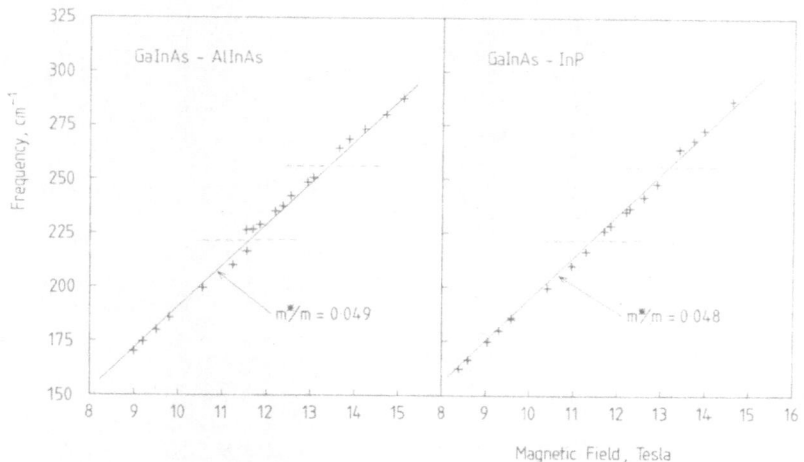

Fig. 8. The frequency dependence of the cyclotron resonance in two
Ga$_{.47}$In$_{.53}$As based heterojunctions.

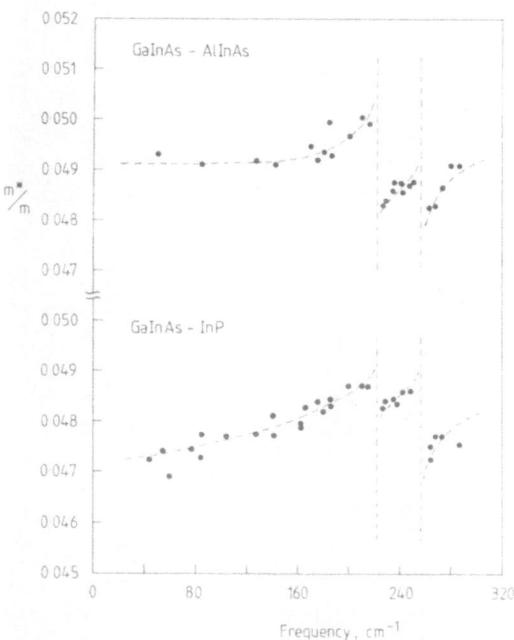

Fig. 9. The frequency dependence of the effective mass as deduced from
the data shown in figure 8. The dashed lines show the
frequencies of the two TO polaron modes.

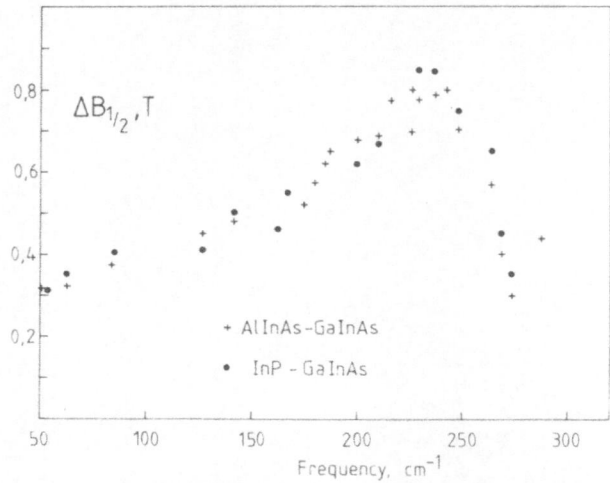

Fig. 10. The frequency dependence of the resonance linewidth from the
samples studied in figures 8 and 9.

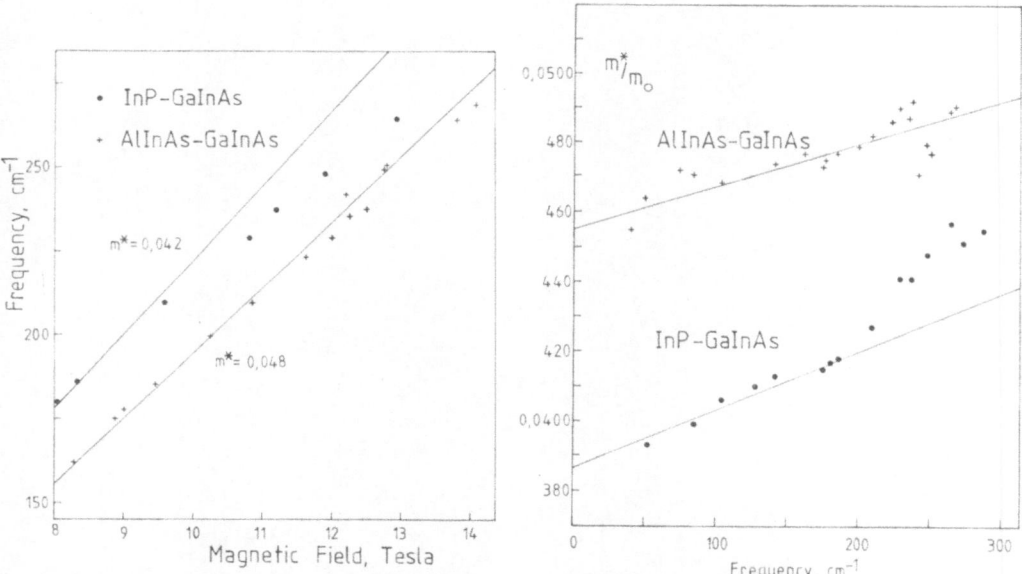

Fig. 11. Frequency dependence of
cyclotron resonance in 150 Å
quantum wells of GaInAs with
InP and AlInAs barriers.

Fig. 12. Effective masses deduced
from the data of figure
11.

the influence of screening, or the different symmetry of the confining potential. Coupling to TO phonons is not observed in bulk material because the only polarisation associated with these modes is at the surface and is insignificant in the bulk case. In an ionic slab whose thickness is much less than the phonon wavelength, transverse vibrations occur at the bulk LO frequency and longitudinal vibrations at the bulk TO frequency (48) and these 'slab modes' have been observed in Raman scattering experiments on superlattices (48). However, the heterojunctions, in which coupling is seen at the TO frequencies, are only a rough approximation to the thin ionic slab, and the observation of slab modes would seem more probable in the superlattice, where coupling at the LO frequencies is seen.

The explanation of this behaviour may again lie in the importance of screening and many body effects. Raman measurements on bulk GaAs (49) have shown that the frequency of the coupled plasmon-optic phonon system approaches the TO frequency for large q-vectors. The LO and TO frequencies differ because of the additional restoring forces due to the polarisation field of the LO phonons, and screening of this field will cause the LO frequency to fall towards the TO phonon value. This idea is supported by the strength of the resonant polaron coupling, which is much less than that predicted theoretically by Das Sarma (3). For a perfectly 2-D system the effective mass splitting at $\omega_c = \omega_{LO}$ was calculated to be

$$\frac{\Delta m^*}{m^*} = (\frac{\pi^{1/2}\alpha}{2})^{1/2} \tag{10}$$

where α has values of approximately 0·02 and 0·03 for the two optic phonon modes in GaInAs. This predicts discontinuities of order 15 − 20%, in contrast to the observed values of 2 − 5%. Some part of this reduction may again be due to the finite extent of the wavefunction (5,6,9). The different behaviour observed in the two multiple quantum well samples studied may be due to rather reduced screening since they had both much lower electron concentrations and much lower mobilities which lead to linewidths which were considerably broader than in the heterojunctions.

A further intriguing possibility is that this behaviour may be associated with a finite electric field present for TO phonons. This idea will be discussed below, when the magnetophonon data has been presented, but is associated with the finite lifetime of optic phonons leading to a finite spatial extent, which may be comparable to the size of the structures studied here.

A further piece of evidence for the existence of resonant polaron coupling closer to TO than LO modes comes from work on surface accumulation layers on $Hg_{.8}Cd_{.2}Te$ (11,20). The system is further complicated by the very narrow band gap of the (Hg,Cd)Te alloy which leads to a highly non-parabolic conduction band and also to the population of a large number of surface electric subbands. Each of these gives a separate cyclotron resonance due to the differing confinement energies. Results from one of these structures are shown in figure 13. Four subband resonances can be followed as a function of frequency, and in addition bulk cyclotron resonance can be seen from the substrate layer. This shows a clear coupling to the LO frequency for the bulk electrons, while the 2-D carriers seem to couple to the TO frequency. The higher mass values correspond to the more tightly bound and heavily populated subbands which seem to show progressively less polaron coupling. The data here are, however, again complicated by the existence of two separate optic phonon branches.

Further complications arise in the interpretation of cyclotron resonance data close to the optic phonon frequencies due to the complicated form of the dielectric function. This can also cause shifts and distortions of the resonances, particularly in multiple layer structures, and when fixed field, variable frequency studies are made. This has been demonstrated by Zeismann et al. (50), who have concluded that no polaron effects are visible in GaSb-InAs quantum wells, where the electrons are bound in the InAs.

While polaron coupling is the result of interactions with virtual optic phonons, real phonon absorption also influences the cyclotron resonance at higher temperatures. The absorption causes scattering which limits the resonance linewidths. Optic phonon scattering also limits the low field mobility, μ, and it has been shown (51) that for short range scattering mechanisms there is a relation between the resonance linewidth ($\Delta B_{1/2}$) and the mobility of the form

$$\Delta B_{1/2} = C (B/\mu)^{1/2} \tag{11}$$

where C is a constant of order $(2/\pi)^{1/2}$. This formula was found to work well for silicon inversion layers (52) and GaAs (53) and GaInAs (54) heterojunctions at low fields. The inclusion of a wavevector dependence into the scattering matrix element, as is the case for phonon scattering mechanisms, may introduce some field dependence into the prefactor C. Calculations of the mobility in 2-D suggest that high purity GaAs/GaAlAs heterojunctions will have optic phonon limited mobilities at temperatures above 60 K (55). Equation (11) has been used (56) to compare the measured linewidths (at B = 6 T) with the calculated zero field mobilities of Walukiewicz et al. (55), as shown in figure 14. Using a value of C = 0·42 gives very good agreement with the calculated values of the optic phonon limited mobility from 300 K to 60 K, however it should be pointed out that the linewidth at high temperatures is independent of magnetic field, so that the value of C cannot be compared directly with the result for short range scattering. At lower temperatures the mobility is limited by both acoustic phonon scattering from the deformation potential and piezo-electric coupling, and by impurity scattering. In this region the results become more sample dependent. The mobilities estimated from equation (11) can be made over an order of magnitude larger than the acoustic phonon limited D.C. mobility, and more than two orders larger than the impurity limit. This is probably due to the failure of equation (11) as the scattering becomes increasingly long range. Screening may also be important in reducing the strength of the acoustic phonon scattering, as has been suggested by some zero field calculations (57,58), and will be still more effective in high magnetic fields as discussed earlier.

4. THE MAGNETOPHONON EFFECT

The magnetophonon effect is the result of a resonant inelastic scattering process in high magnetic fields, usually involving LO phonons. It is characterised experimentally by an oscillatory component of the magnetoresistance, periodic in 1/B, whose amplitude reaches a few per cent of the monotonic component at maximum. The effect was first observed in the magnetoresistance of InSb, and has subsequently been observed in various transport coefficients in a number of different materials, as reviewed by Harper et al. (30), Peterson (59) and more recently by Nicholas (16).

The normal magnetophonon effect arises from the absorption of optic phonons, which are essentially monoenergetic, causing transitions between the Landau levels formed at high magnetic fields. In high fields the

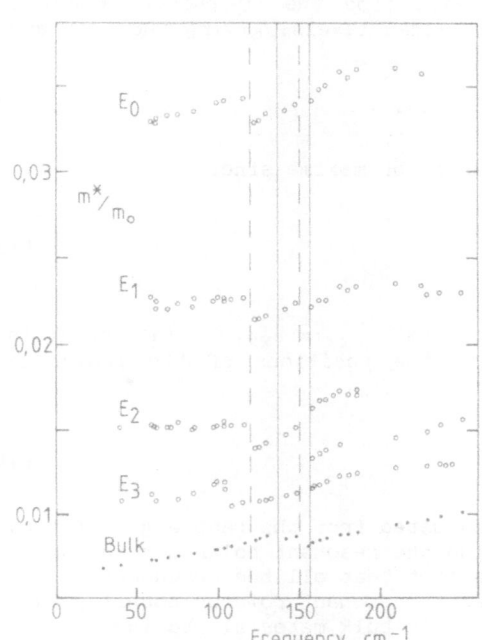

Fig. 13. Plots of the frequency
dependent effective mass in
a $Hg_{.8}Cd_{.2}Te$ surface
accumulation layer. Data are
shown for four bound subbands
and bulk material. The solid
lines show the frequencies of
the LO phonons and dashed
lines are TO models.

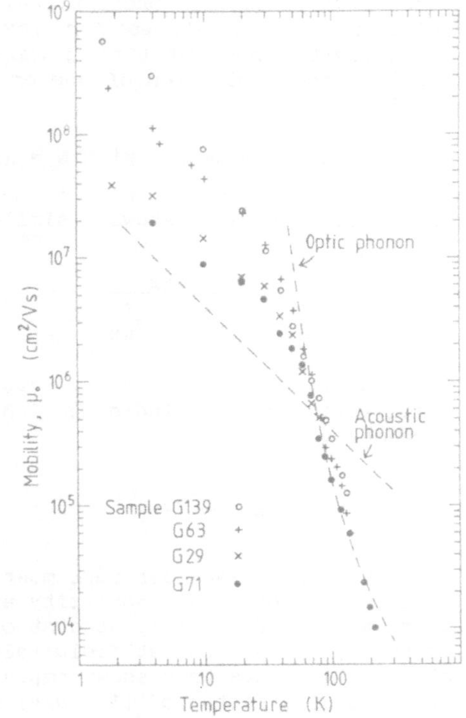

Fig. 14. Temperature dependence of
the mobility deduced from
the resonance linewidth
(eq. 12) for three GaAs/
GaAlAs heterojunctions.
The solid and dashed lines
show the optic and acoustic
phonon limited mobility
(55).

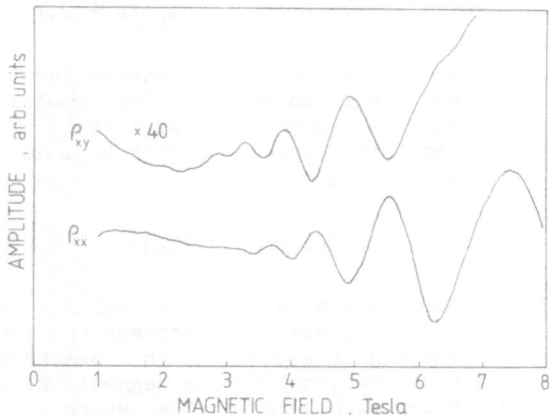

Fig. 15. Second derivative recordings of magnetophonon oscillations in
Hall voltage and resistivity at 150 K.

conductivity, σ_{xx}, is proportional to the scattering rate, and hence to the joint density of states for levels separated in energy by $\hbar\omega_{LO}$. The conductivity thus exhibits a maximum each time the LO phonon energy is equal to the separation of one or more Landau levels, giving the resonance condition:

$$N\hbar\omega_c = \hbar\omega_{LO}, \quad \text{with } N = 1, 2, 3, \ldots \tag{12}$$

The resistivity also shows a similar series of maxima since

$$\rho_{xx} = \frac{\sigma_{xx}}{\sigma_{xy}^2 + \sigma_{xx}^2} \tag{13}$$

and usually $\mu B \gg 1$, $\sigma_{xy} \gg \sigma_{xx}$, so that $\rho_{xx} \propto \sigma_{xx}$. The resulting oscillations are periodic in $1/B$, with the positions of the individual extremes given by

$$NB_N = \frac{m^*\omega_{LO}}{e} \tag{14}$$

where the effective mass m^*_{MPR} must be adjusted from the band edge value to take account of non-parabolicity and also the resonant polaron coupling. This latter effect takes account of the fact that all the resonances occur, by definition, at field values where resonant polaron coupling is strong. This has been shown empirically, in bulk material, to increase the mass by a factor of $(1 + \alpha/4)$ (63). The value for 2-D systems is not accurately known, however, the cyclotron resonance studies of resonant polaron coupling discussed above suggest that this correction is likely to be smaller for 2-D systems than in 3-D, but may be dependent on the carrier concentration, temperature and level broadening in the particular case being studied. In practice the correction is only of order 1% in bulk GaAs, and so should be even smaller in 2-D. A resonable assumption is that the correction is approximately 75% of that for 3-D systems, i.e. a factor of $(1 + \cdot75\alpha/4)$ (14).

The increase in mass due to non-parabolicity can be quite large for a 2-D system. Using the multi-band k.p theories discussed above and equation (8), the effective mass for the magnetophonon resonances is given by:

$$1/m^*_{MPR} = 1/m^*(1 + K_2/E_g[(2L+N+1)\hbar\omega_{LO}/N + 2\langle T_z\rangle]) \tag{15}$$

where L is the quantum number of the initial Landau level and N is the harmonic number of the resonance, as before. For samples with a low electron concentration at high temperatures we have $L \sim kT/\hbar\omega_c$, and for more heavily doped systems, under the conditions $\hbar\omega_c$, $\hbar\omega_{LO} < E_F$, the Fermi energy, equation (15) may be reduced to

$$1/m^*_{MPR} = 1/m^*(1 + K_2/E_g[2E_F + 2\langle T_z\rangle]). \tag{16}$$

In order to detect the usually rather weak oscillatory components of the magnetoresistance, it is necessary to suppress the monotonic magnetoresistance. In the high field regime this may be done by subtraction of a voltage proportional to the magnetic field, or by taking the second derivative of the magnetoresistance, which strongly enhances the rapidly varying oscillatory components. This is mainly done by the use of high pass electronic filters, but has also worked successfully with both field modulation and digital processing. The action of taking the second derivative changes the sign of the oscillations, so that $-d^2R/dB^2$ is displayed to give maxima at the positions of the maxima in the original resistance.

To date 2-D magnetophonon resonance has only been observed in hetero-structures with electrons confined in GaAs or $Ga_{.47}In_{.53}As$. Several workers have made careful studies of silicon inversion layers. However, the many valleyed conduction band minima, lower mobilities and non-polar coupling have meant that any structure is much weaker and it has not been observed experimentally. Searches for magnetophonon structure in narrow gap semiconductors have been hampered by the persistence of Shubnikov-de Haas peaks at high temperatures.

The first observations of 2-D magnetophonon resonance were made in GaAs/GaAlAs heterojunctions and multi-quantum wells by Tsui et al. (60), and were extended to higher magnetic fields by Englert et al. (61) and Kido et al. (62). Recently more accurate studies have been reported by Brummell et al. (14,19). Some typical experimental recordings are shown in figure 15, using the second derivative technique. This shows oscillations in both the resistivity ρ_{xx} and the Hall coefficient ρ_{xy}. The oscillations in the Hall coefficient are in antiphase to those seen in the resistivity and are over an order of magnitude weaker. This is probably due to oscillations in σ_{xx} which are observed in ρ_{xy} due to the tensor nature of the resistivity, as

$$\rho_{xy} = \frac{\sigma_{xy}}{(\sigma_{xx}^2 + \sigma_{xy}^2)} \qquad (17)$$

At high fields $\sigma_{xy} \gg \sigma_{xx}$ so strong resonances in σ_{xx} will lead to weaker resonances in ρ_{xy} in antiphase to the original oscillations.

The temperature dependence of the oscillation amplitudes is determined by two factors. There must be a sufficient population of LO phonons to cause scattering, and so the amplitude decreases at low temperatures, while at higher temperatures the increased broadening of the Landau levels smears out the resonances in the joint density of states. Hence there is a maximum in the amplitude, as shown in figure 16, which occurs around 150 K, depending upon the harmonic index, N. The amplitude also seems to follow a systematic behaviour as a function of electron concentration, as shown in figure 17. This shows data from a number of very high mobility heterojunctions, for the N = 3 resonance at 180 K, which do not appear to be influenced by sample dependent variations such as impurity scattering. This is in distinct contrast to the behaviour in bulk material (63) where the impurity content apparently determines the strength of the magnetophonon structure, even at temperatures where the mobility is phonon limited. Figure 17 thus shows the 'intrinsic' strength of the magnetophonon structure in the 2DEG. The strength of the oscillations falls rapidly as the electron concentration decreases and the system becomes more degenerate. At 180 K we would expect the system to be close to degenerate statistics at approximately 3×10^{11} cm^{-2}, where we have $kT \sim \hbar\omega_c \sim E_F$ (B=0). At this point we might expect screening to assume a more important role, leading to a reduction in the strength of the electron-optic phonon coupling, and the relative importance of transitions occurring between higher Landau levels will increase. It has been suggested both theoretically (6,8,9) and from the experimental studies of cyclotron resonance described above, that screening may reduce polaron coupling. It should be remembered that while the statistics remain non-degenerate there is nearly no change in the relative contribution to the resistivity from the different Landau level transitions, and so the carrier concentration dependence of the amplitude at very low densities

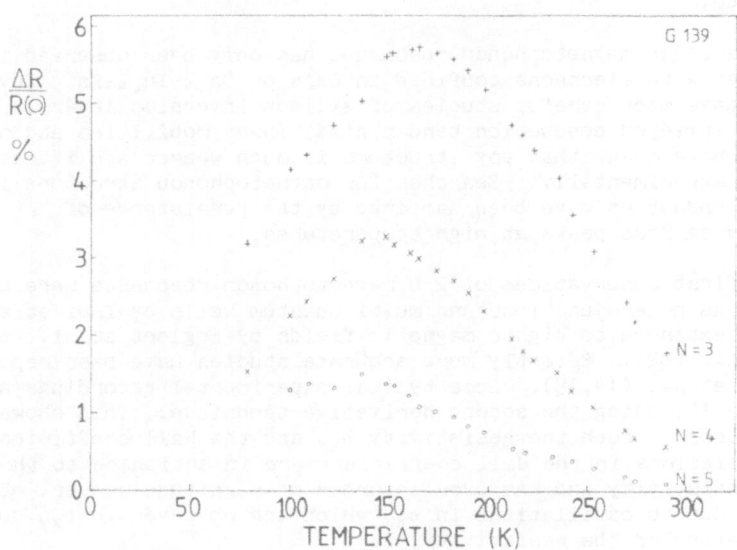

Fig. 16. The temperature dependence of the magnetophonon amplitudes in a
GaAs/GaAlAs heterojunction.

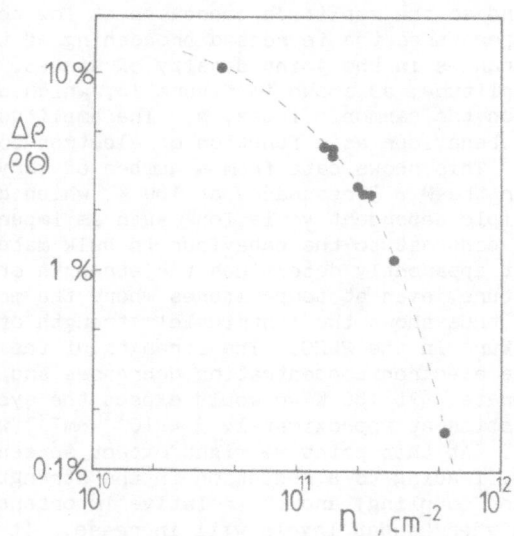

Fig. 17. The carrier concentration dependence of the oscillation amplitude
at 180 K for N = 3. This uses results from a high mobility
GaAs/GaAlAs heterojunction.

can only be associated with some modification of the basic electron-phonon interaction.

In order to deduce the phonon frequencies from the positions of the oscillations it is necessary to use equations (14) and (16), and the most dependable results will be obtained when accurate values of the effective masses are measured directly on the same samples by cyclotron resonance, particularly if this can be measured at the same temperature. The oscillations form an exponentially damped sinusoidal series, and this means that the peak values of resistivity do not correspond exactly with the resonance conditions. This can be easily corrected (16) by using a measured damping factor of the series and tabulated correction factors. Taking these factors into account, and correcting for non-parabolicity and the polaron mass enhancement discussed above, Brummell et al. have calculated accurate values for the optic phonon energy as a function of temperature in GaAs/GaAlAs heterojunctions (14,19). These are shown in figure 18 for one sample as a function of temperature. At a tempearture of 120 K a range of six different samples all gave frequencies between 279 and 286 cm^{-1}. In this study the various corrections to the effective mass and resonance positions were all chosen to overestimate, rather than underestimate the phonon frequency, so this value should be regarded as a lower limit.

The conclusion from these studies is that the phonon interacting with the 2-D electrons has a frequency well below that of the bulk LO phonon, and is in fact closer to the TO frequency. These results seem consistent with the studies of resonant polaron coupling by cyclotron resonance, described above, where coupling close to the TO frequencies has been suggested. One interpretation suggested that strong screening of the polarization field of the LO phonons was important, leading to a reduction in their frequency (17,18,64). However, this seems rather unlikely because the screening should be strongly temperature and electron concentration dependent, due to changes in both the level broadening and carrier statistics (14). Possible interface phonon modes have frequencies of 290 cm^{-1} and 270 cm^{-1}, with the latter having only a very weak oscillator strength (14). The measured phonon frequency actually corresponds most closely with the 'GaAs-like' LO mode in the GaAlAs, and interactions with this phonon have been observed in resonant Raman experiments on GaAs/GaAlAs superlattices (65). However the scattering by the confined modes of the GaAs was stronger, and it seems unlikely that interactions across the interface should dominate over coupling to the GaAs LO phonon. One possibility is that since the measured phonon frequency is near the top of the region where the LO phonon bands of the two materials overlap, modes which are continuous across the interface may exist at these frequencies. Thus there are several possibilities to explain some interaction with a lower frequency mode, with the possibility also of unresolved contributions from both LO and TO phonons.

Tilting the magnetic field direction relative to the surface normal to the heterojunction, through an angle θ, usually has little effect upon magneto-transport phenomena, such as the quantum Hall effect and Shubnikov-de Haas oscillations, other than to demonstrate their two dimensionality. The features are found to depend only upon the perpendicular field component $B\cos\theta$. In contrast something very different happens for magnetophonon resonances in 2-D. Figure 19 shows that the amplitude of the oscillations decreases very rapidly on rotation, with turning by 10° sufficient to halve the amplitude. The small subsidiary wings are due to some beating with bulk magnetophonon resonance from the p-type substrate layer (19). The resonance positions also show an anomalous behaviour, seen in figure 20. Initially the resonances follow a 1/cosθ dependence, but then vary more rapidly, moving to a second 1/cosθ curve with a resonance field ~ 3% higher than would be expected from the θ = 0 value. The change in behaviour corresponds with the rapid decrease

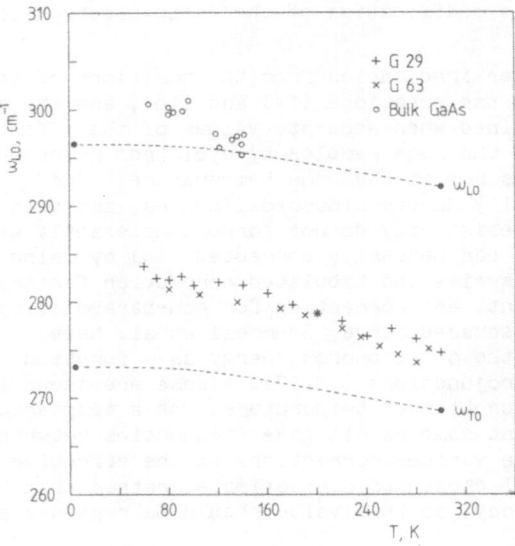

Fig.18.

The temperature dependence of the phonon frequencies in two GaAs/GaAlAs heterojunctions as deduced from magneto-phonon resonance. Some values deduced for bulk GaAs are shown for comparison.

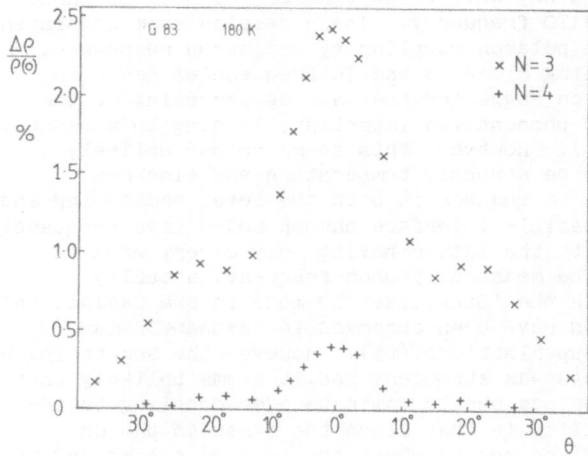

Fig.19.

The orientational dependence of the magnetophonon amplitudes in a GaAs/GaAlAs hetero-junction. θ is the angle between the field direction and the surface normal.

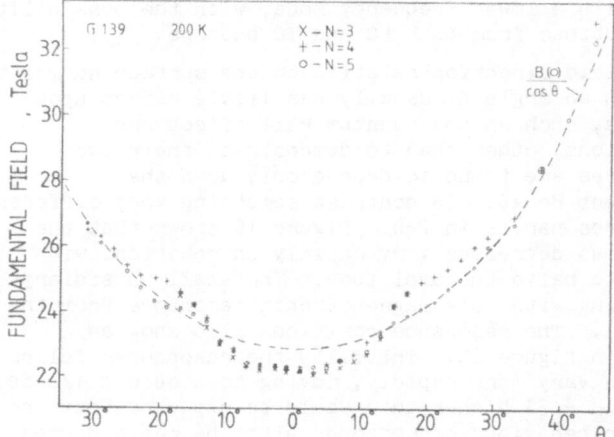

Fig.20.

The orientational dependence of the resonance positions. The dashed curve shows a 2 dimensional behaviour, with a funda-mental field of 22·9 T. The anomalous behaviour at around 20° is the result of the presence of unresolved bulk magnetophonon resonance from the p-type buffer layer.

in amplitude. If we use the high angle results to deduce the phonon frequencies, we find the interesting result that the frequency rises to almost that of the bulk LO phonon. This appears to be a general result for all of the samples studied.

The reason for this behaviour is not clear, but is most probably associated with a mixing of the electric subbands of the heterojunction. This is the major consequence of introducing a magnetic field component along the direction of the interface (66). One possible mechanism for this influence could be a modification of the screening of charge fluctuations in the third dimension. In a perfectly quantized 2-D system the subband wavefunction is unable to distort in order to provide screening of fields in this direction. The mixing of subbands could then produce a large increase in this screening. This would be consistent with a removal of coupling to a phonon with an electric field perpendicular to the interface, such as a TO mode, to allow the expected LO phonon scattering to be observed. The remaining unresolved question posed by both the magnetophonon and resonant polaron cyclotron resonance studies is the origin of the TO phonon coupling, and its dominance over LO phonon absorption. For the LO phonons involved in the magnetophonon resonance the k-vector and electric field lie in the plane of the 2DEG and hence may be very strongly screening in high magnetic fields. One possible mechanism for the appearance of scattering by the TO phonons would be that they have a finite size, due to their finite lifetime, which means that they possess a finite electric field which can couple to the electrons. The phonon 'size' is estimated from the energy uncertainty, and hence the k-vector spread. This gives values of order 30 Å, suggesting fields at least an order of magnitude lower than for the LO phonon. However one polarisation will have a field perpendicular to the interface, which is unscreened, and may also have an additional interaction through its influence on the subband levels. Thus the TO mode may give significant scattering until the subbands are mixed, after which its intensity falls and the strongly screened LO coupling dominates over the weakly screened TO coupling.

Very recently Leadley et al. (67) have reported the observation of hot electron magnetophonon oscillations in GaAs/GaAlAs heterojunctions. These are shown in figure 21 at around 70 K. At low electric fields normal oscillations occur, giving resistivity maxima at resonance, while at higher electric fields the oscillations change sign to give minima. This is thought to be due to resonant cooling of the hot electron gas which causes the mobility to increase, thus giving minima at resonance. The amplitude of the resistivity minima increases very rapidly with increasing electric field, as shown in figure 22. The amplitude is also very strongly temperature dependent, having a peak at around 60 K. This is thought to be due to enhanced electron heating at higher temperatures in the high magnetic fields. At low temperatures the Landau levels are very sharp, and it is difficult for the electrons to move upwards in energy space by hopping from one Landau level to the next. As the temperature rises the levels are broadened and a more rapid flow upwards is possible. Eventually the electron heating is suppressed by the falling carrier mobility, thus giving a peak in amplitude at about the temperature where the dominant phonon scattering process switches over from acoustic to optic modes, as shown in figure 14.

Magnetophonon measurements on GaInAs/InP heterojunctions have been reported by Portal et al. (68), and more recently by Gauthier et al. (24). Typical oscillations are shown in figure 23, which also shows a residual Shubnikov-de Haas peak. In samples with a single occupied subband the oscillations form a single series with a fundamental field,

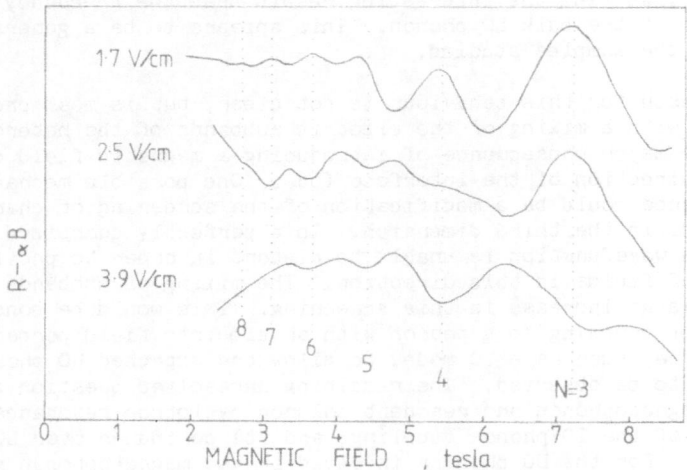

Fig. 21. Normal (1·7 V/cm) and hot electron (3·9 V/cm) magnetophonon resonance in a GaAs/GaAlAs heterojunction at 80 K.

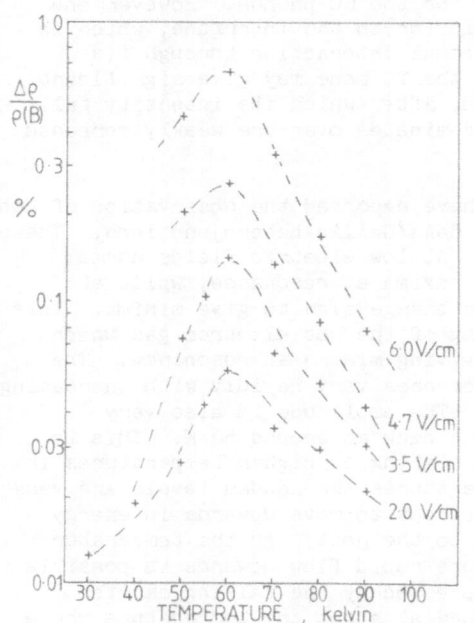

Fig. 22. The temperature dependence of the amplitude of the hot-electron magnetophonon resonance in a GaAs/GaAlAs heterojunction for several electric fields

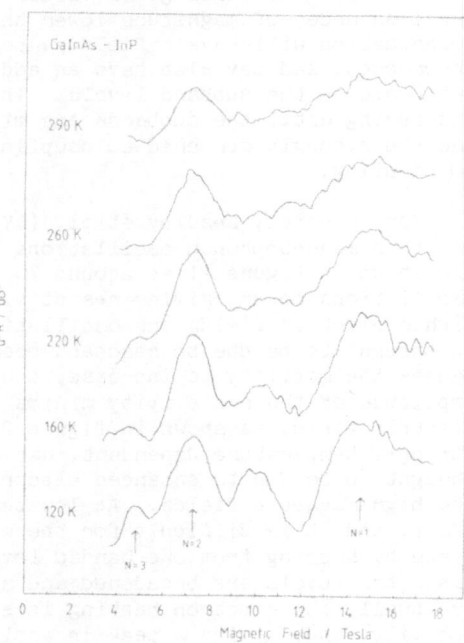

Fig. 23. Temperature dependent magnetophonon resonance in a GaInAs/InP heterojunction.

NB$_N$, of 14·4 T. This corresponds to a phonon energy of approximately 275 cm^{-1}, using the cyclotron resonance results reported above. In samples with more occupied subbands it was found (24) that the fundamental field fell, and at the highest carrier concentrations the optic phonon frequency deduced was as low as 240 cm^{-1}. This latter value is substantially lower than any of the strong 'GaAs-like' optic phonon modes, and it is not clear which phonon is involved in this behaviour.

Magnetophonon oscillations have also been observed in GaInAs/AlInAs heterojunctions (23), where a single series was found with a fundamental field of 12·6 T. The oscillations are two dimensional in character, as shown in figure 24, and the periodicity gives a phonon frequency of 236 cm^{-1}. This is close to the frequency of the 'InAs-like' LO phonons which are present on both sides of the barrier, and lead Brummell et al. (23) to conclude that the dominant optic phonon interaction was very different from the case of GaInAs/InP heterojunction where the 'GaAs-like' modes appear to be stronger. This is consistent with the resonant polaron coupling studied in the same structures, and shown in figure 9. The GaInAs/InP heterojunctions show the strongest mass discontinuity at the 'GaAs-like' mode, while the GaInAs/AlInAs showed much stronger coupling to the 'InAs-like' mode.

Two reports have also appeared recently of the observation of magnetophonon oscillations in GaInAs/InP superlattices (22,68). The samples studied had lower carrier concentrations than in the heterojunctions, ~ 1 × 10^{11} cm^{-2}, and were also studied by cyclotron resonance as described above (figures 11,12). The magnetophonon resonances in these structures are shown in figure 24, and illustrate once again that they are 2-D up to high temperatures. It is also clear that there are two series of resonances present in all three of the samples studied. The series with the higher fundamental field, which is shown by primed indices in figure 25, increases in relative intensity as the layer thickness of GaInAs is reduced: for the 150 Å layers this second series is rather weak compared with the main resonances, while by 80 Å both series are of comparable intensity. The phonon energies were deduced for both series, and correspond to frequencies of around 265 and 350 cm^{-1} respectively. The lower energy phonon is obviously 'GaAs-like', as seen in the GaInAs/InP heterojunctions. The upper phonon energy corresponds to that of the LO phonon in bulk InP, and it is this series of oscillations which increases in intensity as the wells are made thinner. Therefore the electrons bound in the GaInAs are being scattered by phonons from the InP barriers. The exact mechanism and nature of the phonon modes for this process is not clear. However, it has been proposed (69) that a long range polar phonon interaction could occur in silicon MOS devices, due to the interface optic phonon modes of the silicon-silicon dioxide interface. This type of mechanism would seem to be an excellent candidate to explain the scattering due to the InP optical phonons. It has been suggested by Lassnig and Zawadski (70) that interface phonons should be involved, at frequencies intermediate between those of the bulk LO and TO modes in the two materials. They also showed that the strength of the InP mode should increase for thinner layers and that that of the GaInAs modes tends to zero for very thin wells. Other possibilities include the finite penetration of the barrier by the electron wavefunction, and some grading of the interface leading to the existence of some strong InP modes close to the heterojunctions. Another factor aiding the InP phonons is screening. This will reduce the coupling to electrons bound in the wells, by changing the local polarizability, without affecting the interaction once outside the wells.

The complete picture of what is going on for GaInAs heterostructures is thus a rather complex one which appears to depend on both the carrier

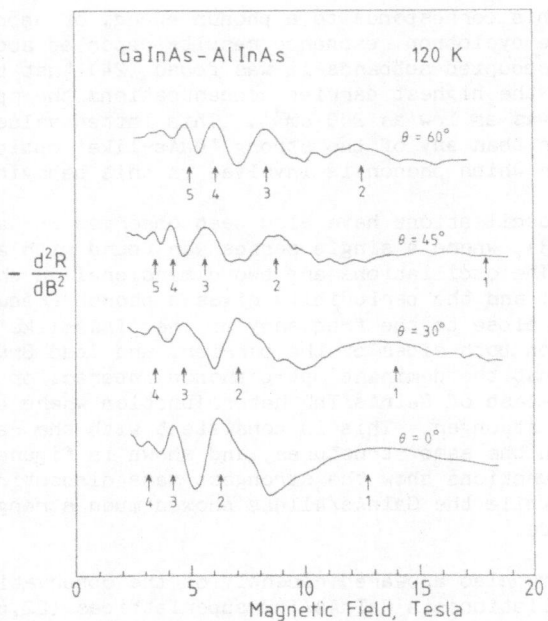

Fig. 24. Experimental recordings of magnetophonon oscillations in a GaInAs-AlInAs heterojunction. The sample is rotated relative to the magnetic field direction.

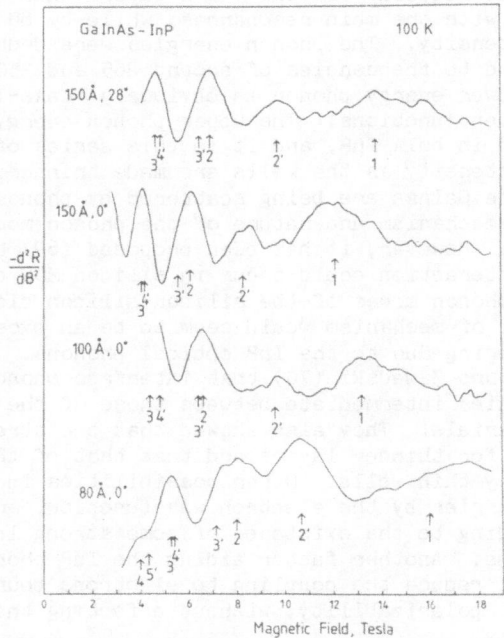

Fig. 25. Experimental recordings of magnetophonon structure in three GaInAsa-InP multi-quantum wells. The dashed peaks arise from absorption of InP phonons.

concentration and the shape of the well in which the carriers are confined. Some evidence (22) and theory (70) suggest that interface phonons may be involved, but this remains to be confirmed.

ACKNOWLEDGEMENT

We would like to thank all the people who have been involved in various aspects of this work including D.F. Barnes, L.C. Brunel, K.Y. Cheng, A.Y. Cho, M.A. di Forte-Poisson, C.T. Foxon, D. Gauthier, J.J. Harris, S. Huant, G.T. Jenkin, K. Karrai, P.K. Knowles, J.C. Portal, M. Razeghi, D.C. Rogers, H. Sigg, J. Singleton and D. Sivco.

REFERENCES

1. F.E. Low and D. Pines, Phys. Rev. $\underline{98}$ 414 (1955).
2. S. Das Sarma and A. Madhukar, Phys. Rev. B$\underline{22}$ 2823 (1980).
3. S. Das Sarma, Phys. Rev. Lett. $\underline{52}$ 859 (1984).
4. F.M. Peeters and J.T. Devreese, Phys. Rev. B$\underline{31}$ 3689 (1985).
5. D.M. Larsen, Phys. Rev. B$\underline{30}$ 4595 (1984).
6. S. Das Sarma, Phys. Rev. B$\underline{27}$ 2590 (1983).
7. R. Lassnig and W. Zawadzki, Surf. Sci. $\underline{142}$ 361 (1984).
8. S. Das Sarma and B.A. Mason, Phys. Rev. B$\underline{31}$ 5536 (1985).
9. H. Sigg, P. Wyder and J.A.A.J. Perenboom, Phys. Rev. B$\underline{31}$ 5253 (1985).
10. M. Horst., U. Merkt and J.P. Kotthaus., Phys. Rev. Lett. $\underline{50}$ 754 (1983).
11. J. Singleton, R.J. Nicholas and F. Nasir, Sol. Stat. Comm. $\underline{58}$ 833 (1986).
12. W. Seidenbusch, B. Lindemann, R. Lassnig, J. Edlinger and E. Gornik, Surf. Sci. $\underline{142}$ 375 (1984).
13. M. Horst, U. Merkt, W. Zawadzki, J.C. Maan and K. Ploog, Sol. Stat. Comm. $\underline{53}$ 403 (1985).
14. M.A. Brummell, R.J. Nicholas, M.A. Hopkins, J.J. Harris and C.T. Foxon, Phys. Rev. Lett. $\underline{58}$ 77 (1987).
15. M.A. Hopkins, R.J. Nicholas, M.A. Brummell, J.J. Harris and C.T. Foxon, to be published in Phys. Rev. B (1987).
16. R.J. Nicholas, Prog. Quantum Electron. $\underline{10}$ 1 (1985).
17. R.J. Nicholas, L.C. Brunel, S. Huant, K. Karrai, J.C. Portal., M.A. Brummell, M. Razeghi, K.Y. Cheng and A.Y. Cho, Phys. Rev. Lett. $\underline{55}$ 883 (1985).
18. L.C. Brunel, S. Huant, R.J. Nicholas, M.A. Hopkins, M.A. Brummell, K. Karrai, J.C. Portal, M. Razeghi, K.Y. Cheng and A.Y. Cho, Surf. Sci. $\underline{170}$ 542 (1986).
19. M.A. Brummell, D.R. Leadley, R.J. Nicholas, J.J. Harris and C.T. Foxon, Surf. Sci. to be published.
20. R.J. Nicholas and H. Sigg, to be published (1988).
21. S. Das Sarma, Phys. Rev. Lett. $\underline{57}$ 651 (1986).
22. J.C. Portal, J. Cisowski, R.J. Nicholas, M.A. Brummell, M. Razeghi and M.A. Poisson, J. Phys. C$\underline{16}$ L573 (1983).
23. M.A. Brummell, R.J. Nicholas, J.C. Portal, K.Y. Cheng and A.Y. Cho, J. Phys. C$\underline{16}$ L579 (1983).
24. D. Gauthier, J.C. Portal, M. Razeghi, D.R. Leadley, M.A. Hopkins, M.A. Brummell and R.J. Nicholas, to be published.
25. H. Frohlich, H. Pelzer and S. Zienau, Phil. Mag. $\underline{41}$ 221 (1950).
26. H. Frohlich, Adv. Phys. $\underline{3}$ 325 (1954).
27. F.C. Brown, in "Polarons and Excitions", ed. C.G. Kupar and G.D. Whitfield, Oliver and Boyd, London, p.323 (1962).
28. D.M. Larsen, in "Polarons in Ionic Crystals and Polar Semiconductors", ed. J.T. Devreese, North Holland, Amsterdam.
29. F.M. Peeters, Wu. Xiaoguang and J.T. Devreese, Phys. Rev. B$\underline{33}$ 4338 (1986).
30. P.G. Harper, J.W. Hodby and R.A. Stradling, Rep. Prog. Phys. $\underline{36}$ 1 (1973).

31. G. Lindemann, R. Lassnig, W. Seidenbusch and E. Gornik, Phys. Rev. B28 4693 (1983).
32. Wu. Xiaoguang, F.M. Peeters and J.T. Devreese, Phys. Rev. B34 2621 (1986).
33. Wu. Xiaoguang, F.M. Peeters and J.T. Devreese, Phys. Stat. Sol. (b) 133 229 (1986).
34. Wu. Xiaoguang, F.M. Peeters and J.T. Devreese, Proc. 7th E.P.S. Conf., Pisa (1987), to be published in Physica .
35. C.T. Foxon, J.J. Harris, R.E. Wheeler and D.E. Lacklison, J. Vac. Sci. Technol. B4 511 (1986).
36. M.A. Hopkins, R.J. Nicholas, P. Pfeffer, W. Zawadzki, D. Gauthier, J.C. Portal and M.A. di Forte-Poisson, Semicond. Sci. and Technol. in press (1987).
37. M. Braun and U. Rössler, J. Phys. C18 3365 (1985).
38. W. Zawadzki and P. Pfeffer, Proc. Int. Conf. on the Application of High Magnetic Fields in Semiconductor Physics, Würzburg 1986, Springer-Verlag to be published.
39. J. Singleton, R.J. Nicholas, D.C. Rogers and C.T. Foxon, Surf. Sci. to be published (1987).
40. T. Ando and Y. Murayama, J. Phys. Soc. Japan 54 1519 (1985).
41. Th. Englert, J.C. Maan, Ch. Uihlein, D.C. Tsui and A.C. Gossard, Sol. Stat. Comm. 46 545 (1983).
42. R. Lassnig and E. Gornik, Sol. Stat. Comm. 47 959 (1983).
43. W. Seidenbusch, E. Gornik and G. Weimann, Physica 134B 314 (1985).
44. M. Brodsky and G. Lucovsky, Phys. Rev. Lett. 21 990 (1968).
45. A. Pinczuk, J.M. Worlock, R.E. Nahory and M.A. Pollack, Appl. Phys. Lett. 33 461 (1978).
46. T.P. Pearsall, R. Carles and J.C. Portal, Appl. Phys. Lett. 42 436 (1983).
47. B.D. McCombe and R.J. Wagner, Adv. in Electronics and Electron Phys. 37 1 and 38 1 (1975).
48. R. Fuchs and K.L. Kliewer, Phys. Rev. 140 A2076 (1965); J.E. Zucker, A. Pinczuk, D.S. Chelma, A.C. Gossard and W. Wiegmann, Phys. Rev. Lett. 53 1280 (1984).
49. A. Mooradian and G.B. Wright, Phys. Rev. Lett. 16 999 (1966).
50. M. Zeismann, D. Heitmann and L.L. Chang, Phys. Rev. B35 4541 (1987).
51. T. Ando and Y. Uemura, J. Phys. Soc. Japan 36 959 (1974).
52. G. Abstreiter, J.P. Kotthaus, J.F. Koch and G. Dorda, Phys. Rev. B14 2480 (1976).
53. P. Voisin, Y. Guldner, J.P. Vieren, M. Voos, P. Delescluse and N.T. Linh, Appl. Phys. Lett. 39 982 (1981).
54. M.A. Brummell, R.J. Nicholas, L.C. Brunel, S. Huant, M. Baj, J.C. Portal, M. Razeghi, M.A. di Forte-Poisson, K.Y. Cheng and A.Y. Cho, Surf. Sci. 142 380 (1984).
55. W. Walukiewicz, H.E. Ruda, J. Lagowzki and H.C. Gatos, Phys. Rev. B30 4571 (1984).
56. M.A. Hopkins, D.F. Barnes, M.A. Brummell, R.J. Nicholas, J.J. Harris and C.T. Foxon, to be published (1988).
57. P.J. Price, Phys. Rev. B32 2643 (1985).
58. B. Vinter, Phys Rev. B33 5904 (1986).
59. R.L. Peterson, in "Semiconductors and Semi-metals", Academic, New York, 10 221 (1975).
60. D.C. Tsui, Th. Englert, A.Y. Cho and A.C. Gossard, Phys. Rev. Lett. 44 341 (1980).
61. Th. Englert, D.C. Tsui, J.C. Portal, J. Beerens and A.C. Gossard, Sol. Stat. Comm. 44 1301 (1982).
62. G. Kido, N. Miura, H. Ohno and H. Sakaki, J. Phys. Soc. Japan 51 2168 (1982).
63. R.A. Stradling and R.A. Wood, J. Phys. C1 1711 (1968).
64. R. Lassnig, Surf. Sci. 170 549 (1986).
65. A.C. Maciel, L.C. Campelo Cruz and J.F. Ryan, J. Phys. C20 3041 (1987).

66. T. Ando, A.B. Fowler and F. Stern, Rev. Mod. Phys. $\underline{54}$ 437 (1982).

67. D.R. Leadley, M.A. Brummell, R.J. Nicholas, J.J. Harris and C.T. Foxon, Proc. 5th Int. Conf. on Hot Carriers in Semiconductors, Boston, 1987, to be published.

68. J.C. Portal, G. Gregoris, M.A. Brummell, R.J. Nicholas, M. Razeghi, M.A. di Forte-Poisson, K.Y. Cheng and A.Y. Cho. Surf. Sci. $\underline{142}$ 368 (1984).

69. K. Hess and P. Vogl, Sol. Stat. Comm. $\underline{30}$ 807 (1979).

70. R. Lassnig and W. Zawadzki, Surf. Sci. $\underline{142}$ 361 (1984).

DENSITY OF STATES AND ELECTRON-PHONON COUPLING IN

TWO-DIMENSIONAL ELECTRON SYSTEMS AT HIGH MAGNETIC FIELDS

J. P. Eisenstein

AT&T Bell Laboratories
600 Mountain Avenue
Murray Hill, NJ 07928

One of the central problems in the two-dimensional electron field[1] is the determination of the Landau level density of states. This quantity plays an essential role in determining almost all physically measurable properties. For example, understanding the striking characteristics of the quantum Hall effect requires the existence of localized electronic states somewhere in the gap region between Landau levels[2]. How many such states are there and how are they distributed? Considerable effort, both theoretical and experimental, has been devoted to this problem over the last several years and much progress has been made. In these lectures I will discuss in detail two sets of experiments on 2D electron systems (2DES) in which the density of states (DOS) is directly observable. The first example will be our measurements of the static magnetization of the 2DES at high magnetic fields[3]. Being an equilibrium thermodynamic variable, the magnetization depends only on the DOS. The strength of the observed oscillations in the magnetic moment as a function of field (deHaas-van Alphen effect) gives a direct measure of the broadening of the Landau levels. Our observations have led to the conclusion, now widely corroborated, that a significant density of states exists in the gap region between Landau levels. The second example concerns our experiments on the thermal conductivity at high magnetic fields of multi-layer heterostructures that contain 2D electron systems[4]. Although these are really phonon transport measurements, the DOS of the 2DES enters through the coupling of the phonons to the electrons. Striking magneto-oscillations are observed in the temperature gradient generated by an applied heat flux. These oscillations are again a direct manifestation of the DOS only they depend on the electron-phonon matrix elements as well.

This paper is organized into five main sections. In the first the basic physics of the 2DES is reviewed. The next section gives a very brief discussion of how such systems are realized in semiconductor heterostructures. In the third and fourth parts of the paper the magnetization and thermal conductivity experiments are described in detail. Both measurements are fitted into the mosaic of related work in the literature although a comprehensive review is not our aim here. The last section summarizes the important findings of both experiments and gives some indication of possible future work.

SECTION I. BASIC PHYSICS

For our purposes a two-dimensional electron (or hole) system is defined as a collection of elementary charges confined near an interface or surface by some potential. The system is sufficiently clean and the length scale sufficiently short that the quantization of the electronic motion perpendicular to the interface is dominant and all the charges reside in the lowest bound state of the potential. The motion in the plane is not quantized in the absence of a magnetic field. In two dimensions the density of plane wave states is a constant $D_0 = m^*/\pi\hbar^2$ including spin degeneracy. For GaAs, with effective mass m^*/m = 0.067 the zero field DOS is $2.8 \times 10^{10} meV^{-1} cm^{-2}$. Typically the first excited state of the confining potential is 20meV or so above the ground state. Therefore roughly 5×10^{11} carriers per cm^2 can be accommodated in the ground state, or subband, before the Fermi level rises above the second subband edge. Population of the higher subbands is tantamount to quasi-three-dimensionality and will not be considered further here.

Application of an external magnetic field perpendicular to the 2D plane drastically alters the DOS. The classical cyclotron motion is completely quantized and the previously uniform DOS is reduced to a sequence of discrete energy levels

$$\varepsilon_j = (j + 1/2)\hbar\omega_c \pm g\mu_B B/2 \qquad j = 0,1,2... \tag{1}$$

where ω is the cyclotron frequency eB/m^*, μ_B the Bohr magneton, and g the g-factor. The small spin splitting of these levels is ignored for now. Each Landau level is highly degenerate since the cyclotron orbit center may be placed anywhere (almost) in the sample area. Assuming the perpendicular field B is in the z-direction and using the vector potential $A = (0,-xB,0)$ the wavefunctions for the in-plane motion are

$$\Phi_{jk} = \exp(iky)\phi_j(x - x_0) \tag{2}$$

where ϕ_j is the simple harmonic oscillator wavefunction

$$\phi_j(x) = c_j \exp(-x^2/2l^2) H_j (x/l) \tag{3}$$

Here c_j is a normalization constant, H_j is a Hermite polynomial, and $l = \sqrt{\hbar/eB}$ is the magnetic length. In this gauge the states are plane waves in the y-direction and are localized in the x-direction to within roughly one magnetic length of the position x_0. The "momentum" k and the coordinate x_0 are related through

$$x_0 = kl^2 \tag{4}$$

Taking an area defined by $0 \le y \le L_y$ and $0 \le x \le L_x$ applying periodic boundary conditions in the y-direction shows that there are $L_x L_y / (2\pi l^2)$ states in each Landau level. Per unit area each level contains

$$N_L = eB/h \tag{5}$$

states. An additional factor of two can be added if the spin splitting is not resolved. Equation 5 is a very general result for 2D systems in a magnetic field and does not depends on any sample parameters (effective mass, for example) or the presence of line broadening.

If we assume that the carrier density N_s in the 2D layer does not change with magnetic field (as is usually the case) then the Fermi level E_F will lie (at T=0) in the jth (j=0,1,2...) Landau level where $j = INT(N_s/(2eB/h))$. As a function of field E_F will exhibit a sawtooth oscillation, making a discontinuous jump each time an integer number of Landau levels are filled. This dependence is illustrated in Fig. 1. If the spin splitting is resolved, additional, albeit smaller, jumps will occur when E_F crosses between the spin sublevels of each Landau level. Defining the filling factor ν as

$$\nu = N_s / (eB/h) \tag{6}$$

implies that for even integral values of ν E_F lies in the orbital gap between Landau levels whereas for odd ν E_F lies in the spin gap of a given Landau level.

In a real system these Landau levels are broadened by various effects including scattering and long range potential fluctuations. The most simple-minded estimate of the linewidth would be $\hbar/\tau$ where τ is the scattering time determined from the zero-field mobility μ via $\tau = m^* \mu / e$. This, however, ignores the singular nature of the density of states which is crucial in calculating the scattering rate at high magnetic field. Ando and Uemura[5] showed how to self-consistently determine (the so-called self-consistent Born approximation) the scattering rate and Landau level linewidth. They found a semi-elliptical lineshape with base half-width Γ given by

$$\Gamma^2 = \frac{2}{\pi} \hbar \omega_c \frac{\hbar}{\tau} \tag{7}$$

for the case of short-range scatterers. Here τ is the mobility lifetime. As will be discussed at length in this paper, significant refinements to Ando and Uemura's approach are required to obtain any agreement with the experiments on 2D electron systems in GaAs/AlGaAs heterostructures. Nevertheless, Eq. 7 serves as a useful starting point. Figure 1 shows how the discontinuities of E_F are smoothed out when broadened Landau levels are present. The result in the figure is calculated via the formula (valid at T=0)

$$N_s = \int_0^{E_F} D(\varepsilon, B)\, d\varepsilon \qquad (8)$$

where $D(\varepsilon, B)$ is a sum over a sequence of gaussian Landau levels each normalized to contain $2eB/h$ states.

The magnetization of the 2DES consists of both an orbital and a spin part. Due to the small effective mass in GaAs, the orbital contribution is much larger than the spin part. The magnetization M is a thermodynamic variable and may be calculated (at T=0) by first determining E_F via Eq. 8 and then calculating the total energy

$$E_{TOT} = \int_0^{E_F} \varepsilon\, D(\varepsilon, B)\, d\varepsilon \qquad (9)$$

and then differentiating, $M = -dE_{TOT}/dB$. At finite temperature the procedure is the same except the free energy must be calculated. In the absence of Landau level broadening M exhibits a sawtooth oscillation around zero similar to E_F except that the amplitude is independent of magnetic field. Numerically the amplitude is $M_0 = N_s A \mu_B^*$ where A is the

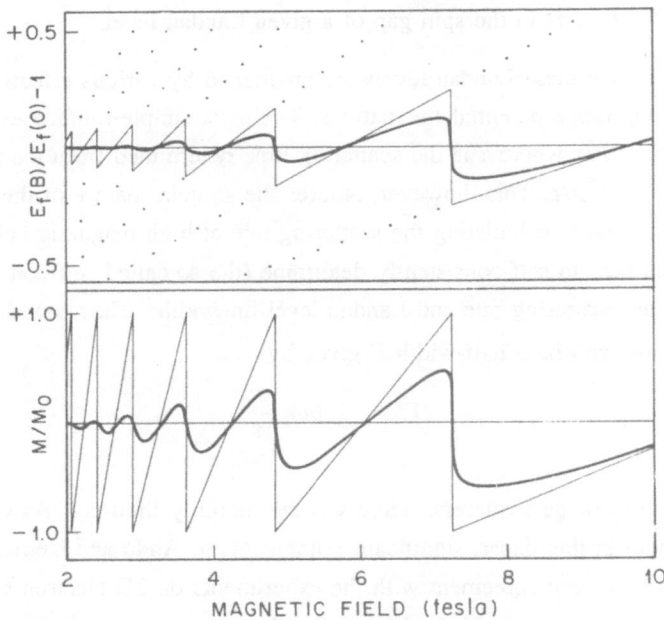

Fig. 1. Top panel: Magnetic field variation of the Fermi level of the 2DEG at T=0 for both the ideal case of unbroadened Landau levels (sawtooth) and for gaussian levels of rms half-width 1.5meV (smooth curve). The dotted lines are the individual Landau levels. Carrier density is $5x10^{11}$ cm^{-2}. Bottom panel: Calculated magnetization for same conditions.

sample area and μ_B^* is the Bohr magneton with the effective electron mass inserted. Broadening the Landau levels attenuates and smooths these deHaas-van Alphen oscillations[6] as shown in Fig. 1.

An important aspect of the Landau level DOS not yet mentioned concerns localization. Essential to our understanding of the quantum Hall effect[2] is the notion that there are two classes of electron states within each Landau level, as far as transport is concerned. It is now widely held that only those states near the center of each Landau level are delocalized and capable of carrying current. States in the flanks of the levels are localized and, at T=O, can pin the Fermi level in a region of zero conductivity. This picture is supported by both the percolation model of conduction[7] in a disordered 2D system and by numerical studies of the localization length[8]. A thermodynamic quantity such as magnetization is not affected by this distinction among states, so long as the measurement proceeds slowly enough that thermal equilibrium is maintained. While this is the case for our magnetization experiment, it is not obviously so in the thermal conductivity studies and there the issue gives insight into the electron-phonon scattering process.

SECTION II. SAMPLES

All of the samples to be discussed in this paper are modulation doped GaAs/AlGaAs heterostructures. Grown by molecular beam epitaxy (MBE) onto <100> oriented GaAs substrates, these structures contain from a few up to hundreds of layers of semiconductors. In a typical multi-quantum well sample thin (~200Å) layers of GaAs are alternated with thicker (~500Å) layers of the alloy semiconductor $Al_xGa_{1-x}As$ where x~0.3. These two materials are lattice matched and thus strain-free epitaxial interfaces result. The AlGaAs has the larger band gap and the Type 1 alignment at the interface results in roughly 65% of the band-gap discontinuity being in the conduction band. Thus both electrons or holes can be confined within the GaAs layer. Carriers are generated by modulation doping[9] whereby a thin central region of the AlGaAs layer is doped, usually with silicon if electron layers are desired. The dopants ionize and the charge is transferred to the smaller band-gap GaAs layer. Spatial separation from the parent donor atoms prevents low temperature freeze-out and enhances the mobility because coulombic scattering is thereby reduced. A carrier concentration of $5x10^{11}$ cm^{-2} is typical. Mobilities in such multi-quantum well (MQW) samples have recently reached the $5x10^5$ cm^2/Vs mark.

Two-dimensionality, as defined in the previous section, is assured by the small quantum well thickness and by limiting the doping to prevent second-subband filling. Furthermore, the AlGaAs thickness and the band offsets can be made large enough that tunnelling of electrons from one GaAs well to another is inhibited. The multi-quantum well samples to be discussed here can always be regarded as stacks of independent 2D electron systems. It is obviously possible to grow structures in which tunnelling between quantum wells is significant and an anisotropic 3D system with Kronig-Penney bands results. The integer quantum Hall effect has even been observed in such 3D structures[10].

The highest mobility 2D electron systems are created in single interface heterojunctions. Here a thick (~1μm) GaAs buffer is grown on the GaAs substrate. Atop this AlGaAs is grown, with doping commencing after a setback of typically 500Å. The structure is usually capped with a thin GaAs layer. Carriers from the ionized dopants transfer both to surface states and into the GaAs buffer. The resulting band bending creates a confined 2DES in a roughly triangular well on the GaAs side. (The cap layer is fully depleted by the Schottky barrier at the surface.) Mobilities as high as $5x10^6$ cm^2/Vs have been obtained in such samples[11]. Although such samples are preferable to MQW's for study of the isolated 2DES, the small total number of carriers makes thermodynamic measurements very difficult.

SECTION III. MAGNETIZATION MEASUREMENTS

We have carried out magnetization measurements using a torsional technique that combines great sensitivity with rejection of much of the unwanted background magnetization arising from sample substrates, holders, etc. As depicted in Fig. 2, a thin torsion fiber made of PtW alloy is held under tension perpendicular to the axis of the superconducting solenoid used to apply the magnetic field. The heterostructure sample is mounted on this fiber in such a way that the normal to the 2D plane is inclined to the applied field by a small angle, typically 15 degrees. Since the orbital moments of the 2DES are constrained to be perpendicular to the 2D plane a torque, $\vec{\mu} \times \vec{B}$ is created along the fiber axis. This torque twists the fiber and is thus detected using a sensitive capacitive technique. Since it is the absolute sample orientation relative to the applied field that is

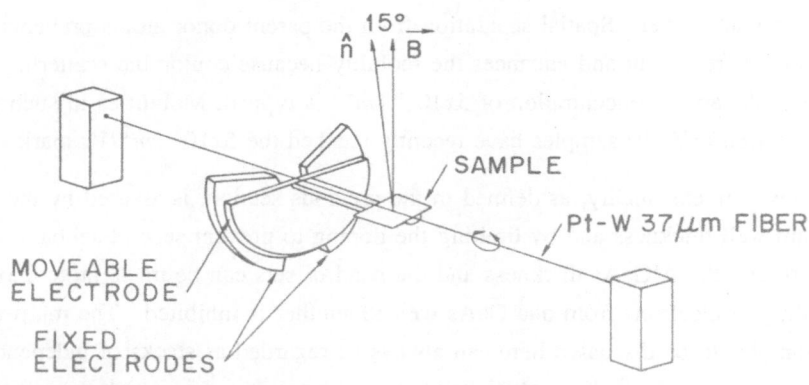

Fig. 2.   Schematic diagram of torsional magnetometer.

measured, the technique is dc. This is limited only by the rate at which the external field is swept. With sufficiently thin torsion fibers and a precision ac bridge for measuring the capacitance, an ultimate magnetization sensitivity of 0.5 pJ/T can be obtained at 5T. This sensitivity has proven sufficient to observe the magnetization of a single layer of electrons and provides excellent signal to noise ratios for multi-quantum well samples. Further, since most background magnetization sources (sample holder, substrate nuclear paramagnetic and orbital diamagnetic moments, etc.) are isotropic, they create no torque and are thereby rejected. There are, however, both geometric effects and genuine anisotropic moments that create background signals and these must be subtracted. In an effort to reduce these effects the samples are usually thinned to at least 50 μm total thickness before the magnetization is measured. Further details of this method have been described previously[12].

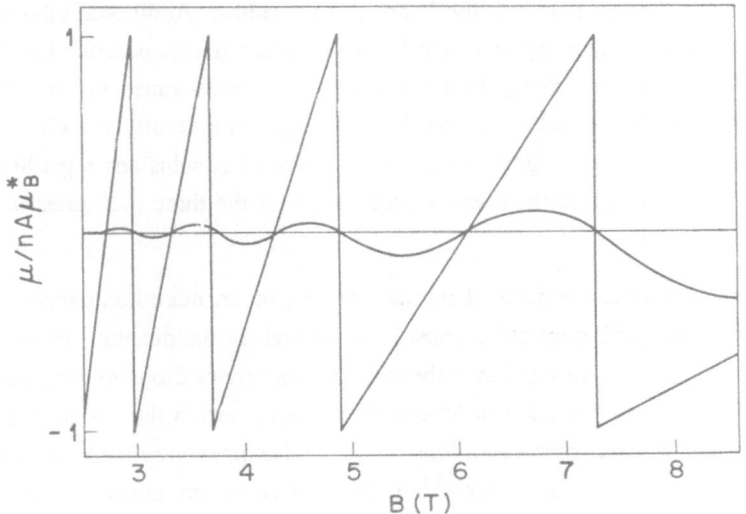

Fig. 3.   Observed and ideal dHvA oscillations for a 172-layer structure.

In Fig. 3 magnetization data from a low mobility (20,000 cm^2/Vs) multi-quantum well sample with 172 layers of 2D electrons is shown. Also shown is the sawtooth magnetization expected from an ideal 2DES of the same density and total 2D area. Comparing the two shows that the data has the correct periodicity and phase to be unambiguously identified with the deHaas-van Alphen (dHvA) effect. It is also clear that there must be significant broadening mechanisms operative to account for the reduced amplitude and lack of discontinuities in the data. It was determined from a careful study (via transport measurements) of the carrier density across the MBE wafer from which this sample was cleaved, that a gradient across the wafer exists. The off-axis molecular beams

Table I. Relevant sample parameters for dHvA experiment. Measured carrier density inhomogeneities are listed.

Sample	Periods	Low-T mobility $(m^2/V/s)$	2D density (cm^{-2})	Total 2D area (cm^2)	rms $\Delta N/N$ (%) In plane	rms $\Delta N/N$ (%) Layer to layer	Measurement Temperature (K)
A	50	8.0	$5.4x10^{11}$	9.9	1.1	<1.8	0.4
B	1	28.5	$3.7x10^{11}$	1.5	1.0	n/a	4.2
C	51	3.9	$5.5x10^{11}$	12.6	0.8	<2.0	0.3

and lack of substrate rotation used during sample growth are responsible for this effect. Such a gradient can be easily incorporated into calculations of the dHvA effect and show, for this sample, that roughly half the observed attenuation of the oscillations arises from the gradient and the rest is due to the Landau level widths. Additional inhomogeneity in the carrier density between layers of the MQW structure is also possible but this too can be assessed approximately using the same transport measurements. For the samples to be described in detail in the rest of this work these large-scale density variations have been measured and found to not significantly alter the general conclusions regarding the Landau level widths. Table I gives the relevant parameters for the three magnetization samples I will discuss in detail.

Before moving to the detailed results from the other, more homogeneous, samples an important non-equilibrium effect must be discussed. In the quantum Hall regime, when a plateau exists in the Hall resistance the diagonal resistivity drops to exceedingly small values[13]. This is because the Fermi level resides deeply within the region of localized electronic states. Under these conditions the 2DES becomes very susceptible to the generation of long-lived eddy currents[14]. These may be set up inductively by simply sweeping the applied magnetic field into the plateau region. These currents are very slow to decay; half-lives exceeding 1000s having been observed in a high mobility MQW structure[15]. Such circulating currents generate magnetic moments which can easily exceed the thermodynamic magnetization of the sample. The effect, though confined to a narrow field range surrounding the quantum Hall plateau center (i.e. integer filling factor), worsens as the temperature is reduced or the field sweep rate is increased. Although we do not understand the large sample-to-sample variation of the strength of the eddy currents, among those of interest here Sample A showed the largest effect. At 0.4K the eddy current magnetic moment was comparable to the dHvA amplitude itself in a narrow range around 5.8T, the center of the $\nu = 4$ filling factor. At lower fields (higher filling factors) the effect was negligible. Data was taken for sweeps up and down in field and then averaged. This removes the eddy current contribution since it reverses sign with sweep direction. For the other samples listed in Table I this procedure was unnecessary.

Magnetization data from samples A and B are shown in Fig. 4. For sample A, a 50-layer multi-quantum well sample the data was taken at 0.4K. A small, linear in field, background magnetization has been subtracted. The data for Sample B, a high-mobility single-interface structure, was taken at 4.2K. A large, temperature-dependent background magnetization, again linear in field, has been subtracted from the data. The source of this

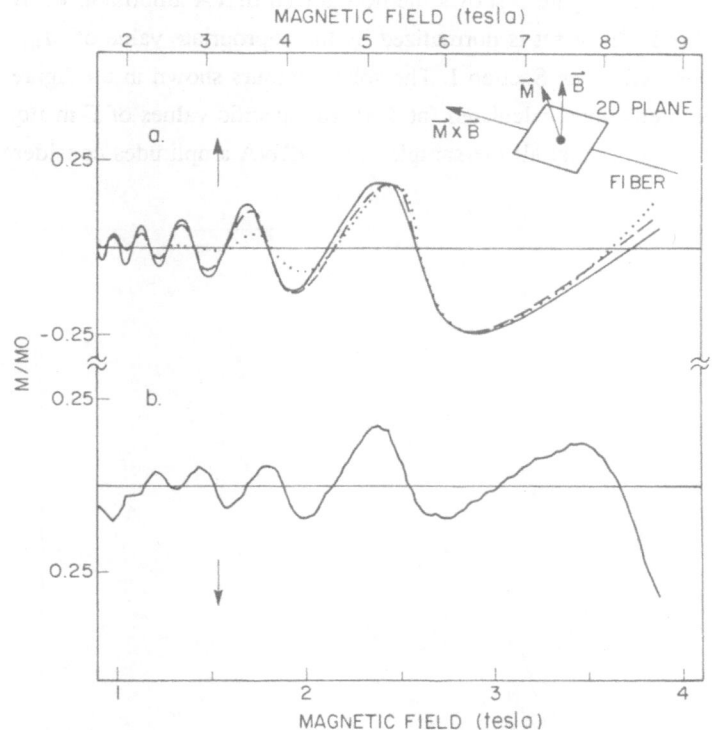

Fig. 4. Normalized magnetization for (a) sample A and (b) sample B. Note different field scales. Dotted and dashed lines are fits: $\Gamma = 2.4meV$ and $\Gamma = (1meV/T^{1/2})B^{1/2}$, respectively. The basic geometry of the magnetization measurement is depicted at top right.

large background, similar to that seen in all single-interface samples, is unknown. Its magnitude and temperature dependence prevented taking reliable data below 4.2K and above 4T. Although the uncertainties of the background subtraction for this sample make interpreting the shape of the oscillations impossible, the field dependence of the dHvA amplitude can be reliably determined and presents valuable information on the Landau level widths[3].

In order to systematically analyze the dHvA data from these samples we have modeled the DOS and then calculated the resultant magnetization and compared it to experiment. A simple sum of gaussian Landau levels, with spin splitting ignored, has been used:

$$D(\varepsilon) = \frac{2eB}{h} \sum_{j=0}^{\infty} \frac{1}{(2\pi)^{1/2} \Gamma} \exp\left[-\frac{(\varepsilon-\varepsilon_i)^2}{2\Gamma^2}\right] \qquad (10)$$

In this definition Γ is the rms half-width of the Landau levels and $\varepsilon_j =(j + 1/2)\hbar\omega_c$ is their mean energy. Figure 5 shows the normalized dHvA amplitude vs. B for all the samples in Table I. Each set is normalized by the appropriate value of M_0, the ideal dHvA amplitude defined in Section I. The solid contours shown in the figure represent the dHvA oscillation envelope calculated (at T=0) for specific values of Γ in Eq. 10. From the figure it is apparent that all the samples show dHvA amplitudes considerably smaller

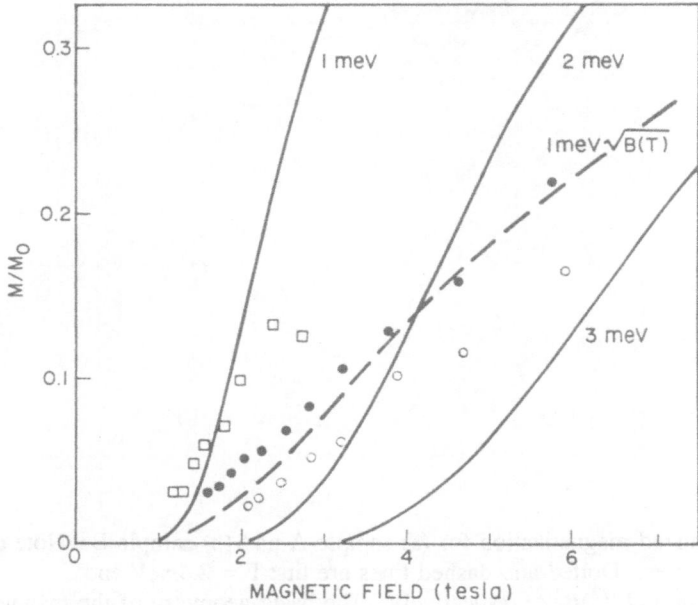

Fig. 5. Normalized dHvA oscillation amplitude vs. field for all three samples listed in Table I: filled circles, sample A; squares, sample B; open circles, sample C. The solid lines are the contours described in the text; the labels give the rms half-widths. The dashed line results from a linewidth proportional to $B^{1/2}$.

than the ideal value and that the apparent Landau level linewidths are in the 1-3 meV range. Also, the low field decay of the amplitudes is always less than that predicted by the model calculations which have assumed a constant linewidth Γ. One way to interpret this is to assume that the effective linewidth is itself dependent on magnetic field. The dashed line in Fig. 5 shows the calculated dHvA envelope under the assumption that Γ is proportional to $\sqrt{B}$. Using a prefactor of $1meV/T^{1/2}$ the calculated magnetization oscillation amplitudes are in much better agreement with the data from sample A than are any of the constant linewidth calculations. Similar results apply to samples B and C.

Figure 4 also shows the calculated dHvA oscillations for both constant and $\sqrt{B}$ linewidth field dependences compared to the data from sample A. In both cases the only fitting parameter is the linewidth, which was adjusted to give a good fit around 6T. As already determined from Figure 5, the field-independent fit decays much too quickly at low field. The $\sqrt{B}$ dependence gives a reasonable fit to both the amplitudes and the general shape of the oscillations. Obviously, one can further "refine" the model DOS to give even better fits. Changes in Landau level shape, field dependence, adding constant backgrounds to the DOS, etc. can all be pursued.

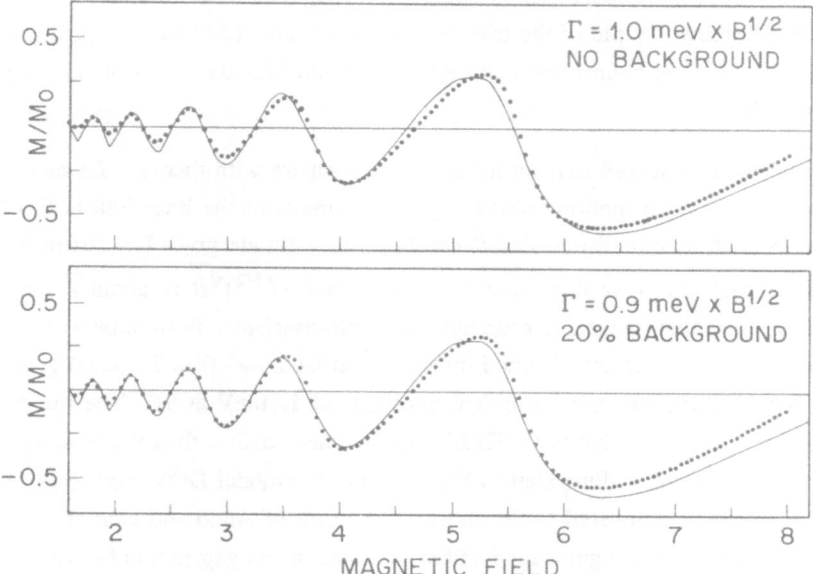

Fig. 6.   Fits (dotted curves) to sample A magnetization data (solid lines) with and without including a flat background in the DOS.

A number of workers[16-19] have found some evidence that a flat background of states exists atop of which the Landau levels sit. If a fraction f of all states are assumed to reside in the background then the degeneracy of each Landau level is taken as 2(1-f)eB/h (including spin degeneracy.) This is an ad hoc assumption since no theoretical model yet predicts such a DOS. In order to facilitate comparison to these other works, most notably the heat capacity studies of Gornik, et al.[16], we have fitted the dHvA data from sample A with and without such a background of states. These two fits are compared in Fig. 6. In the top panel the data is fitted without a background and with a $\sqrt{B}$ field dependence for the linewidth. In the lower panel 20% of the states are assumed to reside in a flat background while the rest are in gaussian Landau levels, again with a $\sqrt{B}$ linewidth. The best-fit linewidths are reduced by about 10% compared to the no-background case. Although the fit is better at the low field end with the background, there does not seem to be a compelling reason to adopt this DOS. The simpler assumption of gaussian levels, with one adjustable width parameter, provides an adequate basis for comparison to theory.

As mentioned previously, sample B is a single interface with mobility roughly 3.5 times higher than the multi-layer sample A. Remarkably, the normalized dHvA oscillations are not much larger. Only a small part of this comes from the higher measurement temperature used for sample B (4.2K compared to 0.4K). At 2T the observed rms half-width, from Fig. 5, is about 1.1meV for sample B. Accounting for the temperature broadening leaves a residual linewidth of about 0.9meV. This is only 25% smaller than that observed in sample A at the same magnetic field. The data from sample C, the lowest mobility sample of the trio, also confirms this rather weak dependence on zero-field mobility. The connection between mobility and Landau level broadening will be returned to below.

How do these observed Landau level widths compare with theory? As mentioned above, the simple-minded mobility scattering time estimate of the linewidth is $\Gamma = \hbar/\tau$. For sample A, with its mobility of $8x10^4$ cm^2/Vs, this estimate gives $\Gamma = 0.2$ meV. The best gaussian lineshape fit for this sample is $\Gamma = (1.0meV/T^{1/2})\sqrt{B}$ or about 2.2meV at 5T, ten times larger than our naive estimate. The self-consistent Born approximation result for short-range scatterers obtained by Ando and Uemura[5] (Eq. 7 above) gives a base half-width for the calculated semi-elliptical lineshape of 1.1meV at 5T. The rms half-width is 1/2 this value, or 0.55meV. This is still 4 times smaller than we observe. This discrepancy is dramatically illustrated in Fig. 7 where the model DOS used to fit the 5T data from sample A is compared to the short-range result of Ando and Uemura. Experimentally one finds a significant number of states in the gap region between Landau levels. This result has by now been widely confirmed. Although the experimental lineshapes appear to scale with field roughly as $\sqrt{B}$ as do the short-range theory results, this area of agreement seems relatively unimportant compared to the large disagreement on the magnitude of Γ. Making the same analysis for samples B and C, with their widely different mobilities, yields a similar level of disagreement between the short-range result and the experimentally observed DOS.

Several other experimental techniques have been applied to DOS determination for the 2D electron gas in heterostructures at high magnetic fields. The specific heat of the 2DES has been measured by two groups[16,20]. Magneto-capacitance measurements, which are sensitive to the DOS under appropriate circumstances, have also been performed[21,22]. Closely related to the DOS is the magnetic field dependence of the Fermi energy. This has been probed with both activated transport experiments[17-19] and direct electro-chemical potential measurements[23,24]. A comparative study of most of these methods has been presented elsewhere[43]. Quantitatively, the precise linewidths obtained from these experiments vary by as much as a factor of two for similar quality samples. There is, however, a systematic way in which this occurs and a plausible explanation will be given below. On the other hand, at the qualitative level there is broad agreement between the various methods. The basic conclusion that an unusually large DOS exists in the gap region between Landau levels is common to all the experiments.

Over the last three years a number of advances have been made in the theoretical understanding of Landau level widths. Important among these are the realizations that transport lifetimes in heterostructures can greatly exceed the actual elastic scattering time[25] and that screening of the impurity potential dramatically affects the DOS[26,27]. Both of these effects result in broader Landau levels than predicted by the short-range result given in Eq. 7.

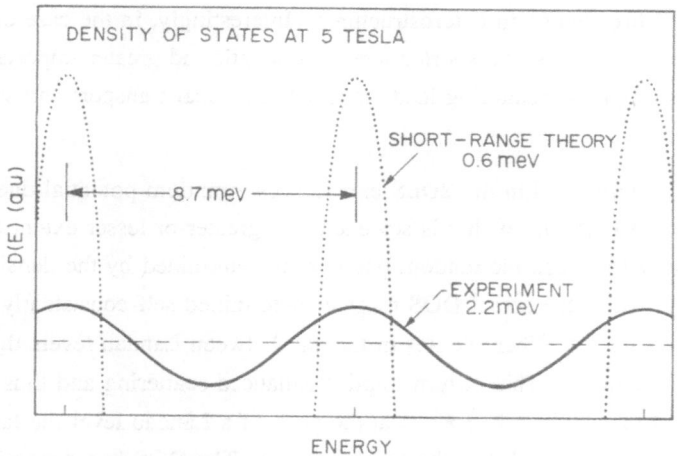

Fig. 7.   Comparison of model DOS (solid line) used to fit the sample A dHvA data at 5T, with the short-range theory. The rms half-widths of the levels are also shown.

The influence of an inhomogeneous carrier concentration has also been investigated[28]. Proper accounting for the associated inhomogeneities in the chemical potential leads to a plausible explanation of the anomalous DOS in the Landau gap. Finally, Luryi has argued[44] that the DOS is considerably more complicated than our simple gaussian model owing to the effects of localization and the so-called "perfect screening" mechanism. Luryi finds additional sharp structures in the DOS atop the broad distribution of localized states in each Landau level. As yet no analytic calculations have given this picture a basis for comparison to experiment.

Das Sarma and Stern[25] have calculated the ratio of the transport lifetime (determined from the mobility at B=0) to the so-called single-particle lifetime. In modulation-doped heterostructures the 2D electron sheet is separated from its parent donor impurities by a spacer, or set-back, layer typically 200Å thick. It has been widely established that coulombic scattering off of these ionized donors is usually the dominant factor in determining the low-temperature mobility. The spatial separation enhances the mobility and simultaneously produces a smooth long-range impurity potential, free of 1/r divergences. The result is a great predominance of small angle scattering, which is inefficient in producing the momentum relaxation required for electrical resistivity. Das Sarma and Stern find that the transport lifetime can exceed the true scattering time, which is the correct indicator of level broadening, by two orders of magnitude. Although the calculations are strictly valid only at zero magnetic field, it is anticipated that the results will influence the broadening of the Landau levels. This has been experimentally verified by examining the magnetic field dependence[29] of the Shubnikov-de Haas oscillations of the diagonal resistivity[20,31]. It has been found that the scattering rate needed to fit the oscillation amplitudes exceeds that estimated from the mobility by as much as a factor of ten. It is likely, therefore, that insertion of the transport lifetime into Eq. 7 underestimates the Landau level broadening in heterostructures. Interestingly, in the case of inversion layers[31] in Si-MOS devices, the shorter screening length and greater importance of short-range interface-roughness scattering leads to nearly identical transport and single-particle scattering times.

A single electron within the 2DES experiences a random potential due to the ionized impurities in the AlGaAs layer that is screened to a greater or lesser extent by the 2DES itself. The ability to screen the random potential is determined by the density of states at the Fermi level. Thus, the actual DOS must be determined self-consistently taking the screening into account[32]. When E_F lies in the gap between Landau levels the DOS is low and screening is reduced. This in turn implies enhanced scattering and thus broader Landau levels. Conversely, when E_F is at the peak of a Landau level the large DOS allows efficient screening and thus the levels narrow. The DOS is a dynamic function of the Fermi level position, the Landau level widths oscillating as E_F moves through the Landau ladder of states. This effect has been quantitatively addressed by Cai and Ting[27] and by Ando and Murayama[26]. Both groups find very strong oscillations of the level widths. When E_F is in the "gap" the levels actually strongly overlap. Although the calculations are only valid assuming such level mixing does not occur, it seems clear that a significant DOS, associated with the longer range components of the random potential, exists between Landau levels. Equation 7, valid in the short-range limit, does not apply. Although these ideas are in qualitative accord with our experimental findings, a quantitative comparison must await calculations that include Landau level mixing.

The discrepancies between the various experiments alluded to above provide evidence, albeit circumstantial, for such a dynamic DOS. Since this has been discussed elsewhere[43], I will mention here only the comparison of the specific heat experiment of Gornik, et al.[16] with our magnetization results[3]. Using a sample roughly comparable in carrier density and mobility to our sample A (see Table I) Gornik, et al. find Landau level widths about half as large as the $\Gamma = (1.0meV/T^{1/2})\sqrt{B}$ result found via the dHvA effect.

A possible explanation for this lies in the self-consistent screening of the impurity potential. At low temperature and high magnetic fields the specific heat is really just a measure of the density of states at the Fermi level. As such it is largest when E_F is at the center of the Landau levels. By contrast, the magnetization is near zero in this situation and is largest when E_F is in the gap between levels. Obviously, when fitting the experimental data from either measurement, the fit is most strongly weighted by those regions of data showing the largest signal. Consequently, the specific heat data emphasizes the peaks in the DOS while the magnetization data highlights the gaps. The narrower Landau levels found via specific heat are therefore to be expected, given the screening effects outlined above. Similar arguments can also explain qualitatively the level width variations found in the other DOS experiments.

SECTION IV. THERMAL CONDUCTANCE MEASUREMENTS

In this section I will discuss our measurements of the dc thermal conductivity at high magnetic fields of GaAs/AlGaAs heterostructures containing 2D electron systems[4]. In these phonon transport experiments the DOS of the 2DES enters through the coupling of the phonons carrying the heat to the 2D electrons residing near the sample surface. Being a transport measurement one cannot unambiguously extract the DOS from the data without detailed knowledge of the electron-phonon scattering matrix elements. On the other hand, these measurements provide a novel new avenue for study of both the high-field DOS and the electron-phonon interaction in 2D systems.

Part of the original motivation for these experiments was to observe the thermal conductivity of the 2DES itself, on top of the much larger contribution of the lattice phonons. In the quantum Hall regime the thermal conductivity tensor of the 2DES is predicted to show novel properties related to the localization of the electronic states[33]. Not surprisingly, these electronic effects will be extremely difficult to observe. A simple estimate is provided by assuming the lattice conduction to be determined by

$$\kappa_l = \frac{1}{3} \, C_{ph} \, c \lambda_{ph} \tag{11}$$

where C_{ph} is the Debye specific heat of the lattice[34], c is an appropriate average phonon velocity in GaAs (~3300m/s) and λ_{ph} is the phonon mean free path. The mean free path (mfp) will be determined by boundary scattering at low temperature. For the electronic contribution we appeal to the Wiedemann-Franz law relating the electronic thermal conductivity to the electrical conductivity

$$\kappa_e = \frac{1}{3} \left[\frac{\pi k_B}{e} \right]^2 T \, \sigma \tag{12}$$

The conductivity σ is conveniently expressed in terms of the 2D carrier concentration N_s and mobility μ as $\sigma = N_s e\mu$. Since $N_s = 5x10^{11} cm^{-2}$ and $\mu = 10^5 cm^2/Vs$ are typical, we expect the electronic thermal conductivity to be approximately $10^{-10} W/K$ for a square sample at 0.5K. Assuming the sample has been thinned down parallel to the 2D plane to a thickness of 50 μm (typical of our experiments) and is much larger in both width and length, we can estimate the boundary scattering phonon mfp to be also about 50 μm (see below, however.) From these numbers we find that the ratio of electronic thermal conduction to lattice conduction is about 10^{-5} at 0.5K. This is a very small value and will be very difficult to observe. Even with a MQW structure with 100 layers of 2D electrons the prospects are not great. The results which we are about to describe are not consistent with observation of the electronic part of the net thermal conductance of the samples. It may be possible by going to lower temperature (since the electronic part increases relative to the lattice contribution as T^{-2}) and using samples of higher mobility to finally detect this small electronic effect.

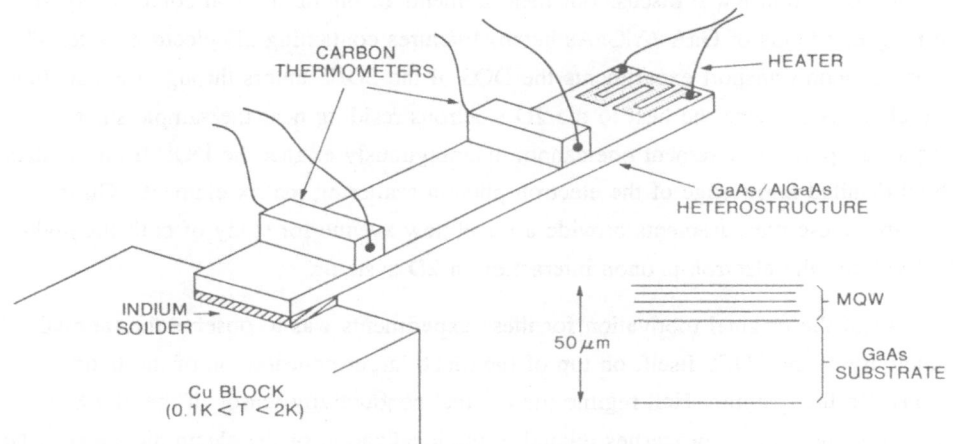

Fig. 8.   Schematic of thermal conductance experimental arrangement.

The experimental arrangement used to measure the thermal conductance of heterostructure samples is shown schematically in Fig. 8. An approximately 1x8mm bar is cleaved from the <100> oriented GaAs MBE wafer and the substrate is then thinned down using a bromine-methanol etch until a total sample thickness of about 50 μm is achieved. This particular etch leaves a very smooth surface. Attached to one end of this bar is a constantan foil strain gauge which serves as a resistive heater exhibiting very little magneto-resistance. The opposite end of the bar is thermally anchored, using indium solder, to a copper block that is attached to the mixing chamber of a dilution refrigerator. A magnetic field of up to 11T can be applied perpendicular to the heat flow direction and the 2D electron layers. To measure the temperature gradient along the bar two small carbon resistance thermometers are employed. These are both cut from the same parent

220-ohm Speer resistor and are therefore extremely well matched in both temperature and magnetic field dependence. This is essential given the small magnitudes of the effects observed. Both resistors are connected into a constant-current ac-differential bridge configuration the output of which yields the temperature difference between the thermometers. The thermometers and the heater are attached, using GE 7031 varnish, to the substrate side of the sample to avoid capacitive effects associated with the 2D electron layers in the samples. The several wires (usually very fine stainless steel or superconducting NbTi alloy) used for the thermometers and heater have negligible thermal conductance compared to the sample itself. No noticeable thermal resistance has been yet found associated with the indium solder joint. The multi-quantum well samples used in this experiment are described in Table II.

Table II. Sample parameters for thermal conductance experiment.

Sample	Length[a] (mm)	Width (mm)	Thickness (μm)	Number of Periods	2D density (cm^{-2})	Low-T mobility ($m^2/V{\cdot}s$)
1	1.8	1.0	50	172	$8.9x10^{11}$	2.0
2	3.7	1.0	48	50	$5.1x10^{11}$	8.0

a. Thermometer spacing, center-to-center

The thermal conductance of sample 2 at zero magnetic field is shown in Fig. 9a. The data is consistent with a power law T^n with $n = 2.65$. This is close to the T^3 dependence expected for boundary scattering of phonons. The mean free path of the phonons may be calculated from Eq. 11 and the sample geometry using published values[34] for the lattice specific heat of GaAs (we ignore the differences due to the small amount of the alloy AlGaAs present in the heterolayers.) The results are displayed in Fig. 9b. These mean free paths are of order 1mm, considerably larger than the 48 μm thickness of the sample. Assuming the phonon-surface collisions are diffuse (i.e. absorption followed by random thermal re-emission) one can calculate the expected mfp. Due to the large width-to-thickness ratio of the samples the correct diffuse mfp exceeds the thickness considerably. Unlike in a cylinder, a phonon moving between parallel plane boundaries has a large chance of travelling more than the plane separation before striking the boundaries. Wybourne, et al.[35] have calculated the appropriate mfp for samples with aspect ratios similar to ours. Applying their results to our geometry yields an expected diffuse mfp of 153 μm. The observed mfp exceeds this by as much as 10 times. The surfaces must therefore reflect the phonons specularly much of the time.

Assuming that the temperature gradient along the sample is uniform one can estimate the degree of surface specularity. Again following Wybourne, et al.[35] our data show an apparent specularity (fraction of surface collisions which are specular) ranging from 75% around 1K to over 90% below 0.2K. Klitsner and Pohl[36] have pointed out that such estimates are likely to be inaccurate for highly specular surfaces. The source of the problem lies in those regions of sample surface on which the thermometers are mounted. These regions are almost undoubtedly diffuse and the mfp will be close (modulo a factor

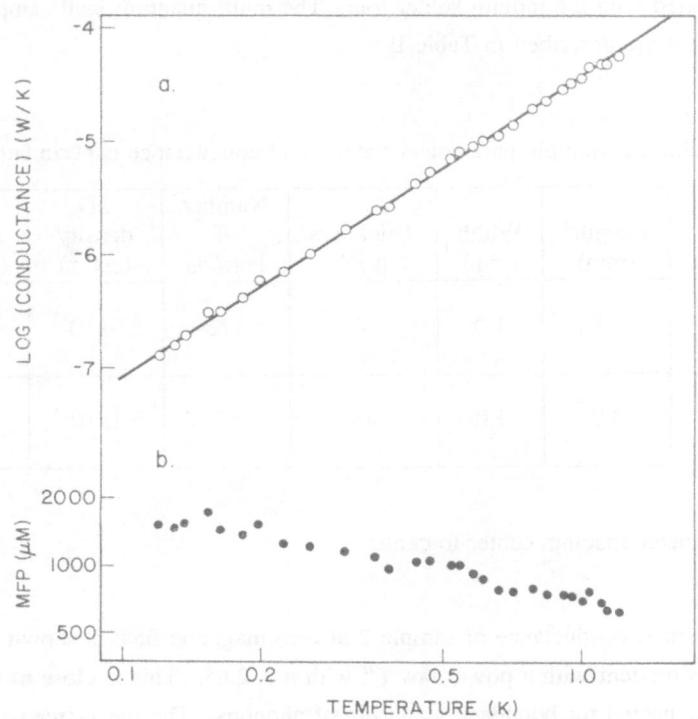

Fig. 9.   a. Thermal conductance at zero magnetic field of sample 2 vs. temperature. Solid line represents $T^{2.65}$ power law.

           b. Apparent mean free path deduced from B=0 conductance data on sample 2.

of 2 since only one surface is involved) to the diffuse estimate given above. The temperature gradient under the thermometers will be greater than that in the region between them. Our data support this picture since, for a given heat flux, the temperature differences between each thermometer and the copper mounting block are not in proportion to their distance from the block. While geometrically we expect about a 5:1 ratio for sample 2, the observed ratio is only 2.5:1. If the sample were perfectly specular between the thermometers and diffuse under them, the ratio would be 2:1. As will be noted below the small magneto-oscillations in the temperature gradient, arising from the electron-phonon coupling are in the proper ratio, indicating a homogeneous effect. A reliable estimate of the true surface specularity in the region between thermometers requires resorting to numerical methods[37]. For our purposes it is sufficient to note that the surfaces are highly specular, probably even more so than the simple estimate just given indicates.

Application of a magnetic field perpendicular to the heterostructure layers causes oscillations in the observed temperature difference between the two thermometers. Data for both sample 1 and 2, under conditions of constant applied heat flux, are shown in Fig. 10. In both cases the figure includes a large offset; these oscillations are only a few percent of the total temperature difference between the thermometers. The arrows in both

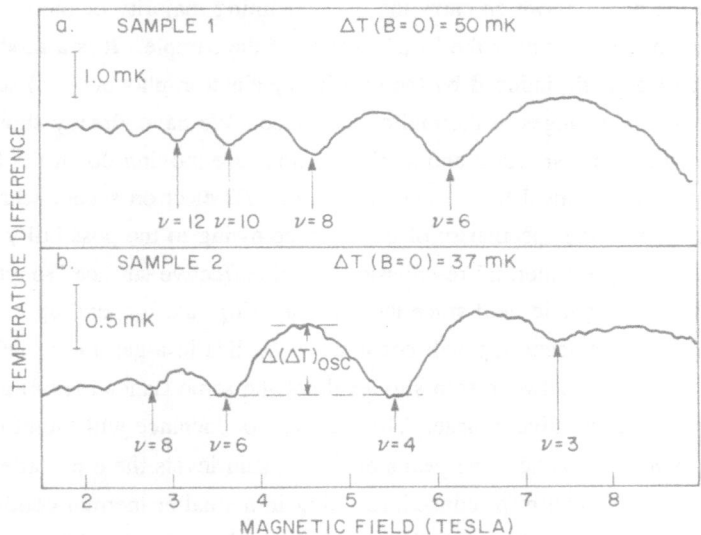

Fig. 10. a. Thermal conductance oscillations at 0.7K for sample 1. Note the large
vertical offset. Arrows indicate the "cusps" at which E_F lies in the Landau gap.
b. Thermal conductance oscillations at 0.35K for sample 2. The arrow at $v = 3$
represents E_F in the spin gap of the j=1 Landau level. The definition of
$\Delta(\Delta T)_{osc}$ is shown.

data sets in the figure indicate integral values of the filling factor v. The 1/B periodicities of these filling factors are consistent with the known 2DES carrier densities in each sample. The indexings shown are the only reasonable ones possible. The data dramatically show the influence of the 2DES on the bulk thermal conductance of modulation-doped heterostructures.

The cusp-like features occurring at the integral filling factors are local minima in the temperature difference and therefore correspond to local maxima in the thermal conductance of the samples. Since these occur at even integral filling factors (except one, $v = 3$) the Fermi level is in the gap between Landau levels. In this situation the direct electronic contribution to the thermal conductivity should be smallest. Conversely, at the broad maxima in temperature difference, and consequent minima in thermal conductance, E_F is near the peaks of the Landau levels where the DOS is largest. Here one expects the electronic thermal conductance to be large. Thus, the observed thermal conductance oscillations, while obviously related to the 2DES, are exactly out of phase with what one would expect from the direct electronic contribution. In addition, as we will discuss further below, the oscillation amplitudes remain a roughly fixed fraction of the total temperature difference in the range 0.15K < T < 1.2K. As noted above, this too is inconsistent with a simple electronic thermal conductance channel in parallel with the much larger phonon channel.

The magneto-oscillations observed in the thermal conductance appear to arise from the coupling of the phonons which carry the overwhelming majority of the heat and the 2D electron systems residing near the "top" surface of the sample. It is a modulation of the phonon mean free path, induced by the oscillatory electron-phonon (e-p) scattering rate, that generates the changes in thermal conductance. We have already shown that the sample surfaces are highly specular and so the phonons are moving down the bar in a way analogous to light in an optical fiber. The presence of 2D electron sheets near one sample surface slightly reduces the specularity of that surface owing to the possibility for phonon absorption (and subsequent thermal re-emission). This effective surface "roughness" may be modulated by the magnetic field since the e-p scattering rates depend upon the DOS at the Fermi level. Reduced scattering will occur when E_F lies in a gap in the DOS (either between Landau levels or between spin sub-levels of the same Landau level) and so the net phonon mfp will be relatively large. The thermal conductance will therefore also be large. At fields where E_F is near the peaks of the Landau levels the e-p scattering rate will be large and the phonon mfp reduced, resulting in a smaller thermal conductance. This mechanism has the same phase as the observed results.

The minimum at $v = 3$ seen in the data for sample 2 displayed in Fig. 10 arises because E_F resides in the spin gap of the j=1 Landau level. The higher electron mobility of this sample results in the narrower Landau levels needed to resolve the high field spin doublets. In this connection it also appears that it is the total DOS at E_F that is being sampled, without distinction between "localized" or "extended" electron states. This conclusion is based on the generally smooth field dependence of the thermal conductance, in sharp contrast to the orders-of-magnitude variation of the electrical conductivity characteristic of the quantum Hall regime[13]. The spin features seen in electrical transport (on a sample from the same MBE wafer as sample 2) are much stronger than the weak cusp seen at $v = 3$ in thermal conductance. We speculate that the fundamental electron-phonon processes occur on a length scale (typical phonon wavelength ~1000Å at 0.5K) that is small compared to the dominant localization lengths.

The strength of the e-p coupling responsible for the thermal conductance magneto-oscillations can be determined from the temperature dependence of the oscillation amplitude. As defined in Fig. 10 this is defined as the maximum change in the observed temperature drop between two of the adjacent cusp-like minima. Assuming that the e-p scattering rate is much smaller when E_F is in a Landau gap than when at the peak of the Landau levels, the oscillation amplitude measures the maximum e-p scattering rate for E_F in the center of a particular Landau level. This follows from a few simple assumptions. In addition to Eq. 11 for the thermal conductivity, we assume that the net phonon lifetime may be written as

$$\tau_{ph}^{-1} = \tau_o^{-1} + \tau_{ep}^{-1}(B) \tag{13}$$

where $\tau_{ep}^{-1}(B)$ is the e-p rate at magnetic field B and τ_o^{-1} is some residual, magnetic field-independent phonon scattering rate. Assuming that the temperature gradient in the region between the thermometers is uniform and that $\tau_{ep}^{-1}(B)$ is much smaller at fields where E_F is in a gap (i.e. the "cusps" in the conductance) than at the peaks of the Landau levels we find

$$\Delta(\Delta T_{osc}) = \{3QL/(AC_{ph}c^2)\}\ (\tau_{ep}^{-1})_j \tag{14}$$

In this equation $(\tau_{ep}^{-1})_j$ is the maximum electron-phonon scattering rate with E_F in

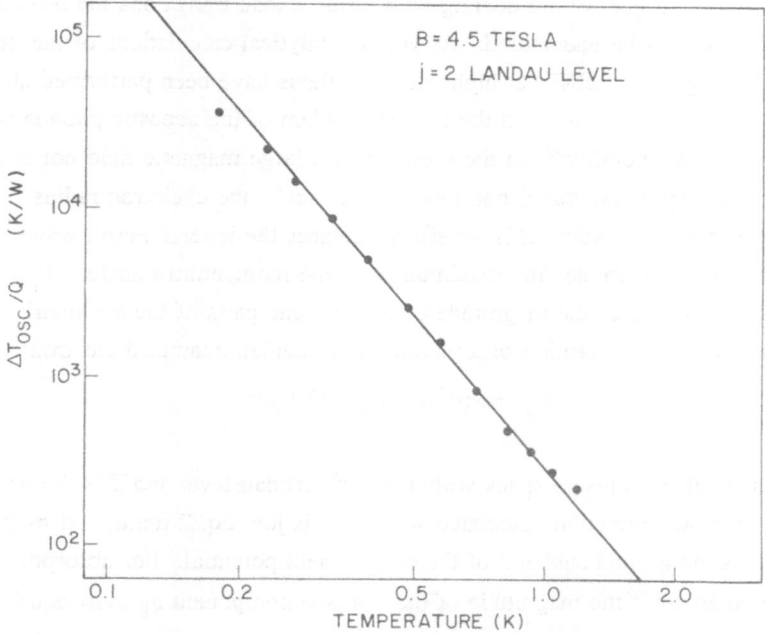

Fig. 11. Temperature dependence of $\Delta(\Delta T)_{osc}$ for sample 2 at 4.5 tesla in the j=2 Landau level. Solid line is a T^{-3} power law.

the jth Landau level, Q is the applied heat flux, and A and L are the sample cross-sectional area and thermometer spacing respectively. We have assumed that the only significant electronic transitions are those within a given Landau level. Inter-Landau level transitions are being ignored. This is an excellent assumption since the phonon energies (of order $k_B T$) are negligible compared to the splittings between Landau levels.

Figure 11 shows the temperature dependence of $\Delta(\Delta T_{osc})/Q$ for sample 2. These data represent the e-p scattering rate at 4.5 tesla with E_F in the middle of the j=2 Landau level. The data are consistent with a T^{-3} power law over the temperature range displayed. Since the lattice specific heat scales as T^3 we can conclude from Eq. 14 that $(\tau_{ep}^{-1})_2$ is roughly temperature independent under these particular conditions. Numerically these data give $\tau_{ep} \sim 20\mu s$ or, equivalently, a mean free path for electron-phonon scattering of approximately 7cm. Now, the mean distance a phonon must travel before striking one surface of the sample is just the diffuse boundary scattering mfp of 153 μm discussed above. Since sample 2 has 50 layers of electrons the probability for absorption of a phonon by any one of these layers is approximately $4x10^{-5}$. This result has been very recently supported by ballistic phonon absorption measurements on yet another sample from the same MBE wafer[38].

Several distinct effects must be included in any detailed calculation of the electron-phonon scattering rate in modulation doped heterostructures. Both piezoelectric and deformation potential scattering channels are operative and screening of the e-p interaction should be included. Since we are dealing with intra-Landau transitions the broadening of the Landau levels must be understood. As yet no analytical calculations of the lifetime of phonons interacting with a 2DES at high magnetic fields have been performed although considerable work has been done on the related problem of the acoustic phonon scattering limit to the zero-field mobility[39]. In the presence of a large magnetic field not only does the DOS become highly structured but a new length scale, the cyclotron radius $l = (\hbar/eB)^{1/2}$ becomes relevant. This, in effect, replaces the inverse Fermi wavevector which, at zero field, determines the maximum in-plane momentum transfer. It is the product $q_\| l$ that determines the magnitude of the in-plane parts of the e-p matrix elements. The matrix element for absorption of a phonon by a Landau quantized electron is

$$M_{j\vec{q}} = <jk' |\exp(i\vec{q} \cdot \vec{r})| jk> \tag{15}$$

where $|jk>$ and $|jk'>$ represent states within the jth Landau level and $\vec{q}$ is the phonon wavevector. The wavefunction associated with $|jk>$ is just Eq. 2 multiplied by $\chi_0(z)$ which describes the ground subband of the confinement potential. For absorption of a phonon of momentum $\vec{q}$ the magnitude of the in-plane component $q_\|$ must equal the difference in the "momenta" associated with the plane wave part of the Landau level wavefunctions given in Eq. 2

$$k' = k + q_\| \tag{16}$$

From Eq. 4 it is apparent that this absorption involves a spatial jump for the electron. We anticipate that this jump

$$\Delta x = (k' - k)l^2 = q_{\parallel}l^2 \qquad (17)$$

cannot greatly exceed the magnetic length or the wavefunctions will not overlap sufficiently. This gives an intuitive criterion for a large transition probability:

$$q_{\parallel}l \sim 1 \qquad (18)$$

Phonons of wavevector much in excess of l^{-1} will not be absorbed. Evaluating Eq. 15 gives, to within a constant and an overall phase factor,

$$M_{j\vec{q}} = g(q_z)\exp(-q_{\parallel}^2 l^2/4) \, L_j \, (q_{\parallel}^2 l^2/2) \qquad (19)$$

The gaussian factor contains the intuitive criterion of Eq. 18 and the Laguerre polynomial $L_j(x)$ represents an additional oscillatory dependence on $q_{\parallel}l$. The function $g(q_z)$ depends on the perpendicular component of phonon momentum and the subband wavefunction $\chi_0(z)$

$$g(q_z) = \int |\chi_0(z)|^2 \, \exp(iq_z z)dz \qquad (20)$$

This function falls off when the z-projection of the phonon wavelength becomes much shorter than the z-extent of the subband wavefunction.

To actually calculate the lifetime of a phonon of momentum q one must account for emission as well as absorption, square the matrix elements, multiply by the q-dependent coupling strengths[40] associated with the type of scattering (piezoelectric or deformation potential, transverse or longitudinal modes), and incorporate screening[41]. Also the phase space must be worked out subject to energy conservation and Fermi statistics. It is in the sum over final and the average over initial electron states that the Landau level density of states enters.

Correlation effects associated with inhomogeneous broadening may play a role in determining what is the proper measure of the Landau level width[42]. Obviously, if the "jump" distance (Eq. 17) is small compared to the correlation length for inhomogeneities then the effective linewidth needed for energy conservation may be much smaller than that measured by a bulk thermodynamic probe like magnetization. A similar situation occurs in the "vertical" transitions of cyclotron resonance. Generally, CR linewidths are smaller than those determined by true thermodynamic DOS probes.

Although the full numerical calculations outlined have not yet been performed, there are some interesting aspects of the matrix elements which may already have shown up in the experiments. As Eq. 19 shows there is rich structure in the $q_\parallel l$ dependence of the matrix elements around $q_\parallel l \sim 1$. Since in our experiments the dominant phonon wavelengths ($\lambda/2\pi \sim 100\text{Å}/T$) are comparable to the magnetic length ($l \sim 250\text{Å}/B^{1/2}$) the product $q_\parallel l$ is of order unity and can be scanned, via temperature or field over a wide range. Although some of the complex structure of the matrix elements will wash out upon angular averaging over a thermal distribution of phonons, we anticipate that some of the $q_\parallel l$ dependence will remain and show up in the magneto-oscillation amplitude.

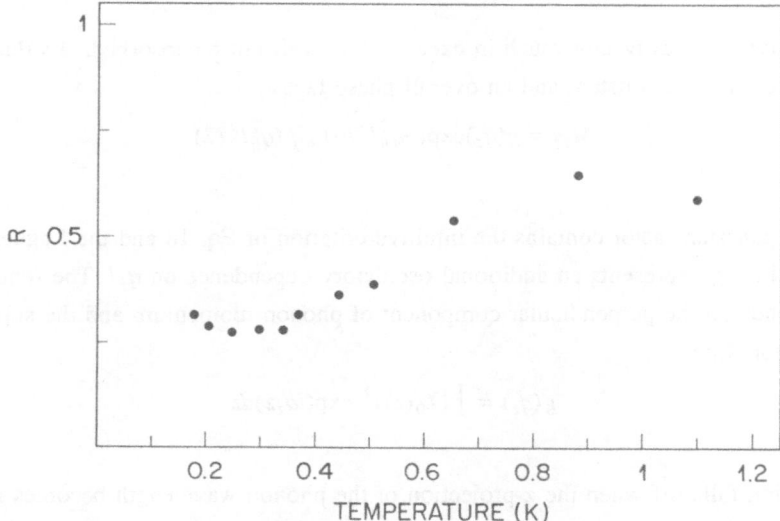

Fig. 12. Ratio R of $\Delta(\Delta T_{osc})$ for the j=3 level at 3.1 T to the j=2 level at 4.5T in sample 2.

Possible evidence for such effects is contained in Fig. 12. There the ratio R of $\Delta(\Delta T_{osc})$ for the j=3 oscillation at 3.1T to the j=2 oscillation at 4.5T is plotted vs. temperature for sample 2. Surprisingly, the data show the j=3 amplitude decreasing relative to the j=2 amplitude as the temperature is reduced. This is in contradiction to the simple argument that reducing the thermal broadening should cause lower field magneto-oscillations (as in the Shubnikov-de Haas effect) to become more prominent. A possible, though not yet proven, explanation lies in the matrix elements just discussed. Beyond slightly different magnetic lengths (146 vs. 121Å), the j=3 and j=2 matrix elements contain different order Laguerre polynomials in $q_\parallel l$. The j=3 matrix element has three zeroes whereas the j=2 element has only two. As a function of $q_\parallel$ the oscillations of the two matrix elements are approximately out of phase. Rudimentary calculations averaging the matrix elements over a Planck distribution of piezoelectrically scattered screened transverse phonons reveals relative scattering rates for the j=2 and 3 levels that qualitatively agree with the data in Fig. 12.

SECTION V. CONCLUSIONS

Although fundamentally different measurements, the magnetization and thermal conductance experiments just described both highlight the detailed structure of the density of states of 2D electron systems at high magnetic field. The magnetization, being a thermodynamic variable, directly gives the DOS without complicating effects arising from either localization or scattering matrix elements. Our studies have revealed an unexpectedly large DOS in the "gap" region between the Landau levels as well as an apparent $B^{1/2}$ dependence to the actual Landau level widths. These effects have been corroborated by a number of independent experiments[16-24]. Considerable theoretical advances have been made as well. The crucial effect of modulation doping on the ratio of transport to single-particle lifetimes has been elucidated[25]. The strong oscillations of the Landau level widths arising from the self-consistent screening mechanism have also been investigated[26,27] as have the effects of inhomogeneities[28]. While not yet fully determined the high field DOS in heterostructures is much better understood than it was just a few years ago. We anticipate additional information will be obtained from planned magnetization measurements on single interface structures with mobilities exceeding $10^6 cm^2/Vs$. A tantalizing possibility remains in observing the magnetization in the fractional quantum Hall regime.

The thermal conductance studies provide a look not only at the DOS but also at the complex matrix elements for electron-phonon scattering in 2D systems at high fields. This is an almost untouched area. While the density of states is obviously involved in the e-p scattering rate, the question of how this DOS is related to that determined by a bulk thermodynamic measurement remains. The electron-phonon scattering proceeds on the length scale of the cyclotron radius and it is not clear how correlations and long-range potential fluctuations are involved. It appears that further measurements of the thermal conductance and of ballistic phonon absorption[38] will succeed in revealing some of the intricate matrix element structure arising from the Landau level wavefunctions. The possibility of studying the coupling of phonons to the correlated many-body ground state of the fractional quantum Hall effect is very enticing.

ACKNOWLEDGEMENTS

It is a pleasure to thank H. L. Stormer and V. Narayanamurti for many enlightening discussions throughout the course of the experiments described in these lectures. I am indebted to A. C. Gossard, A. Y. Cho, C. W. Tu and W. Wiegmann for growing the MBE heterostructures without which the results described here would not exist. Many thanks are due to K. Baldwin, A. Savage, and M. A. Chin for expert and extensive technical help.

REFERENCES

1. T. Ando, A. B. Fowler and F. Stern, Rev. Mod. Phys. *54*, 437 (1982).
2. R. B. Laughlin, Phys. Rev. B *23*, 5632 (1981).
3. J. P. Eisenstein, H. L. Stormer, V. Narayanamurti, A. Y. Cho, A. C. Gossard and C. W. Tu, Phys. Rev. Lett. *55*, 875 (1985).
4. J. P. Eisenstein, A. C. Gossard and V. Narayanamurti, to be published in Proceedings of Seventh International Conference on Electronic Properties of Two-Dimensional Systems, Santa, Fe, 1987.
5. T. Ando and Y. Uemura, J. Phys. Soc. Jpn., *36*, 959 (1974).
6. W. Zawadzki, in *Two-Dimensional Systems, Heterostructures, and Superlattices*, eds. G. Bauer, F. Kuchar and H. Heinrich, Springer Series in Solid State Sciences, vol. 53 (Springer-Verlag, New York, 1984).
7. R. F. Kazarinov and S. Luryi, Phys. Rev. B *25*, 7626 (1982).
8. H. Aoki and T. Ando, Phys. Rev. Lett. *54*, 831 (1985).
9. R. Dingle, H. L. Stormer, A. C. Gossard and W. Wiegmann, Appl. Phys. Lett. *7*, 665 (1978).
10. H. L. Stormer, J. P. Eisenstein, A. C. Gossard, W. Wiegmann and K. Baldwin, Phys. Rev. Lett. *56*, 85 (1986).
11. J. H. English, A. C. Gossard, H. L. Stormer and K. W. Baldwin, Appl. Phys. Lett. *50*, 1826 (1987).
12. J. P. Eisenstein, Appl. Phys. Lett. *46*, 695 (1985).
13. D. C. Tsui, H. L. Stormer and A. C. Gossard, Phys. Rev. B *25*, 1405 (1982).
14. J. P. Eisenstein, H. L. Stormer, V. Narayanamurti and A. C. Gossard, Superlatt. and Microstruc. *1*, 11 (1985).
15. J. P. Eisenstein, unpublished.
16. E. Gornik, R. Lassnig, G. Strasser, H. L. Stormer, A. C. Gossard and W. Wiegmann, Phys. Rev. Lett. *54*, 1820 (1985).
17. E. Stahl, D. Weiss, G. Weimann, K. v. Klitzing and K. Ploog, J. Phys. C *18*, L783 (1985).
18. D. Weiss, E. Stahl, G. Weimann, K. Ploog and K. v. Klitzing, Electronic Properties of Two-Dimensional Systems, ed. T. Ando (Elsevier 1986) p. 285.
19. H. P. Wei, A. M. Chang, D. C. Tsui and M. Razeghi, Phys. Rev. B *32*, 7016 (1985).
20. J. K. Wang, J. H. Campbell, D. C. Tsui and A. Y. Cho, Bull. Am. Phys. Soc. *32*, 462 (1987).
21. T. P. Smith, B. B. Goldberg, P. J. Stiles and M. Heiblum, Phys. Rev. B *32*, 2696 (1985).
22. V. Mosser, D. Weiss, K. v. Klitzing, K. Ploog and G. Weimann, Sol. State Commun. *58*, 5 (1986).
23. R. T. Zeller, F. F. Fang, B. B. Goldberg, S. L. Wright, and P. J. Stiles, Phys. Rev. B *33*, 1529 (1986).
24. D. Weiss, V. Mosser, V. Gudmundsson, R. R. Gerhardts and K. v. Klitzing, Sol. St. Commun. *62*, 89 (1987).
25. S. Das Sarma and Frank Stern, Phys. Rev. B *32*, 8442 (1985).
26. T. Ando and Y. Murayama, J. Phys. Soc. Jpn. *54*, 1519 (1985).
27. W. Cai and C. S. Ting, Phys. Rev B *33*, 3967 (1986).
28. Vidar Gudmundsson and Rolf R. Gerhardts, Phys. Rev B *35*, 8005 (1987).
29. T. Ando, J. Phys. Soc. Jpn. *37*, 1233 (1974).
30. M. Paalanen, D. C. Tsui and J. C. M. Hwang, Phys. Rev. Lett. *51*, 2225 (1983).
31. J. P. Harrang, R. J. Higgins, R. K. Goodall, P. R. Jay, M. Laviron, and P. Delescluse, Phys. Rev. B *32*, 8126 (1985).
32. T. Ando, J. Phys. Soc. Jpn. *43*, 1616 (1977).

33. Herbert Oji, Phys. Rev. B 29, 3148 (1984).
34. T. C. Cetas, C. R. Tilford and C. A. Swenson, Phys. Rev. 174, 835 (1968).
35. M. N. Wybourne, C. G. Eddison and M. J. Kelly, J. Phys. C 17, L607 (1984).
36. T. Klitsner and R. O. Pohl in "Phonon Scattering in Condensed Matter V", Springer Series in Solid-State Sciences, vol. 68, eds. A. C. Anderson and J. P. Wolfe, Springer-Verlag, Berlin 1986.
37. J. E. VanCleve, T. Klitsner and R. O. Pohl in "Phonon Scattering in Condensed Matter V", Springer Series in Solid-State Sciences, vol. 68, eds. A. C. Anderson and J. P. Wolfe, Springer-Verlag, Berlin 1986.
38. N. C. Jarosik, J. P. Eisenstein and M. A. Chin, unpublished.
39. For example, Peter J. Price, Surf. Sci. 143, 145 (1984) and W. Walukiewicz, H. E. Ruda, J. Lagowski and H. C. Gatos, Phys. Rev. B 30, 4571 (1984).
40. P. J. Price, Annals of Physics 133, 217 (1981).
41. P. J. Price, J. Appl. Phys. 53, 6863 (1982).
42. M. Lax and J. P. Eisenstein, to be published
43. J. P. Eisenstein, Physica Scripta $T14$ (1986).
44. S. Luryi in "High Magnetic Fields in Semiconductor Physics", ed. G. Landwehr, Springer Series in Solid State Sciences, vol. 71 (Springer-Verlag 1987).

THE QUANTUM HALL EFFECT: FROM INTEGRAL TO FRACTIONAL, FROM

TWO DIMENSIONAL TO THREE DIMENSIONAL

H.L. Stormer

AT&T Bell Laboratories
Murray Hill, New Jersey 07974
U.S.A.

This lecture reviews some recent developments in the experimental aspects
of the quantum Hall effect (QHE) such as the observation of the integral
QHE in quasi three-dimensional systems, the magnetic field dependence of
the energy gaps in the fractional QHE and the absence of even-denominator
quantization in the lowest Landau level.

INTRODUCTION

The occurrence of exact integral quantization in a process as
"chaotic" as electrical transport in macroscopic, inevitably disordered
specimens unaffected by the irregularities of their geometrical shape and
untidiness of their electrical contact pads must come as a surprise. The
sudden unfolding of an abundance of rational quantum numbers from
similarly inconspicuous materials, seemingly uncorrelated with any
inherent counting scheme, represents a true enigma. Yet, these are
exactly the hallmarks of the integral[1] and fractional[2] quantum Hall
effects. Their startling properties have attracted the attention of
experimentalists and theorists alike.

The past years have seen a tremendous amount of activities in
precise experimental measurements establishing the accuracy of the
quantization and in probing deeper into the distribution of electronic
states in high magnetic fields while complex theories and extensive
numerical schemes have been developed to assess the origin of the
experimental observations. Most developments until 1986 are
authoritatively reviewed in a recent monograph[3].

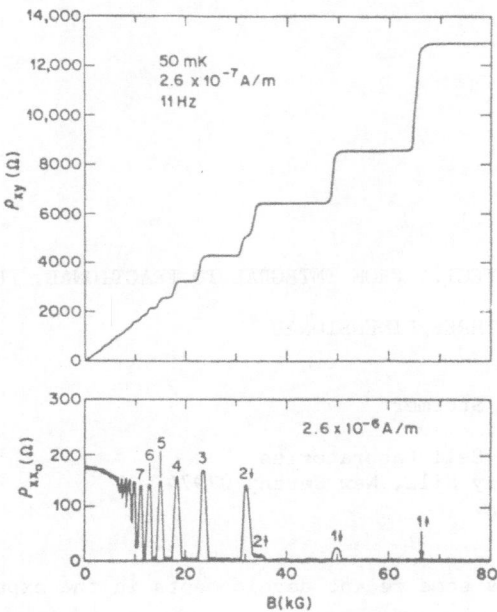

Fig. 1. The integral quantum Hall effect (IQHE)[6]. Diagonal resistivity
(ρ_{xx}) and Hall resistance (ρ_{xy}) are shown.

These lecture notes describe more recent experimental progress. In
the integral quantum Hall effect the focus rests on the discovery of
quantization in systems that are not two-dimensional[4]. The section on
the fractional quantum Hall effect concerns itself with recent, more
systematic experiments on extremely high-quality two-dimensional (2D)
electron systems[5] which allow contact to be made with some of the
theoretically derived quantities.

INTEGRAL QUANTUM HALL EFFECT

The essence of the experimental observations termed the IQHE are
quickly stated. At low temperatures and in high magnetic fields the Hall
resistance ρ_{xy} of a two-dimensional electron system is quantized to $\rho_{xy} =$
h/ie^2, i = 1,2,... to an accuracy as high as a few parts in 10^8, Fig. 1.[6]
Quantization in ρ_{xy} and concomitant loss of resistivity ρ_{xx} are
intimately associated with the singular nature of the density of states
of a two-dimensional system in a magnetic field, B.[7] Under its
influence, the continuous energy spectrum of the electronic system
transforms into a sequence of highly degenerate Landau levels separated

by the cyclotron energy $\hbar\omega_c$ each having a degeneracy of $S = 2eB/h$. Lifting of the twofold spin degeneracy results in a sequence of pairs of singularities of degeneracy $s = eB/h$. For a fixed electron density n and a particular field value B this sequence of levels is occupied to a filling factor $\nu = n/s$.

Since electronic transport is governed by the density of states at the fermi energy, intuition tells that the transport coefficients at integral filling factor $\nu = i = 1,2,3...$ may be unique. In its vicinity the characteristic features of the IQHE develop: the Hall resistance forms an extended plateau which is quantized to $\rho_{xy} = (h/e^2)/i$ and the diagonal resistivity ρ_{xx} approaches zero over the same range. The IQHE appears as a result of the stringent quantization condition of noninteracting charged particles in a two-dimensional system[8] leading to a highly singular density of states. The finite range over which this quantiztion extends is explained by localization of states induced by disorder at the fringes of the Landau levels.

In this standard derivation of the IQHE two-dimensionality of the electronic system is pivotal. The singularities in the density of states require the lack of dispersion parallel to the field direction as is the case in two-dimensional systems.

Quantum Hall Effect in Quasi Three-Dimensional Systems

Over the years the IQHE has been observed in a variety of structures[9,10] that deviate from the simple one-layer systems of the early work. Nevertheless, all those structures consist of electronically strictly two-dimensional systems lacking any dispersion and, therefore, conduction in the direction normal to the layers. Hence, they must be regarded as a stack of independent quantized Hall resistors connected in parallel. The question arises whether strict two-dimensionality is indeed a prerequisite for the IQHE. This case was tested in a system which is electrically three-dimensional by virtue of being a good, although anisotropic conductor in all spatial dimensions.[4,11]

The structure chosen was a GaAs/AlGaAs superlattice with highly penetrable barriers. Two identical samples were grown via MBE on a semi-insulating substrate for in-plane transport and on a n^+-substrate for normal transport. The dimensions of the superlattice are illustrated in Fig. 2(a). Measurements of the Hall density and Hall mobility at 4.2K yielded $n_H = 2.1 \times 10^{17}$ cm^{-3} and $\mu_H = 6400$ cm^2/Vsec, assuming a thickness of 30 x (188Å + 38Å) from TEM measurements.

Heterostructures fabricated from GaAs/(AlGa)As are well understood and well represented by a simple square-well potential.[12] The miniband structure and wavefunction of the electronic system can be calculated using a Kronig-Penney model with the parameters listed in Fig. 2(b). From the calculations, the dispersion relation shown in Fig. 2(b) is obtained. Fig. 2(a) also shows the z-dependence of the wavefunction for $k_z = 0$. The variation from maximum in the well to minimum in the barrier is less than a factor of 4, demonstrating the high degree of transparency of the barriers.

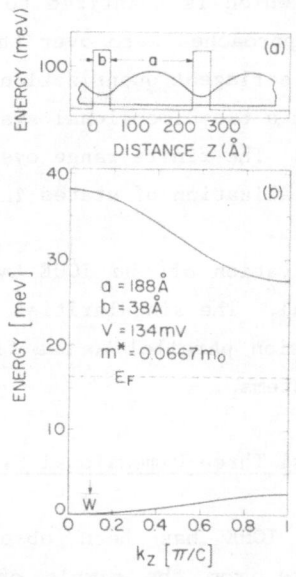

Fig. 2. (a) Kronig-Penney model and wavefunction for $k_z = 0$ for the GaAs/(AlGa)As superlattice employed. (b) Dispersion relation in the z direction calculated with the parameters shown in (a) (c = a + b).

Fig. 3 shows contours of constant energy in the k_z, k_{xy} plane for various energies up to $E_F = 16.4$ meV. The three-dimensionality of the electronic system is evident. In comparison, a strictly two-dimensional system is represented in such a plot as a set of straight lines parallel to the k_z axis, indicating the lack of dispersion in the z direction.

Measurements of the conductivity of the system showed that the superlattice not only conducted in the plane ($\sigma_{xx} = 210 \ \Omega^{-1} \ cm^{-1}$) but also perpendicular to it ($\sigma_{zz} = 0.12 \ \Omega^{-1} \ cm^{-1}$) demonstrating clearly its quasi-three-dimensional nature. The anisotropy of $\alpha \sim 10^3$ represents an

upper limit since the z-measurement is highly affected by contact
resistances. Beyond measurements of the conductivity to support the
three-dimensionality of the electronic system, it proved feasible to
determine directly the characteristic shape of the fermi surface shown in
Fig. 3 by derivative magneto-transport. The magneto-oscillations shown
in Fig. 4(a) indicate a beating pattern between two different
oscillations, close in frequency. A node is clearly visible around
$B \sim 2.3 T$ which establishes the difference between belly and neck orbit to
be $\Delta A = 1.1 \times 10^{11}$ cm^{-2} in close agreement with the calculated value of
1.4×10^{11} cm^{-2}. The non-vanishing conductivity in the z-direction at
zero-field and the observation of separated belly and neck orbits

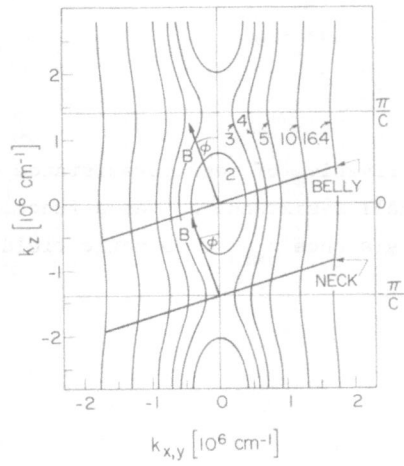

Fig. 3. Contours of constant energy (E = 1,2,3..., 16.4 meV) in the
k_{xy}-k_z plane for the band structure of Fig. 2(b). Belly and
neck represent the two extremal orbits of the fermi surface.

establishes convincingly the three-dimensionality of the electronic
system. Fig. 4b and 4c show low-temperature data on ρ_{xy} and ρ_{xx}. At
$B \sim 9 T$, the diagonal resistivity develops a clear zero resistance state
and ρ_{xx} is activated with a rather small activation energy of $\Delta/2 = 0.13$
mV, Fig. 5. Concomitant with the loss of resistivity the characteristic
plateau of the QHE appears and converges towards a value of $\rho_{xy} = h/48e^2$
to 5 parts in 10^5. These features are exactly the same as those observed
in the traditional, purely two-dimensional electron systems. This
indicates that two-dimensionality is not a prerequisite for the
observation of the QHE. The non-zero perpendicular conductivity,
parallel to the magnetic field turns out to be also field dependent. As
Fig. 6 shows, σ_{zz} oscillates in phase with ρ_{xx} and tends toward $\sigma_{zz} \to 0$ at
a field position where ρ_{xy} approaches a plateau and ρ_{xx} vanishes.

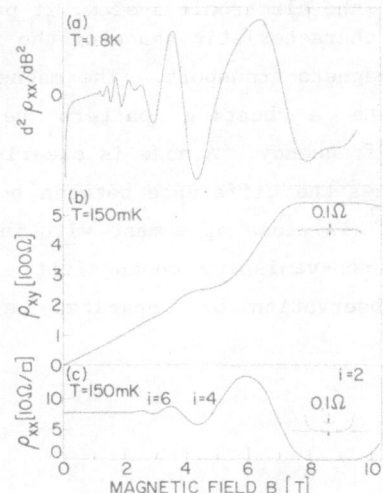

Fig. 4.   (a) Second derivative of magnetoresistance ρ_{xx} vs magnetic
field.   (b) Hall resistance ρ_{xy} as a function of magnetic field.
(c) Magnetoresistance ρ_{xx} vs magnetic field.

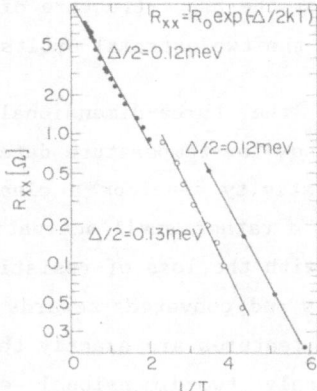

Fig. 5.   Magnetoresistance in the i = 2 minimum of ρ_{xx} versus inverse
temperature.   Data are taken on 3 different specimens in 2
different systems.

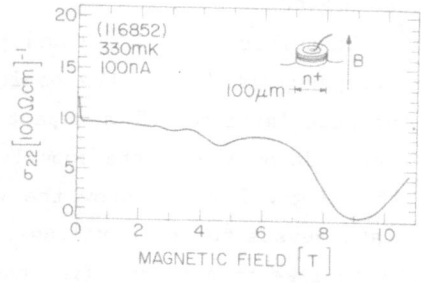

Fig. 6. Conductivity σ_{zz} normal to the layers of the superlattice as a function of magnetic field. This sample was grown on a conducting substrate.

These findings suggest that in the quantized state the resistivity and the conductivity tensor assume the form

$$\rho = \begin{bmatrix} 0 & \dfrac{h}{ie^2} & 0 \\[2mm] -\dfrac{h}{ie^2} & 0 & 0 \\[2mm] 0 & 0 & \rho_{zz} \to \infty \end{bmatrix} \qquad \sigma = \begin{bmatrix} 0 & -\dfrac{ie^2}{h} & 0 \\[2mm] \dfrac{ie^2}{h} & 0 & 0 \\[2mm] 0 & 0 & \sigma_{zz} \to 0 \end{bmatrix}.$$

One can develop an intuitive picture which is able to explain some of the experimental facts.[13] In an ideal two-dimensional system in a high magnetic field along z, the Landau levels consist of δ-functions separated by the cyclotron energy $\hbar\omega_c$. In a three-dimensional system, each quantized state in the plane is associated with a range of k_z, each having a slightly different energy. Therefore, each Landau level develops into a band. Since the z motion is not affected by the magnetic field, the shape of each band is field independent and reflects the one-dimensional density of states of the k_z dispersion in the absence of a field. Its width reproduces the zero-field miniband width W of Fig. 2b. In high magnetic fields, when $\hbar\omega_c$ exceeds W, the density of states again exhibits gaps, as in the ideal two-dimensional case, and the condition for the observation of a IQHE is fulfilled.

However, the quantum number i = 48 does not seem to have anything in common with characteristic numbers of the system. From the above model one would expect that with a total thickness L = jc there exist j different k_z states for each state in the plane. With a level degeneracy of d = eB/h in the x-y plane, a total of jeB/h states exist in each

Landau band, and standard arguments used for the strictly two-dimensional case yield $\rho_{xy} = h/jie^2$. Therefore, with i = 2 and j = 30, one expects $\rho_{xy} = h/60e^2$. This discrepancy is due to the depletion of several top and bottom layers of the superlattice. This aspect can be investigated by fabricating a gate electrode on top of the superlattice allowing us to vary the depletion depth. Figs. 7 and 8 show the variation of ρ_{xx} and ρ_{xy} in the i=2 IQHE state versus the gate voltage. Large oscillations in ρ_{xx} and concomitant step-like transitions from one quantized value to the next are being observed when the sample is depleted to increasing depth. A similar pattern results when a backside bias is applied to the substrate side. The quantum numbers vary in steps of two from i = 48 to i = 40 consistent with a reduction in the number of participating layers. From the periodicity of the superlattice, we expect the peaks in ρ_{xx} (the best defined structures) to appear at gate voltages

$$V_K + \overline{V} = \frac{e}{2\varepsilon\varepsilon_0} Nc^2[k - k_0 + \phi]^2 , \quad k, k_0 = 0, 1, 2, \ldots; \ 0 \leq \phi \leq 1 .$$

Here, $\overline{V}$ (a negative voltage) is the built-in surface potential measured from E_F, N is the 3D density, c is the period of the superlattice, k_0 is the unknown number of initially depleted layers and ϕ is a phase factor close to 1/2. The voltage differences then are linear in k:

$$V_{k+1} - V_k = \frac{e}{2\varepsilon\varepsilon_0} Nc^2[2(k - k_0 + \phi) + 1] .$$

Such an assessment leads to Fig. 9 which demonstrates the expected linearity and yields c = 207Å and k_0 = 4. The periodicity c is in good agreement with the 226Å found in TEM. Since 4 layers seem to be depleted from the substrate side of the structure in agreement with a simple

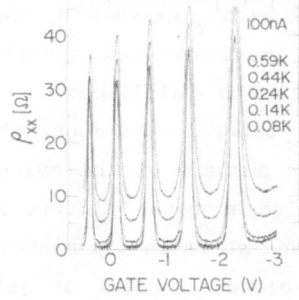

Fig. 7. Variation of ρ_{xx} at i = 2 minimum as a function of gate voltage.

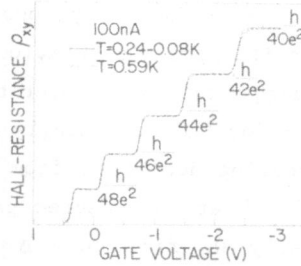

Fig. 8. Variation of ρ_{xy} at i = 2 plateau as a function of gate voltage.

calculation assuming the fermi energy is trapped in the center of the Cr-doped GaAs substrate band gap.

It is striking that the transitions between subsequent states are so well defined and that a model assuming sequential depletion of individual layers seems to account for the observations. In a superlattice such a distinction between individual layers is no longer possible due to strong interlayer tunneling. Indeed, the transition between subsequent quantized states are not due to sequential depletion of layers but are transitions at which the whole bulk participates.

In the z-direction the superlattice represents a periodic one-dimensional system with finite boundaries. Since the barriers are very transparent all electronic states within each subband extend all the way across. On decreasing the distance between the boundaries the uppermost state of each band rearranges dramatically. From being an extended state it rather abruptly accumulates in the vicinity of the boundary, and its energy is rapidly increasing. These states are sequentially peeled off the continuum, accumulate at the boundary and cross the fermi level

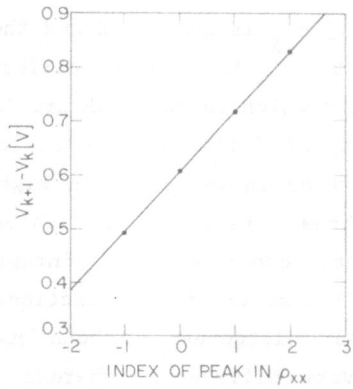

Fig. 9. Voltage difference between neighboring peaks of Fig. 7 versus peak index.

whenever the confinement is reduced by one period. It is this periodicity which is responsible for the behaviour shown in Fig. 7 and 8. This intuitive model is supported by theoretical calculations[14] which also provide an explanation for the vanishingly small activation energy. From an ideal model, neglecting depletion from the surfaces, we arrive with $B \sim 9T$ and with $W \sim 2.5$ meV at an expected gap energy of $\Delta = \hbar\omega_c - W \sim 12$ meV, while experimentally we find only $\Delta \sim 0.26$ meV. In general, broadening of the Landau level (here Landau band) is assumed to be responsible for such a reduction. From a mobility of $\mu = 6400$ cm^2/Vsec one deduces a lifetime broadening of $\Delta E \sim \hbar/\tau \sim 3$ meV, which accounts only in part for the small activation energy. The numerical calculations show clearly that in a system with finite depletion the probability of finding one of the accumulated states close to the fermi level is exceedingly high, resulting in a much reduced apparent energy gap. This fermi level tracking may be responsible to a large degree for the small activation energies observed.

In conclusion, the transport experiments on highly transparent GaAs-(AlGa)As superlattices with quasi three-dimensional transport behavior demonstrate convincingly that two-dimensionality of the electronic system is not a prerequisite for the observation of the IQHE. It is evident that there exists a much wider class of materials which are capable of exhibiting this intriguing quantum phenomenon.

FRACTIONAL QUANTUM HALL EFFECT

Phenomenologically, the fractional quantum Hall effect (FQHE), Fig. 10, resembles very much the integral quantum Hall effect (IQHE). Once again the Hall resistivity ρ_{xy} is quantized and the diagonal resistivity ρ_{xx} vanishes concomitantly. The crucial difference is found in the associated quantum numbers which in the FQHE are rational numbers $f = p/q$ occurring in the vicinity of fractional Landau level occupancy $\nu = p/q$. While the IQHE is understood in terms of the highly singular density of states of a two-dimensional system in a high magnetic field and the existence of exceptional conditions at integral filling of these singularities, no such linkage exists at fractional filling. Indeed, in spite of the superficial relatedness of both quantum Hall effects the underlying physics is very different. Whereas the IQHE rests on the quantization conditions for degenerate, non-interacting carriers in the presence of a magnetic field, the FQHE is a many-particle effect, and the result of strong carrier-carrier correlation.[3] It is presently

understood in terms of a novel electronic quantum fluid which exists
exclusively at primitive odd-denominator fractions $\nu = 1/q$ and $1 - 1/q$.
An analytic expression for the groundstate of a 2D system in the extreme
quantum limit at rational filling factor was proposed by Laughlin.[15]
This many-particle wavefunction,

$$\psi(z_1, z_2 \ldots z_N) = C_i \prod_{i<j}^{N} (z_i - z_j)^q \exp\left[-\frac{1}{4} \sum_k^N |z_k|^2 \right] ,$$

with built-in pair correlation, presently forms the basis for most
theoretical models of the FQHE. $z = x - iy$ denotes the position of each
electron in units of the magnetic length $\ell_o = (eB/h)^{\frac{1}{2}}$ and C is a
constant. The exponent q is an integer and related to the filling factor
by $\nu = 1/q$.

Laughlin's wavefunction has the following properties:

1)   It describes a state only at filling factor $\nu = 1/q$, where q
      is an integer. Assuming electron/hole symmetry, a case can

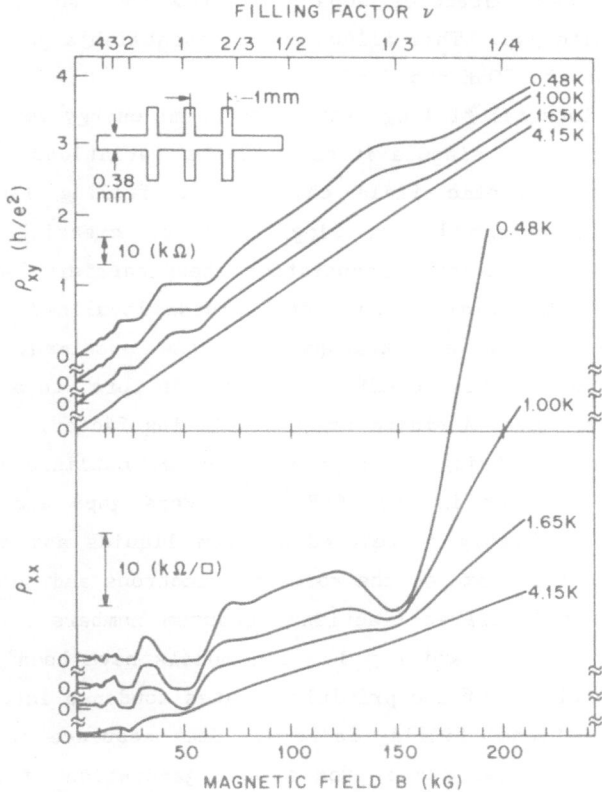

Fig. 10.   First indications[2] of quantum Hall effect at fractional Landau
level filling. Minima develop in ρ_{xx} at $\nu \sim 1/3$ and $\nu \sim 2/3$.
concomitmant with the development of plateaus in ρ_{xy}.

also be made for $v = 1 - 1/q$.

2)   It is antisymmetric only for odd q hence only odd denominators are allowed.

3)   Its pair correlation function suggests it to be a novel quantum-fluid rather than a Wigner solid for $q \lesssim 10$.

4)   The elementary excitations are separated from the groundstate by a finite gap.

5)   These quasi-particle excitations have fractional charge $e^* = e/q$.

6)   The quantum-fluid is incompressible and has no low-lying excitations. Hence, it flows with no dissipation at T = 0.

7)   For $q \gtrsim 10$, the quantum liquid is expected to crystallize into a Wigner solid.[16]

Central to the observation of a QHE at fractional filling factor is the existence of quasi-particles separated from the many-particle ground state by a finite gap. This allows one to establish a phenomenological analogy between the IQHE and FQHE.

At exact integral filling factor the fermi energy resides in the gap between Landau level singularities. Slight variations of the carrier density or the magnetic field changes the filling factor promoting electrons to the next higher singularity or creating holes in the singularity below. At low concentration these carriers become localized due to residual potential fluctuations. These localized carriers do not contribute to the current transport in the specimen, leading to a constancy of the transport coefficients and, in turn, to a plateau in ρ_{xy} and vanishing ρ_{xx} over a finite range of filling factor.

An equivalent string of arguments can be outlined explaining the appearance of plateaus in the FQHE.[15,17] Here gaps are the result of condensation into highly correlated quantum liquids and quasi-particles of fractional charge take on the roles of electrons and holes.

In order to interpret fractional quantum numbers other than those primitive ones $f = 1/q$ and $f = 1 - 1/q$ models have been developed[18,19] where quasi-particles of the primitive states condense into new states at different but rational filling factors. Each sequence of such daughter states becomes parental states for a next generation of ground states. The conditions that a higher order state is favored is given by a continued fraction

$$\nu = \cfrac{1}{q + \cfrac{\alpha_1}{p_1 + \cfrac{\alpha_2}{p_2 + \ldots}}}$$

where q is odd, the p_i's are even and α_i is 0 or ± 1.

Theoretical models based on Laughlin's wavefunction are greatly augmented by numerical few-particle calculations which lent strong support to the analytical results.

Recent Experiments on the Fractional Quantum Hall Effect

Early experiments on the FQHE were largely concerned with its unique qualitative characteristics and its phenomenology.[20-22] With the advent of two-dimensional systems of increasingly higher quality, it has now become possible to conduct more systematic studies intended to discern between various theoretical concepts and to make direct contact with theoretically predictable quantities.

Observations of a phenomenon as fragile as the FQHE with a characteristic energy scale of a few degrees K require materials in which the carrier scattering time τ is exceedingly long so that homogeneous broadening of the electronic states $\hbar/\tau$ is negligible. Modulation-doped GaAs-(AlGa)As heterostructures have been very successful in this respect. They presently provide the best material system in which to study the FQHE. Over the past decade mobilities, $\mu = em^*/\tau$ (m^* = effective carrier mass), of modulation-doped materials have consistently improved[23] reaching $\mu = 5 \times 10^6$ cm^2/Vsec, see Fig. 11. With these extremely high quality two-dimensional systems, it is becoming possible to attempt quantitative experiments on the FQHE.

Activation Energies. In analogy to the IQHE where activation energy measurements of ρ_{xx} (or σ_{xx}) reveal the size of the associated Landau gap $\hbar\omega_c$, activation energy measurements on the FQHE provide a measure for the energy gap above the condensed ground state. Such experiments have recently been performed by various groups,[5,24-26] mostly on the strongest of the FQHE states at $\nu = 1/3$ and 2/3. Fig. 12 summarizes data from systematic experiments at the Francis Bitter National Magnet Lab in fields up to 30T. The crosses are from earlier samples[25] of mobility 0.4 $\times 10^6$ cm^2/Vsec $\lesssim \mu \lesssim 1 \times 10^6$ cm^2/Vsec whereas the solid circles[5] are from the presently highest mobility specimen of $\mu = 5.0 \times 10^6$ cm^2/Vsec. In all cases the energy gap was obtained from fitting the temperature

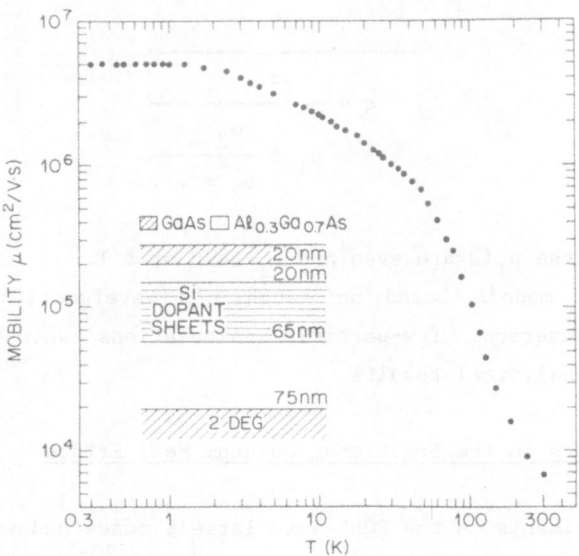

Fig. 11. Temperature dependence of mobility in modulation-doped GaAs-(AlGa)As heterojunctions with areal density $n = 1.6 \times 10^{11}$ cm^{-2}. The low temperature mobility reaches $\mu = 5 \times 10^6$ cm^2/Vsec.

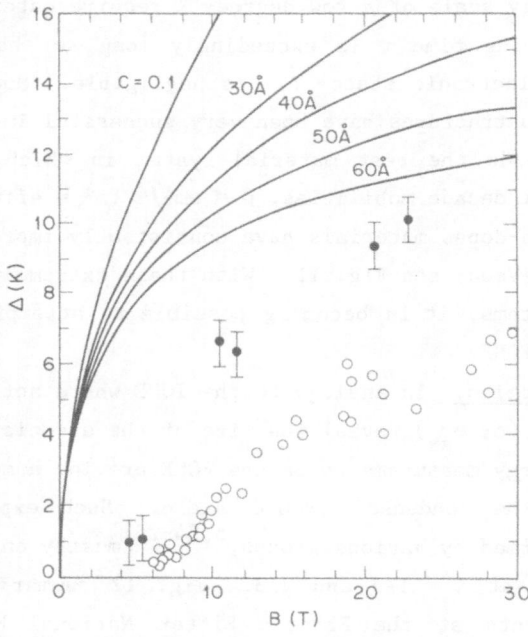

Fig. 12. Results of activation energy measurements in the FQHE. Open circles indicate a collection of data from samples with $0.4 \times 10^6 \lesssim \mu \lesssim 1 \times 10^6$ cm^2/Vsec. Circles are from $\mu = 5 \times 10^6$ cm^2/Vsec sample. Curves are described in text.

dependence of ρ_{xx} at the minimum to an expression $\rho_{xx} = \rho_o \exp(-\Delta/2T)$.
Fig. 12 demonstrates clearly a correlation between activation energy and
sample quality. It also contains four theoretical dependences for Δ.
The line designated $C = 0.1$ represents the result, $\Delta = C\, e^2/\varepsilon \ell_o$ of most
numerical calculations[27-30] on the size of the field dependent energy gap
for the $\nu = 1/3$ state. These idealized calculations neglect finite
thickness of the wavefunction perpendicular to the plane and mixing of
higher order Landau levels.[31] Line $C = 0.1$ also neglects the contribution from a roton-minimum[32] in the dispersion relation of the quasi-
particles which, if assumed to contribute to transport experiments,
reduces C by about 20%. The lines designated 30Å through 60Å contain
corrections for finite wavefunction thickness.[33] The differences between
theory and experiments are considerable. In particular, the set of lower
mobility data clearly indicates a finite threshold for Δ of about 6T and
obviously does not follow the $\sqrt{B}$ behavior expected from the ideal theory.
Furthermore there exists a large discrepancy between theory and experiment regarding the magnitude of Δ. Even with an estimated wavefunction
extent of 100Å these discrepancies cannot be resolved, particularly in
the low field region. They are presently believed to result from a
broadening of the quasi-particle band due to disorder leading to a
roughly field independent reduction of the gap size. Such models can
successfully account for the threshold behavior.[34,35] They are supported
by our findings of increasing Δ with increasing mobility as seen in Fig.
12. Yet, even at mobilities as high as 5×10^6 cm^2/Vsec, theories
neglecting disorder are unable to describe the experimental data.
Although there exist only four data points for the high mobility samples,
it appears that this specimen again has a field threshold. Its gap
dependence is approximately described by a wavefunction extent of 40Å and
a fixed level broadening of about 6K (vertical shift by 6K). As sample
mobilities continue to rise one will be able to realize experimental
conditions which approach the idealized theoretical model structures and
allow for direct comparison. At the same time one would hope that theory
can incorporate disorder in a quantitative way with no adjustable
parameters.

Odd-Denominator Rule. The observation of exclusively odd-denominators in the FQHE has been termed the odd-denominator rule. It represents
an important experimental fact which any emerging theory has to explain.
Laughlin's wavefunction and the derived hierarchy of ground states
describe states at exclusively odd-denominator filling factor. This
restriction is a direct consequence of the requirement for anti-symmetry
of the wavefunction. However, this cause/effect relationship cannot be

taken as a justification for the odd-denominator rule. Since the theory does not contain the filling factor as a continuous parameter it cannot make statements about the absence of condensed ground states at even-denominator fractions or, at least, their absence in transport experiments.

During the past years there have been repeated indications for the existence of even denominator fractions in the FQHE.[36,37] Minima in ρ_{xx} were observed at $\nu \sim 3/4$ and $\nu \sim 1/2$ and $\sim 3/2$. More recently similar structure became apparent in the second Landau level[35,38] at $\nu = 2 + 1/2$, $\nu = 1/4$ and $\nu = 2 + 3/4$. None of these even-denominator fractions are yet confirmed in the sense that quantization in the Hall resistance to $\rho_{xy} = (h/e^2)/f$ has been observed with f being an even-denominator fraction. In order to scrutinize the odd-denominator rule, experiments were conducted in the extreme quantum limit $\nu < 1$ of extremely high mobility samples in the vicinity of $\nu = 1/4$, 1/2 and 3/4. Fig. 13 through 15 show data on ρ_{xx} for three different speciments. Fig. 13, from one of our best samples exemplifies the complexity into which the FQHE has developed in recent years. The last integral Hall plateau exists at $\nu = 1$. None of the structures at higher fields are expected in the noninteracting electron model of the IQHE. Minima at filling factors as high as $\nu = 6/13$ are being resolved and quantization in ρ_{xx} to better than 1% has now been observed for quantum numbers as high as f = 5/7.

At present in the extreme quantum limit structure in ρ_{xx} is apparent at

$$\nu = 1/3,\ 2/3$$
$$\nu = 1/5,\ 2/5,\ 3/5,\ 4/5$$
$$\nu = 2/7,\ 3/7,\ 4/7,\ 5/7$$
$$\nu = 4/9,\ 5/9$$
$$\nu = 4/11,\ 5/11,\ 6/11$$
$$\nu = 6/13$$

For all highlighted fractions ρ_{xy} is quantized to the associated quantum number to better than 1%.

Inspecting the vicinity of $\nu = 1/2$ in Fig. 13 one notices a broad basin which was earlier considered as indicating the beginning of a minimum in ρ_{xx} similar to those developing at odd-denominators. With increasing sample quality a vast number of formerly unobservable odd-denominator fractions were resolved without much development of $\nu = 1/2$. The shape of the broad depression stands in sharp contrast to pronounced minima in ρ_{xx} at odd-denominator filling which develop plateaus in ρ_{xy} as they become more distinct. The structure around $\nu = 1/2$ was never found to be associated with any inflection in ρ_{xy}. To

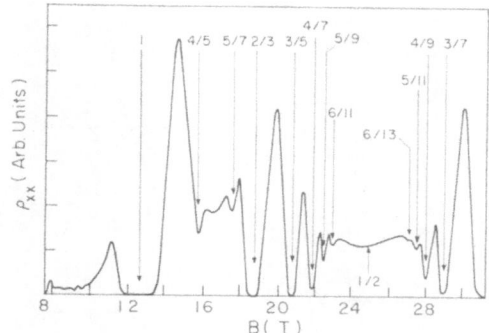

Fig. 13. ρ_{xx} data of high-mobility sample in the extreme quantum limit $\nu < 1$. Fractional filling factors are indicated.

the contrary, the Hall effect assumes the straight classical line in its vicinity. With two sequences of the odd-denominator hierarchy $\nu = (m+1)/(2m+1)$ and $\nu = m/(2m+1)$, $m = 1,2,3...$ expanding towards $\nu = 1/2$ one wonders whether the weak depression around $\nu = 1/2$ may not be caused by higher level daughter states of the same sequences that blend into one

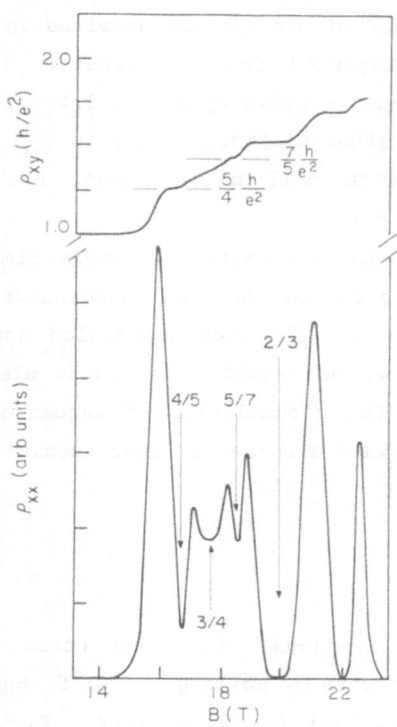

Fig. 14. Diagonal resistivity and Hall resistance data in the vicinity of $\nu \sim 3/4$.

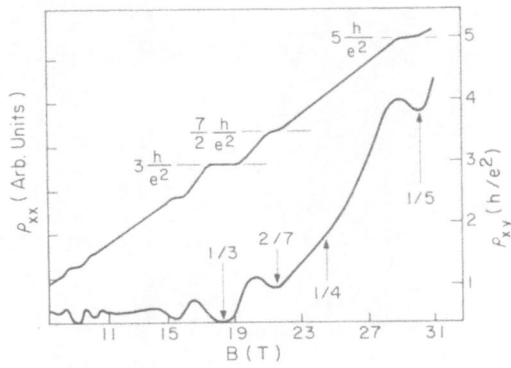

Fig. 15. Diagonal resistivity and Hall resistance data in low-density
sample in the vicinity of $\nu \sim 1/4$.

another. Such an observation can now also be made in the vicinity of
$\nu = 3/4$, Fig. 14. With the discovery of $\nu = 4/5$ and $5/7$ states and
associated ρ_{xy} quantization the first representatives of sequences
converging towards the next even-denominator fraction ($\nu = 3/4$) have been
resolved. They are given by $\nu = (3m+1)/(4m+1)$ and $\nu = (3m-1)/(4m+1)$.
Finally, the neighborhood of $\nu = 1/4$ was examined in a much lower density
sample, Fig. 15. As expected from the general electron-hole symmetry
within each Landau level the vicinity of $\nu = 1/4$ reflects the vicinity of
$\nu = 3/4$. In spite of rather distinct minima in ρ_{xx} around $\nu = 2/7$ and
$\nu = 1/5$ accompanied with Hall quantization, no signs of a similar
phenomenon appear at $\nu = 1/4$.

From our studies in the extreme quantum limit we conclude that
presently there is no evidence for even-denominator quantization in the
lowest Landau level $\nu < 1$. The wide depression apparent in ρ_{xx} at $\nu = 1/2$, $1/4$ and $3/4$ may be the result of enhanced electron correlation in
their vicinity due to the accumulation of sequences of odd-denominator
fractions converging toward even-denominator accumulation points.

ACKNOWLEDGEMENTS

Most of the work reported in these notes result from a very
enjoyable collaboration with my colleagues, G. S. Boebinger, A. M. Chang,
J. P. Eisenstein, D. C. Tsui and R. Willett. Kirk Baldwin's excellent
technical support was invaluable. The extremely important high-quality
MBE materials were provided by A. Y. Cho, J. H. English, A. C. Gossard,
J. C. M. Hwang and G. Weimann. A large portion of the experiments were
performed at the Francis Bitter National Magnet Lab, Cambridge, Mass.

REFERENCES

1. K. v. Klitzing, G. Dorda and M. Pepper, Phys. Rev. Lett. 45:494 (1980).

2. D. C. Tsui, H. L. Stormer and A. C. Gossard, Phys. Rev. Lett. 48:1559 (1982).

3. The Quantum Hall Effect, R. E. Prange and S. M. Girvin, eds., (Springer-Verlag, New York, 1987).

4. H. L. Stormer, J. P. Eisenstein, A. C. Gossard, W. Wiegmann and K. Baldwin, Phys. Rev. Lett. 56:85 (1986).

5. R. Willett, H. L. Stormer and D. C. Tsui, Proc. of the VII Intl. Conf. on the Electronic Properties of Two-Dimensional Systems (EP2DS-VII), Santa Fe, N.M. July 1987.

6. M. A. Paalanen, D. C. Tsui and A. C. Gossard, Phys. Rev. B25:5566 (1982).

7. T. Ando, A. Fowler and F. Stern, Rev. Mod. Phys. 54:437 (1982).

8. R. B. Laughlin, Phys. Rev. B23:5632 (1981).

9. T. Haavasoja, H. L. Stormer, D. J. Bishop, V. Narayanamurti, A. C. Gossard and W. Wiegmann, Surf. Sci. 142:294 (1984).

10. M. Razeghi, J. P. Duchemin, J. C. Portal, L. Dmowski, G. Remeni, R. J. Nicholas and A. Briggs, Appl. Phys. Lett. 48:721 (1986).

11. H. L. Stormer, J. P. Eisenstein, A. C. Gossard, K. W. Baldwin, J. H. English, Proc. 18th Intl. Conf. Phys. Semicon., Olaf Engstrom, ed., (World Scientific, Singapore, 1987) p. 385.

12. For example, L. Esaki in Proc. of the 17th Intl. Conf. on the Physic. of Semiconductors, San Francisco, 1984, J. D. Chadi and W. A. Harrison, eds., (Springer, New York, 1984) p. 473.

13. M. Ya. Azbel, Phys. Rev. B26:3430 (1982).

14. S. E. Ulloa and G. Kirczenow, Phys. Rev. Lett. 57:2991 (1986).

15. R. B. Laughlin, Phys. Rev. Lett. 50:1395 (1983).

16. P. K. Lam and S. M. Girvin, Phys. Rev. B30:473 (1984).

17. B. I. Halperin, Helv. Phys. 56:75 (1983).

18. F. D. M. Haldane, Phys. Rev. Lett. 51:605 (1983).

19. B. I. Halperin, Phys. Rev. Lett. 52:1583 (1984).

20. H. L. Stormer, A. Chang, D. C. Tsui, A. C. Gossard and J. C. M. Hwang, Phys. Rev. Lett. 50:1953 (1983).

21. E. E. Mendez, M. Heiblum, L. L. Chang and L. Esaki, Phys. Rev. B28:4886 (1983).

22. A. M. Chang, P. Berglund, D. C. Tsui, H. L. Stormer and J. C. M. Hwang, Phys. Rev. Lett. 53:997 (1984).

23. J. H. English, A. C. Gossard, H. L. Stormer, and K. W. Baldwin, Appl. Phys. Lett. 50:1826 (1987).

24. A. M. Chang, M. A. Paalanen, D. C. Tsui, H. L. Stormer and J. C. M Hwang, Phys. Rev. B28:6113 (1983).

25. G. S. Boebinger, A. M. Chang, H. L. Stormer and D. C. Tsui, Phys. Rev. Lett. 55:1606 (1985).

26. J. Wakabayashi, S. Kawaji, J. Yoshiro and H. Sakaki, J. Phys. Soc. Jap. 55:1357 (1986).

27. D. Levesque, J. J. Weiss and A. H. MacDonald, Phys. Rev. B30:1056 (1984).

28. F. D. M. Haldane and E. H. Rezayi, Phys. Rev. Lett. 54:237 (1985).

29. R. Morf and B. E. Halperin, Phys. Rev. B33:2221 (1986).

30. W. P. Su, Phys. Rev. B32:2617 (1985).

31. D. Yoshioka, J. Phys. Soc. Jap. 53:3740 (1984).

32. S. M. Girvin, A. H. MacDonald and P. M. Platzman, Phys. Rev. Lett. 54:581 (1985).

33. F. C. Zhang and S. Das Sarma, Phys. Rev. B33:2903 (1986).

34. A. H. MacDonald, K. L. Liu, S. M. Girvin and P. M. Platzman, Phys. Rev. B33:4014 (1986).

35. A. Gold, Europhys. Lett. 1:241 (1986).

36. G. Ebert, K. v. Klitzing, J. C. Maan, G. Remenyi, C. Probst, G. Weimann and W. Schlapp, J. Phys. C. 17:L775 (1984).

37. R. G. Clark, R. J. Nicholas, A. Usher, C. T. Foxon and J. J. Harris, Surface Science 170:141 (1986).

PHOTONS, ROTONS AND FRACTIONALLY-CHARGED VORTICES IN THE

QUANTUM HALL EFFECT

S. M. Girvin*

Surface Science Division
National Bureau of Standards
Gaithersburg, MD 20899 USA

INTRODUCTION

The fractional quantum Hall effect (FQHE) is a remarkable many-body phenomenon occurring in the two-dimensional electron gas (2DEG) at low temperatures in a high magnetic field. As we learned in the lecture by Robert Laughlin at this conference, the experimental manifestations of the effect follow from the existence of an excitation gap and are quite similar to those of the integer case, but the gap arises for rather different reasons: The FQHE is intrinsically a many-body phenomenon associated with Coulomb correlations, while the integer effect is primarily a one-body effect associated with localization. The goal of this lecture is to explore further the FQHE with emphasis on the nature of the collective excitation modes and the origin of the gap. It turns out that a remarkable amount of progress can be made on this problem by drawing on analogies with superfluidity in ^{4}He films.

Let us begin our explorations by reviewing the generic features of the two types of collective excitations in many-particle systems. The first type is the hydrodynamic mode characterized by $\omega\tau \ll 1$ where ω is the mode frequency and τ is a collision time for the particles. Well-defined hydrodynamic modes exist at long wavelengths in many systems despite the damping one might expect from the short collision time of the particles. This is because of the existence of conservation laws for particle number, energy and momentum[1]. A prototypical example would be sound waves in liquid argon. The speed of sound can be computed knowing the compressibility. The latter is typically computed by assuming that collisions are so frequent that the system remains in local equilibrium at all times and so one may just use the adiabatic compressibility of the equilibrium state.

The second regime for collective modes is collisionless and characterized by $\omega\tau \gg 1$. A prototypical example would be sound waves in crystalline sapphire at low temperatures. In this case the crystal translation symmetry causes the frequent collisions among the particles to develop a coherence which in turn leads to a well-defined set of normal modes. Here τ represents the lifetime of the normal mode (due to weak anharmonicities, say) rather than the characteristic collision time for

*Present and permanent address: Department of Physics, Swain Hall West, Indiana University, Bloomington, IN 47405 USA.

individual particles. When expressed in terms of the normal-mode
coordinates, the Hamiltonian becomes that for a set of nearly independent
harmonic oscillators. Recognizing this fact, it is then very easy to pass
to the case of quantum mechanics and realize that this type of collective
excitation has the characteristic of being an eigenstate of the system
Hamiltonian.

SUPERFLUID HELIUM

We are now ready to introduce the our main paradigm, superfluid liquid
helium. A liquid of massive particles like argon is not very quantum-
mechanical[2] and hence is best described classically or semiclassically. If
we insist on describing it by its quantum Hamiltonian, we will find a huge
density of low-lying single-particle excited states with very little
collective nature. We must rely on a *statistical* description to obtain the
hydrodynamic modes. Unlike liquid argon, helium is highly quantum-
mechanical (due to the light mass). It turns out that there is a very *low*
density of excited states and these are collective density-wave eigenstates
of the system Hamiltonian. In this sense, the collective excitations in
liquid helium are analogous to the long-lived phonons in sapphire. It is
this unusual smallness of the density of excited states which accounts for
the superfluidity in ^{4}He. Unlike the liquid, sapphire has a shear strength
and so supports transverse as well as longitudinal phonons (and hence is not
a superfluid).

Having addressed the general question of collective modes we can now
ask how all of this applies to the 2DEG. We are dealing with particles in
the extreme quantum limit but there are two properties not found in helium:
the electrons are charged (and coupled to a magnetic field) and they are
fermions rather than bosons. The effect of the magnetic field is to induce
a Lorentz force perpendicular to the particle velocities which allows the
existence of a quasi-transverse mode even if the plasma is fluid. This so-
called lower-hybrid mode[3] is the analog of transverse phonons in a solid.
The upper-hybrid mode is quasi-longitudinal and occurs near the classical
cyclotron frequency[4]. Thus we expect to see not one, but two modes, which
will be eigenstates of the quantum Hamiltonian. As we shall see later, the
magnetic field has the additional effect of ameliorating some of the
difficulties associated with Fermi statistics and so, except for the fact
that there are two modes, the results are very analogous to superfluid
helium. Let us put off discussion of the role of the Fermi statistics until
we have reviewed the physics for the boson case of helium.

Feynman has argued[5] on quite general grounds that because of the Bose
statistics, there is a remarkably low density of excited states above the
ground state of helium. The only low-lying excited states are collective
phonon-like density oscillations. There are no low-lying single-particle
excitations, a fact which, as mentioned above, helps account for the
remarkable transport properties of the superfluid.

Following Feynman, let us assume that we know the exact ground state
wave function $\Psi(r_1, \ldots, r_N)$ and investigate the properties of a variational
excited state representing a density wave at wave vector k:

$$\Phi_k = N^{-1/2} \rho_k \Psi(r_1, \ldots, r_N) ,$$

(1)

where ρ_k is the Fourier transform of the particle density:

$$\rho_k \equiv \sum_{j=1}^{N} e^{ik \cdot r_j} = \int d^3 r \; e^{ik \cdot r} \; \rho(r) \; . \tag{2}$$

To see that Φ_k represents a density-wave excited state, consider some particular configuration of the particles $\{r_1,\ldots,r_N\}$. If this configuration is relatively uniform in density then ρ_k will be very small (for $k \neq 0$). This will make $|\Phi_k|^2$ small and hence the system has a low probability of being found in this configuration. If on the other hand the density for this configuration is modulated at wave vector k, then ρ_k and hence $|\Phi_k|^2$ will be correspondingly large, meaning that this is a likely configuration. We can make an analogy with the simple harmonic oscillator by viewing ρ_k as a generalized coordinate for the system. The exact wave function of the first excited state of an oscillator is equal to the ground-state wave function multiplied by the coordinate: $\psi_1(x) \sim x \, \psi_0(x)$. Consistent with this interpretation is the fact that for an elastic continuum, eq.(1) gives the exact phonon excited states.

The variational estimate for the excited-state energy is

$$\Delta(k) = f(k)/s(k) \; , \tag{3}$$

where

$$f(k) = N^{-1} \langle \Psi | \rho_k^\dagger \, (H - E_0) \, \rho_k | \Psi \rangle \; . \tag{4}$$

with H being the Hamiltonian, and E_0 the ground state energy. The norm of the state is given by

$$s(k) = N^{-1} \langle \Psi | \rho_k^\dagger \, \rho_k | \Psi \rangle \; . \tag{5}$$

In principle, eqs.(4-5) require us to evaluate 10^{23}-dimensional integrals. Compounding this difficulty is the fact that we do not actually know the ground state wave function Ψ as we originally assumed at the beginning of this discussion. Fortunately two miraculous simplifications occur. The first is that $s(k)$ is the mean square density fluctuation at wave vector k which is just the static structure factor directly measured in neutron scattering experiments[5,6]. Hence we can obtain $s(k)$ from experiment. The second is that $f(k)$ can be rewritten as a double commutator[6]:

$$f(k) = \frac{1}{N} \langle \Psi | \rho_k^\dagger \, [H, \rho_k] | \Psi \rangle = \frac{1}{2N} \langle \Psi | [\rho_k^\dagger, [H, \rho_k]] | \Psi \rangle = \frac{\hbar^2 k^2}{2m} \; . \tag{6}$$

The last equality above follows from the fact that only the kinetic energy fails to commute with the density. Thus $f(k)$ is nothing more than the oscillator strength whose universal value is given by the oscillator strength sum rule[6]. Substituting eq.(6) into eq.(3) yields the famous Bijl-Feynman formula[5,6]:

$$\Delta(k) = \frac{\hbar^2 k^2}{2m} \frac{1}{s(k)} \; . \tag{7}$$

The essence of this expression is that it gives dynamical information, namely the excitation energy, solely in terms of the static properties of the ground state. We can interpret eq.(7) as saying that the collective excitation energy is just the single-particle energy $\hbar^2 k^2/2m$ with a mass renormalization factor $s(k)$ due to correlations in the particle motion.

The Bijl-Feynman expression works rather well in describing the experimentally determined collective mode in helium. Because $s(k)$ vanishes linearly with k for small k, so does $\Delta(k)$ and the slope of Δ vs. k gives the speed of sound. At intermediate values of k, $s(k)$ has a peak associated with the characteristic nearest-neighbor distance among particles in the liquid (short-range solid-like order). This leads to a dip in the value of $\Delta(k)$ known as the roton minimum. These qualitative features appear in the experimental spectrum although the predicted roton energy is about a factor of two too large[5,6]. Improvement of the variational wave function by inclusion of so-called 'backflow' corrections eliminates this difficulty[5,6]. It will turn out that these corrections are unnecessary in the FQHE[7].

An alternative interpretation of eq.(7) is that this expression is equivalent to making the single-mode-approximation (SMA). That is, we are assuming that there is a single excited state (namely $\rho_k|\Psi\rangle$) which absorbs all of the oscillator strength. This works well in the case of a bose-condensed system because there are no low-lying single-particle excitations to steal oscillator strength from the collective mode. This interpretation will prove useful in applying analogous ideas to the collective modes in the FQHE.

PHONONS AND ROTONS IN THE FRACTIONAL QUANTUM HALL EFFECT

Now that we have some understanding of how helium works, let us consider the case of the two-dimensional electron gas in a high magnetic field. As remarked earlier, we expect to see two modes characteristic of a charged plasma in a field. Before finding these we must dispose of the question of the Fermi statistics. We know that for jellium (in the absence of a magnetic field) there is a large density of particle-hole-pair excited states associated with the existence of a sharp Fermi surface[6,8]. These single-particle excitations form a continuum because they can have any kinetic energy down to zero (for $k < 2k_F$). In addition there is a collective mode which is just the well-known plasma oscillation. In general we do not expect the SMA to work well for Fermi systems because the single-particle continuum steals oscillator strength from the collective mode. However in the limit of $k \to 0$, there is no phase space for the particle-hole pairs and the plasmon absorbs 100% of the oscillator strength. Hence the SMA is exact in this limit. In the limit of large k the SMA puts the collective mode at the centroid of the (very broad) single-particle continuum. Thus the SMA is not totally unreasonable for Fermi systems and indeed has proved to be a useful approximation for estimating electron gas correlation self-energies[9,10]. We do not expect however that the SMA will work quite so nicely as it does for the case of bosons.

What happens when we introduce a magnetic field into the problem? The most important thing is that the kinetic energy is 'quenched'. That is, the single-particle eigenstates become Landau levels[11] with the nth level having energy $E_n = \hbar\omega_c(n+1/2)$. In addition to defining a characteristic energy, the magnetic field also defines a characteristic length $\ell = (\hbar c/eB)^{1/2}$. In the extreme quantum limit where the lowest Landau level has fractional filling factor v, the average particle spacing is on the order of $\ell/v^{1/2}$. In the limit $\hbar\omega_c \gg v^{1/2}e^2/\epsilon\ell$ (where ϵ is the dielectric constant of the medium) we expect the kinetic energy to dominate over the Coulomb energy and hence to be able to classify the excitations according to their kinetic energy. Because the kinetic energy comes in discrete lumps of $\hbar\omega_c$, there is no possibility of a single-particle continuum. This destruction of the Fermi surface by the magnetic field thus removes the difficulties associated with the Fermi statistics and renders the problem much more like the boson case. The excitations which carry an electron from the zeroth to the first Landau level have energy near $\hbar\omega_c$ and correspond to the cyclotron or upper-hybrid mode of plasma physics. Excitations to still higher Landau levels will be harmonics of this. The cyclotron mode is discussed in some detail in the lectures by Catherine Kallin[12] at this conference. The intra-Landau-level excitation corresponds to the magnetophonon or lower-hybrid mode and is the primary low-energy mode of interest in the FQHE.

The upshot of all this is that we can make a very simple and useful generalization of Feynman's SMA approximation in which we postulate there to be one collective mode *per Landau level*. To accomplish this we consider the following variational wave function for the nth collective mode with energy near $n\hbar\omega_c$:

$$\Phi^n_k = N^{-1/2} \, \rho_{k;n0} \, \Psi \; . \tag{8}$$

This is analogous to the helium state in eq.(1) except that we have replaced ρ_k by $\rho_{k;n0}$. The latter is that projection of the density operator which takes particles from the zeroth to the nth Landau level. It is clear that Φ^n_k is an eigenstate of the kinetic energy with eigenvalue $n\hbar\omega_c$. Our remaining task is to show that it is also very nearly an exact eigenfunction of the full Hamiltonian. It has been demonstrated[13] for the case of a full Landau level that Φ^1_k is identical to the magnetoplasmon/exciton mode obtained by summation of ladder diagrams[14]. For partially filled Landau levels, Φ^1_k is a non-perturbative improvement on the ladder diagram method[13]. We will concentrate here on the intra-Landau-level mode. Let us simplify the notation in (8) for the case $n = 0$ by writing

$$\Phi_k = N^{-1/2} \, \bar{\rho}_k \, \Psi \; , \tag{9}$$

where $\bar{\rho}_k \equiv \rho_{k;00}$. One might expect that it is rather difficult to compute $\bar{\rho}_k$. It turns out however that a remarkable simplification occurs because of the analyticity properties of the wave functions in the lowest Landau level. In the symmetric gauge the mth lowest-Landau-level single-particle eigenstate of the kinetic energy is:

$$\phi_m(z) = (2\pi 2^m m!)^{-1/2} \, z^m \; ; \qquad m = 0,1,2,3,\ldots \tag{10}$$

where $z \equiv x+iy$ is the complex number representing the particle position in the two-dimensional plane. The ϕ_m form a ladder of harmonic oscillator states with angular momentum given by the oscillator quantum number m. Normally the expression in eq.(10) would include a gaussian factor but it is convenient in the present context to lump this into the integration measure so that the inner product of two states becomes

$$\langle n|m\rangle = \int d^2z \, \exp(-|z|^2/2) \, \phi^*_n(z) \, \phi_m(z) \; . \tag{11}$$

where the magnetic length has been set to unity. The advantage of this formulation is that the eigenfunctions are now analytic in their argument. Indeed, since all states are degenerate in the lowest Landau level, *any* analytic expression yields an eigenfunction[15,16]. This is one of the key features underlying the ground state wave function discussed in the lecture by Robert Laughlin at this conference.

We may take advantage of this analyticity by using the following

$$ik \cdot r = \frac{i}{2}(k^* z + k z^*) \qquad (12)$$

where k and z are complex numbers representing the vectors k and r respectively. We see immediately that the operator z is already projected onto the subspace of the lowest Landau level since it maps analytic functions into analytic functions. Indeed z is the raising operator for the ladder of oscillator states[15,16]. It is straightforward to verify that for any states $|n\rangle$ and $|m\rangle$ in the Hilbert space of analytic functions, the operator z^* satisfies[15,16]

$$\langle m | z^* | n \rangle = \langle m | 2 \frac{d}{dz} | n \rangle \qquad (13)$$

so that the projection of z^* is

$$\overline{z^*} \equiv z^\dagger = 2 \frac{d}{dz} \, . \qquad (14)$$

Because $z^\dagger$ and z fail to commute, it turns out to be necessary to normal-order all the derivatives to the left when doing the projection[16]. Using these results it is now easy to obtain the projected density operator:

$$\overline{\rho}_k = N^{-1} \sum_{j=1}^{N} \exp\left[- ik \frac{d}{dz_j}\right] \exp\left[-i \frac{k^* z_j}{2}\right] \, . \qquad (15)$$

When projecting the Hamiltonian we can ignore the kinetic energy since it is a constant in the lowest Landau level. The potential energy is

$$V = \frac{1}{2} \int \frac{d^2 q}{(2\pi)^2} \, v(q) \sum_{i \neq j} \exp[iq \cdot (r_i - r_j)] \, . \qquad (16)$$

where $v(q)$ is the Fourier transform of the interaction potential. This yields the projected Hamiltonian

$$\overline{V} = \frac{1}{2} \int \frac{d^2 q}{(2\pi)^2} \, v(q) \, (\overline{\rho}_q^\dagger \overline{\rho}_q - \rho e^{-q^2/2}) \, . \qquad (17)$$

where ρ is the mean density. Making use of the commutation relation [which follows directly from eq. (15)]

$$[\overline{\rho}_k, \overline{\rho}_q] = (e^{k^* q/2} - e^{k q^*/2}) \overline{\rho}_{k+q} \qquad (18)$$

it is straightforward to evaluate the projected oscillator strength

$$\overline{f}(k) = \frac{1}{2N} \langle \Psi | [\overline{\rho}_k^\dagger, [\overline{V}, \overline{\rho}_k]] | \Psi \rangle \qquad (19)$$

to obtain:

$$\bar{f}(k) = \frac{1}{2} \sum_q v(q)(e^{kq^*/2} - e^{k^*q/2})$$

$$\times \ [\bar{s}(q)e^{-k^2/2}(e^{-k^*q/2} - e^{-kq^*/2}) + \bar{s}(|k+q|)(e^{+k^*q/2} - e^{+kq^*/2})] \ . \quad (20)$$

where $\bar{s}(k)$ is the projected static structure factor:

$$\bar{s}(k) = \ N^{-1} \ \langle\Psi|\bar{\rho}_k^\dagger\bar{\rho}_k|\Psi\rangle = s(k) - (1 - e^{-k^2/2}) \ . \quad (21)$$

Note that the ordinary structure factor $s(k)$ is non-zero if the lowest Landau level is filled whereas $\bar{s}$ vanishes because it is not possible to create any excitations within a filled level.

The analog of the Bijl-Feynman expression is

$$\Delta(k) = \bar{f}(k)/\bar{s}(k) \ . \quad (22)$$

Clearly evaluation of this expression is somewhat more complex than for the helium case but the spirit of the result is still the same. Eq.(22) gives a dynamical quantity solely in terms of the static properties of the ground state. The reason for the extra complication is that the intra-Landau-level portion of the oscillator strength is non-universal and depends on the details of the particle interactions. On the other hand this is just what we expect is necessary to set the scale of the collective mode energy. Since the kinetic energy has been quenched, Δ must be on the scale of the potential energy.

Let us begin our analysis of eq.(22) by examining the small k limit. The *total* oscillator strength (intra- and inter-Landau level) is $\hbar^2k^2/2m$ independent of the magnetic field. In the limit of small wave vector, Kohn's theorem[17] tells us that the cyclotron mode occurs at exactly ω_c and saturates the oscillator strength sum. Hence the projected oscillator strength given in eqs.(19-20) must vanish *faster* than k^2, and in fact it is easy to show that $\bar{f}(k) \sim k^4$. It can also be shown[7] for any homogeneous, isotropic ground state in the lowest Landau level that $\bar{s}(k) \sim k^4$. Thus eq.(22) predicts the existence of a finite excitation gap for *any* system with a liquid ground state. For the specific case of Laughlin's ground state one can take advantage of the two-dimensional one-component plasma (2DOCP) analogy[15] to make a statement about $\bar{s}$ at small wave vectors. We see from eq.(21) that the gap condition on $\bar{s}$ is equivalent to the following condition on the small-k behavior of s:

$$s(k) = k^2/2 + \ldots \quad (23)$$

This vanishing of the mean-square density fluctuations at large length scales is characteristic of the long-range forces in the 2DOCP. Eq.(23) is equivalent to the perfect screening sum rule[7,18,19] for the 2DOCP and in fact from the compressibility sum rule for the 2DOCP we obtain the exact coefficient of the next term for Laughlin's state[7,18,19]:

$$\bar{s}(k) = \frac{1-v}{8v} \ k^4 \ . \quad (24)$$

It is interesting that there is this deep connection between the incompressibility of the classical analog plasma (which describes the particle distribution only for the ground state) and the incompressibility of the quantum system (existence of an *excitation* gap).

In order to investigate $\Delta(k)$ beyond the $k \to 0$ limit we need to know the static structure factor $s(k)$. Unlike the case of helium this is not available experimentally since there are too few electrons in a typical inversion layer to make the appropriate measurement. Hence, until experimentalists can find a way around this difficulty, we are forced to rely on a model of the ground state. Fortunately Laughlin's ground state wave function has proven to be very accurate[20] and the static structure factor has been previously obtained by Monte Carlo simulation of the associated 2DOCP[21,7]. Using these results, eq. (22) has been evaluated[7] for $v = 1/3$, $1/5$ and $1/7$. The predicted excitation gap in units of the Coulomb energy scale $e^2/\epsilon\ell$ is shown in Fig. 1. As can be seen there is a significant energy gap at zero wave vector, consistent with the previous discussion. In addition there is a 'magnetoroton' minimum at finite k

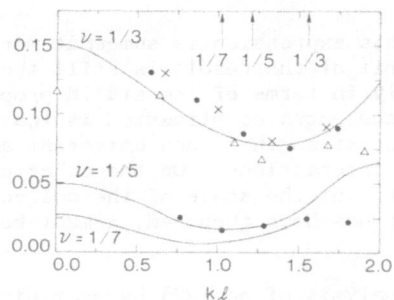

Fig. 1.   Comparison of SMA prediction of collective mode energy
           $vs.$ dimensionless wave vector $k\ell$ for $v = 1/3$, $1/5$, and $1/7$
           (solid lines) with small-system numerical results. Crosses
           indicate ($N = 7$, $v = 1/3$) spherical system[20,24], triangles
           ($N = 6$, $v = 1/3$) hexagonal unit cell[20,24]. Solid dots are
           for ($N = 9$, $v = 1/3$) and ($N = 7$, $v = 1/5$) spherical
           systems[25].

analogous to the roton minimum in superfluid helium. Just as in helium this is associated with a peak in the static structure factor at the wave vector corresponding to the average nearest-neighbor spacing. This can be verified by noting that the position of the minimum is close to the value of the first reciprocal lattice vector for the Wigner crystal at the same density (see Fig. 1). The rapid drop in the gap at the magnetoroton minimum as the filling factor is decreased represents the 'soft mode' which is the precursor to the Wigner crystallization transition which is known[21,22] to occur near filling factor $v = 1/7$. Below about $v = 1/7$ a variational ground-state wave function representing the Wigner crystal can be constructed which is lower in energy than the Laughlin liquid. This is not to be confused with the fact that the state described by the Laughlin wave function itself changes from liquid to solid[15] near $v = 1/65$. What we are saying here is that below about $v = 1/7$ the ground state is no longer described by Laughlin's wave function.

We now have a nice physical interpretation of the nature of the collective excitations in the FQHE at small and intermediate wave vectors. For small wave vectors the collective mode is a quasi-transverse magnetophonon analogous to the phonon in superfluid helium except that it has a 'mass' (gap) associated with the incompressibility of the FQHE state. Chui[23] has interpreted this as saying that the Laughlin state has a shear strength even though it is a liquid. At intermediate wave vectors the collective mode softens and becomes a magnetoroton analogous to the roton in helium. The softening is related to incipient Wigner crystal formation.

Having obtained this physical interpretation of the collective mode dispersion, we now need to examine the validity of the SMA on which our results are based. Fig. 1 shows data points from numerical calculations by Haldane and Rezayi [20,24] and Fano, et al.[25] for the exact excited-state energies for small clusters of particles. These are seen to agree rather well with the excitation energies predicted using the SMA. This comparison suffers from some minor technical complications however. The SMA results are based on the static structure factor for the Laughlin state in an infinite system, whereas the numerical results are for the excited states above the exact ground state for a small spherical or planar system (with periodic boundary conditions). A better comparison has been performed by Haldane and Rezayi who computed the SMA prediction for $\Delta(k)$ using the exact small-cluster ground state. Actually what they did was compute the fraction of the total intra-Landau level oscillator strength absorbed by the lowest excited state at each wave vector. For a considerable range of wave vectors around the magnetoroton minimum, this fraction was very close to unity showing that the SMA is virtually exact[20].

For small wave vectors the numerical studies suggest that the lowest mode does not fully saturate the f-sum rule[20]. It is unclear at present whether this is an artifact of the small cluster size or an indication of a breakdown of the SMA. Because of the peculiar dispersion relation for the collective mode, it happens that for $v = 1/3$, two rotons with oppositely directed wave vectors form a low-momentum excitation which is approximately degenerate with $\Delta(0)$. Mixing of these two types of excitations in the form of a two-roton bound state could reduce the oscillator strength of the lowest state. This is a topic which deserves further investigation.

At very large wave vectors we expect the SMA to break down because it makes no sense to have a density wave with a wavelength shorter than the typical particle spacing. What happens is that the nature of the collective mode crosses over from being a density wave to being a quasiexciton. The density wave is a coherent particle-hole pair excitation, whereas the quasiexciton is a bound pair of Laughlin's fractionally-charged quasiparticles (See Kallin and Halperin[14], Laughlin[19] and Girvin et al.[7].)

In summary we now have a good physical picture of the nature of the collective mode based on an analogy with superfluid helium and we have quantitatively accurate eigenfunctions and eigenvalues for the excitations. We have made only comparison to numerical 'experiments' rather than real devices because we do not have an accurate theory of the effects of disorder. There are no experiments which can directly measure $\Delta(k)$ as yet although attempts are under way at AT&T, Murray Hill[26]. The only number presently available is the activation energy for conduction $E_a = \Delta(\infty)/2$ which measures the asymptotic value of the quasiexciton energy. It turns out that $\Delta(\infty)$ is quite close to the value of the gap at the roton minimum. The experimental number is about a factor of four too small[27]. However a factor of two can be explained from the softening of the effective Coulomb potential due to the finite thickness of the inversion layer charge distribution. The remaining factor of two is presumably due to disorder. Recent experiments[28] on a sample of ultra-high mobility ($\sim$5×10^6 cm^2/V-sec) yield a gap of $\Delta = 6.2$ K for $v = 2/3$ at $B = 9.8$ T and $\Delta = 10$ K for $v = 1/3$ at $B = 13.9$ T. The activation energy is in good agreement with the theoretical prediction using a finite thickness correction but with no disorder correction[28].

Using these results of the SMA theory it is straightforward to compute various useful quantities such as the microwave absorption[29], the dielectric function and the static susceptibility of the FQHE liquid state. For example, the (intra-Landau level portion of the) static susceptibility is given by[7]

$$\chi(q) = - 2 \frac{\bar{s}(q)}{\Delta(q)} \; . \tag{25}$$

The quantity $\alpha \equiv -\chi/2$ is plotted in Fig. 2 for various filling factors. Notice that the susceptibility is sharply peaked at the wave vector corresponding to the magnetoroton and that the magnitude of this peak diverges as the filling factor is reduced. This fits in nicely with our physical picture of the roton minimum being a soft-mode precursor of the Wigner crystal instability. The wavelength at which the liquid is most susceptible to perturbation is the one at which crystallization occurs. The results of Fig. 2 also explain the extreme sensitivity of the FQHE state to disorder at low filling factors.

QUANTIZED VORTICES

It is now appropriate to ask ourselves whether there exists in the FQHE any analog of the quantized vortices found in superfluid helium. The answer turns out to be yes: Laughlin's fractionally-charged quasiparticles[15] are precise analogs of quantized vortices[18]. The density-wave excitations we have been considering so far are coherent superpositions of particle-hole pair excitations of the form

$$\Phi_k = \sum_{j=1}^{N} e^{i\theta(r_j)} \; \Psi(r_1, \ldots, r_N) \; . \tag{26}$$

where the phase angle is given by

$$\theta(r_j) = k \cdot r_j \; . \tag{27}$$

A vortex is a rather different beast. It is a *many-particle* excitation of the form[5]

$$\Phi(R) = \prod_{j=1}^{N} e^{i\varphi(r_j - R)} \; \Psi(r_1, \ldots, r_N) \tag{28}$$

where φ is the azimuthal angle for particle j relative to the vortex center R. Notice that the vortex state has the property that if any particle is moved around the vortex, the phase of the state winds up by an extra 2π relative to the phase change in the ground state. These leads to azimuthal currents and a net circulation around the vortex center just as we expect classically[5]. In order to relate this to the FQHE let us switch from vector to complex-variable notation for the particle coordinates:

$$\Phi(Z) = \prod_{j=1}^{N} e^{i\varphi(z_j - Z)} \Psi(z_1, \ldots, z_N) = \prod_{j=1}^{N} e^{i\,\text{Im}\,\ln(z_j - Z)} \Psi(z_1, \ldots, z_N) \ . \quad (29)$$

This is easily projected onto the Hilbert space of analytic functions with the natural replacement of the imaginary part of the logarithm by the full function

$$i\,\text{Im}\,\ln(z_j - Z) \longrightarrow \ln(z_j - Z) \quad (30)$$

which yields

$$\Phi(Z) = \prod_{j=1}^{N} (z_j - Z)\, \Psi(z_1, \ldots, z_N) \ . \quad (31)$$

This is precisely Laughlin's variational wave function for the fractionally-charged quasihole[15]. The quasielectron state[15] is obtained by the adjoint operator as defined in eq. (14).

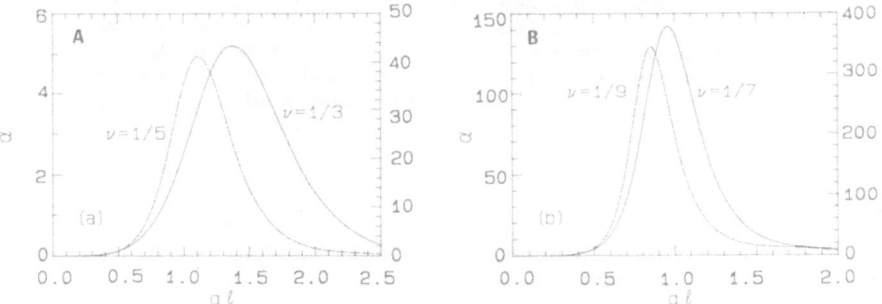

Fig. 2. Susceptibility[18] parameter α vs. dimensionless wave vector $q\ell$. (a) $\upsilon = 1/3$ (scale on left), $\upsilon = 1/5$ (scale on right); (b) $\upsilon = 1/7$ (scale on left), $\upsilon = 1/9$ (scale on right).

Just as is the case for helium, dragging any particle around the vortex yields an extra 2π of phase. Unlike the helium case however, the vortices in the FQHE carry fractional charge and isolated vortices cost only a finite energy to produce. These two facts account for the lack of a finite temperature transition in which the dissipation drops to zero as happens for superfluidity. Instead the dissipation is thermally activated and falls continuously to zero with the number of activated free vortices. To understand this let us recall that the current is proportional to the gradient in the extra phase associated with the vortex[5]. For the case of helium this falls off slowly like

$$J \sim \nabla\varphi = \frac{\hat{\varphi}}{r} \quad (32)$$

so that the total kinetic energy of the vortex diverges logarithmically with the size of the system. Hence isolated vortices can not exist below the Kosterlitz-Thouless temperature[30,31]. Isolated vortices *can* exist in a superconducting film because the particles are charged and couple to the vector potential. This changes the current operator from p to $p + e\mathbf{A}/c$ and the circulating currents can adjust themselves self-consistently to produce a quantized flux in the vortex. The vector potential term in the current then exactly cancels the gradient term, the current is screened out away from the vortex core and the vortex carries a finite energy. This is the origin of flux quantization in superconductors. This effect can *not* occur in the FQHE because the number of carriers is too low to produce any significant flux. There is however the large vector potential associated with the applied field. It turns out that if all the particles move in or out just the right amount, the change in the vector potential which they see cancels the divergent gradient energy. Thus we obtain a quantized *charge* for the vortex (quasiparticle) rather than a quantized flux[18]. The net effect is as if the binding of a fractional charge (proportional to the vortex winding number) to the vortex screens out the circulating currents and yields a finite vortex energy. This is discussed in more detail in ref. (18) and will be mentioned again in my second lecture.

SUMMARY

I have tried to demonstrate that the FQHE shows some deep analogies with superfluidity and superconductivity. Dissipationless particle transport is associated with a very low density of excitations above the ground state. These excitations take the form of magnetophonons and magnetorotons. The Laughlin quasielectron and quasihole are direct analogs of left- and right-handed quantized vortices respectively. In addition to a nice physical picture of the collective excitations, these analogies have yielded a quantitatively very accurate theory of the gap dispersion based on the single-mode approximation. The SMA works well because the magnetic field ameliorates the difficulties normally associated with Fermi statistics.

ACKNOWLEDGMENTS

This lecture is based on work done in collaboration with A. H. MacDonald and P. M. Platzman.

REFERENCES

1. D. Forster,"Hydrodynamic Fluctuations, Broken Symmetry, and Correlatio Functions," Benjamin, Reading, MA (1975).
2. Note however that the effective interaction potential to be used in th classical calculation must determined quantum mechanically.
3. P. M. Platzman and P. A. Wolff, "Waves and Interactions in Solid State Plasmas," [Solid State Physics Supplement 13], Academic, New York (1973).
4. In three dimensions the longitudinal mode occurs at a higher frequency because of the long range of the Coulomb interaction.
5. R. P. Feynman, "Statistical Mechanics," Benjamin, Reading, MA (1972).
6. G. D. Mahan, "Many Particle Physics," Plenum, New York (1981).
7. S. M. Girvin, A. H. MacDonald and P. M. Platzman, Phys. Rev. Lett. 54:581 (1985); Phys. Rev. B 33:2481 (1986).
8. D. Pines, "Elementary Excitations in Solids," Benjamin, Reading, MA (1964).
9. B. I. Lundqvist, Phys. Kondens. Mater. 6:206 (1967).
10. A. W. Overhauser, Phys. Rev. B 3:1888 (1971).
11. R. E. Prange, Chapter 1, in: "The Quantum Hall Effect", R. E. Prange and S. M. Girvin, ed., Springer-Verlag, New York (1987).
12. C. Kallin, Chapter 9, in: "Interfaces, Quantum Wells and Superlattices," C.Richard Leavens and Roger Taylor, eds., Plenum, New York (1988).

13. A. H. MacDonald, H. C. A. Oji and S. M. Girvin, Phys. Rev. Lett. 55:2208 (1985).
14. Yu. A. Bychkov, S. V. Iordanskii and G. M. Eliashberg, Pis'ma Zh. Eksp. Teor. Fiz. 33:152 (1981) [JETP Lett. 33:143 (1981)]; C. Kallin and B. I. Halperin, Phys. Rev. B 30:5655 (1984).
15. R. B. Laughlin, Phys. Rev. Lett. 50:1395 (1983); Chapter 7, in: "The Quantum Hall Effect", R. E. Prange and S. M. Girvin, ed., Springer-Verlag, New York (1987).
16. S. M. Girvin and Terrence Jach, Phys. Rev. B 29:5617 (1984).
17. W. Kohn, Phys. Rev. 123:1242 (1961).
18. S. M. Girvin, Chapter 9, in: "The Quantum Hall Effect", R. E. Prange and S. M. Girvin, ed., Springer-Verlag, New York (1987).
19. R. B. Laughlin, Physica 126B: 254 (1985).
20. F. D. M. Haldane, Chapter 8, in: "The Quantum Hall Effect", R. E. Prange and S. M. Girvin, ed., Springer-Verlag, New York (1987).
21. D. Levesque, J. J. Weis and A. H. MacDonald, Phys. Rev. B 30:1056 (1984).
22. P. K. Lam and S. M. Girvin, Phys. Rev. B 30:473 (1984); 31:613(E) (1985).
23. S. T. Chui, Phys. Rev. B 34:1409 (1986).
24. F. D. M. Haldane and E. H. Rezayi, Phys. Rev. Lett. 54:237 (1985).
25. G. Fano, F. Ortolani and E. Colombo, Phys. Rev. B 34:2670 (1986).
26. H. Störmer , private communication.
27. A. M. Chang, Chapter 6, in: "The Quantum Hall Effect", R. E. Prange and S. M. Girvin, ed., Springer-Verlag, New York (1987).
28. R. Willett, H. L. Störmer and D. C. Tsui, unpublished.
29. P. M. Platzman, S. M. Girvin and A. H. MacDonald, Phys. Rev. B 32:8458 (1985).
30. J. V. José, L. P. Kadanoff, S. Kirkpatrick and D. R. Nelson, Phys. Rev. B 16:1217 (1977).
31. S. K. Ma, "Statistical Mechanics," World Scientific, Singapore (1985).

OFF-DIAGONAL LONG-RANGE ORDER IN THE QUANTUM HALL EFFECT

S. M. Girvin*

Surface Science Division
National Bureau of Standards
Gaithersburg, MD 20899 USA

INTRODUCTION

We saw in my previous lecture on collective excitations that it was quite useful to study analogies between the fractional quantum Hall effect (FQHE) and superfluidity and superconductivity in films. Superfluids and superconductors are characterized by a spontaneously broken gauge symmetry and an associated order parameter. The purpose of this lecture is to investigate the question of whether or not there exists an order parameter which describes the FQHE state, possibly associated with some type of symmetry breaking[1-8].

The close analogy with superfluidity[9] is reinforced by the correlated ring-exchange theory of Kivelson et al.[10] and related work[11-13]. In this picture, the FQHE is viewed as arising from singular contributions to the free energy due to the existence of arbitrarily large exchange rings. D. H. Lee et al.[13] have been able to reconcile the concept of ring exchanges on large length scales with Laughlin's (essentially exact[8,14]) ground-state wave function which focuses primarily on the short-range behavior of the two-particle correlation function. They do this by deriving Laughlin's wave function directly from the ring exchange picture. This ring-exchange picture is analogous to Feynman's[15] theory of the lambda transition to superfluidity in ^{4}He.

In helium the existence of arbitrarily large exchange rings signals the onset of off-diagonal long-range order[16] (ODLRO) associated with the broken gauge symmetry and we may reasonably ask if an analogous concept is relevant here. In a superfluid or superconductor the broken gauge symmetry is represented by the fact that the phase of the order parameter is well-defined. The conjugate quantity to the phase is the particle number[17] which is ill-defined (fluctuating). It is clear that this can not be the case here because one of the hallmarks of the FQHE is the fact that the particle density is ever more sharply defined as the length scale increases (i.e., the 1/3 state has filling factor exactly 1/3). Indeed as we shall see below, there is a sense in which the FQHE is dual to superconductivity: the particle number is exactly defined and it is the phase which is completely uncertain. Associated with this interchange of phase and particle number

---

*Present and permanent address: Department of Physics, Swain Hall West, Indiana University, Bloomington, IN 47405 USA.

will be a mixing up of charge and flux in the Landau–Ginsburg description of the FQHE state.

REVIEW OF OFF–DIAGONAL LONG–RANGE ORDER IN SUPERFLUIDS

Superfluids and superconductors are characterized by the existence of off-diagonal long-range order[16]. To see what this means consider the one-body density matrix defined by

$$\rho(z,z') = \frac{N}{Z}\int d^2z_2 \ldots d^2z_N \; \Psi^*(z,z_2,\ldots,z_N) \; \Psi(z',z_2,\ldots,z_N) \; . \tag{1}$$

where Z is the norm of the ground state wave function Ψ. Notice that the integration is over all the coordinates except the first. If $z = z'$ we just obtain the expectation of the particle density at position z. If $z \neq z'$ we can interpret this as giving the overlap of two different $(N-1)$-particle states parametrized by two different positions for the extra (first) particle: one state with the first particle at position z and one with the first particle at position z'. The second-quantization representation of the density matrix is:

$$\rho(z,z') = \langle 0|\psi^\dagger(z')\psi(z)|0\rangle \tag{2}$$

where ψ is the particle destruction operator. In general we expect $\rho(z,z')$ to decay to zero for large separations of its arguments. However in systems which exhibit ODLRO the density matrix obeys[16]:

$$\lim_{|z-z'| \to \infty} \rho(z,z') \neq 0 \; . \tag{3}$$

This can occur for instance in three-dimensional superfluids and superconductors. In two dimensions the situation is, as we shall see, more complex and $\rho(z,z')$ always goes to zero. However in the superfluid state (i.e., below the Kosterlitz-Thouless[18,19] temperature) $\rho(z,z')$ decays only slowly (algebraically). I shall loosely (but not strictly correctly) also refer to this as ODLRO.

ODLRO is a signature of Bose–Einstein condensation. To see this, consider expanding the particle destruction operator in eq.(2) in some complete set of states $\{\phi\}$:

$$\psi(z) = \sum_m \phi_m(z)b_m \tag{4}$$

where b_m is the destruction operator for the mth state. If condensation occurs so that some particular state ϕ has macroscopic occupation M, then to a good approximation the density matrix factors into the form[16]:

$$\rho(z,z') = M \; \phi^*(z')\phi(z) \; . \tag{5}$$

This is sometimes interpreted as saying that the quantum operator $\psi(z)$ in eq.(2) may be replaced by the c-number order parameter $\phi(z)$. The state of lowest possible energy has a spatially uniform order parameter which therefore yields ODLRO:

$$\rho(z,z') = M/V \tag{6}$$

where V is the system volume.

In the previous lecture on collective excitations we discussed vortices and the extra phase winding associated with them. If a vortex occurs in the superfluid condensate at position Z then the order parameter obeys

$$\phi(z) \sim e^{\pm i\varphi}(z-Z) \tag{7}$$

where φ is the azimuthal angle of z relative to Z and the $\pm$ sign determines the handedness of the vortex. Notice that the presence of a vortex changes the phase but not the magnitude[20] of $\rho(z,z')$. Thermal averaging over the random positions of the vortex will cause $\rho(z,z')$ to phase-average to zero thereby destroying the ODLRO. Recall however that in the previous lecture[21] we found that isolated vortices in two dimensions have divergent energy and so they must occur in "neutral" vortex-antivortex pairs. At low temperatures, these pairs are tightly bound and can cause only weak (convergent) phase fluctuations. These cause the density matrix to decay, but only algebraically. At the Kosterlitz-Thouless temperature[18,19], the vortices unbind through a cooperative screening effect and $\rho(z,z')$ begins to decay exponentially so that ODLRO is destroyed.

ORDER IN THE FQHE

Using the convention of my previous lecture[21], the FQHE ground-state wave function must be analytic in its arguments. Following the clear and very informative discussion by Halperin[22], let us imagine freezing the positions of all the particles except the first. This defines an effective wave function $\chi(z_1)$ parametrized by the positions of the other particles. χ is entire (i.e., analytic throughout the complex z_1 plane). Hence it is uniquely defined by the set of its zeros $\{Z_j; j=1,L\}$:

$$\chi(z_1) = A \prod_{j=1}^{L} (z_1 - Z_j) \tag{8}$$

where A is independent of z_1. (Note that A and the positions of the zeros do depend on $\{z_2,\ldots,z_N\}$.) Restoring the ubiquitous gaussian factor, the probability density for particle one is (with the magnetic length set to unity)

$$\rho(z_1) = |\chi(z_1)|^2 \, e^{-|z_1|^2/2} \, . \tag{9}$$

Using Laughlin's plasma analogy[14] we can rewrite this as

$$\rho(z_1) = |A|^2 \, e^{-\beta\Phi(z_1)} \tag{10}$$

where $\beta \equiv 2/m$ is the "temperature"[23] and

$$\Phi(z_1) = m \sum_{j=1}^{L} \left[-\ln|z_1 - Z_j| + \frac{1}{4} |z_1|^2 \right] \tag{11}$$

is the "electrostatic potential" experienced by particle one. We see that the zeros of the wave function look like 2d Coulomb charges (logarithmic potential) which repel the particle (a quantum particle has a low probability of being found near a zero in its probability amplitude). The $|z_1|^2/4$ term is the electrostatic potential from a compensating uniform background charge of density $\rho_0 = 1/2\pi$ (the magnetic length has been taken to be unity). Now if $\rho(z_1)$ is to be roughly uniform and not confined to the origin by the gaussian term in eq.(9), the density of point charges (zeros) must compensate the background. To summarize: avoiding the kinetic energy of the higher Landau levels enforces analyticity which in turn requires that the wave function have a more or less uniform density ρ_0 of zeros. Halperin[22] has observed that Laughlin's wave function makes optimum use of these zeros by placing them directly on the other particles:

$$\Psi_m(z_1,\ldots,z_N) = \prod_{i<j} (z_i - z_j)^m .$$

(12)

If m zeros are attached to each particle, then the particles see each other as point Coulomb charges of charge m and the particle density will be pinned at $\rho = 1/2\pi m$ (filling factor $\upsilon = 1/m$) by the "charge neutrality" requirement of the two-dimensional one-component plasma (2DOCP)[14,22]. The strong correlations in the 2DOCP analog plasma account for the good variational energy of the Laughlin state[8,14,22] (the particles avoid each other).

We saw in the previous lecture[21] that if the particles repel each other as 2d Coulomb charges, then the density fluctuations are strongly reduced at long wavelengths so that the structure factor vanishes

$$s(k) = \frac{1}{2} k^2 + \ldots$$

(13)

in just such a way as to guarantee the existence of a finite excitation gap, at least within the single-mode-approximation[9] (SMA). *The unique type of ordering which controls the FQHE is thus the binding of the zeros of the analytic wave function to the particles.* Our goal is to try to relate this ordering to the type of ODLRO in superfluids which was discussed above.

SINGULAR GAUGE TRANSFORMATIONS AND "ANYONS"

To make further progress let us reexamine eq.(8) and recall that each particle sees its zeros not only as Coulomb charges, but also as vortices, that is, the phase of χ winds by 2π as z_1 circles a zero. Thus if we evaluate the density matrix in eq.(1) we can *not* find ODLRO because there are free vortices running around which cause the phase cancellations discussed previously. It is even worse than being above the Kosterlitz-Thouless[18,19] temperature in a helium film because all the vortices have the *same* sign. [This can be tolerated because of the background charge (vorticity) density from the gaussian term in Ψ. It is quite analogous to having a bucket (or rather film) of helium rotating at astrophysical speeds so that the required net circulation is carried by m vortices for every helium atom.] This situation results in $\rho(z,z')$ falling off *much faster than exponential*. In fact for any uniform state in the lowest Landau level one can show that[9]

$$\rho(z,z') = \upsilon \, g(z,z') = \frac{\upsilon}{2\pi} e^{-\frac{1}{4}|z-z'|^2} e^{\frac{1}{4}(z^*z'-zz'^*)} ,$$

(14)

where g is the free-particle propagator. The density matrix falls off as a gaussian on a length scale given by the magnetic length. This is precisely the same effect that occurs in superconducting films in the presence of a magnetic field[24].

We see from eq. (14) that there is definitely no ODLRO of the ordinary variety present. This is consistent with our previous observation that the particle density is sharply defined so the phase can not be. Nevertheless there is a peculiar type of order hidden in the density matrix. For reasons that will become clear below, it is useful to consider the singular gauge field $\mathcal{A}_j$ defined by

$$\mathcal{A}_j(z_j) = \frac{\lambda\Phi_0}{2\pi} \sum_{i \neq j} \nabla_j \, \text{Im} \, \ln(z_j - z_i) \, , \tag{15}$$

where $\Phi_0 = hc/e$ is the flux quantum and λ is a real constant. This gauge field has been used in the study of "anyons" (which are objects with arbitrary fractional statistics[25,26]) and in particular in the study of fractional charges in the FQHE[27]. It corresponds to the vector potential that would be included in the Hamiltonian if each particle had attached to itself a solenoid carrying $\lambda/2$ flux quanta[28]. If two such composite objects are exchanged then in addition to the usual statistics phase factor (+1 for bosons, -1 for fermions), there will be an extra Bohm-Aharonov phase $\exp(i\pi\lambda)$. Thus choosing λ to be an odd integer maps bosons into fermions and vice versa.

Now consider making a singular gauge transformation by adding this vector potential to the Hamiltonian. This is of course not a true gauge transformation since a flux tube is attached to each particle. However if we take $\lambda = m$ to be an integer then the only effect is to change the phase of the wave function:

$$\Psi_{new} = \exp\left[-im \sum_{i < j} \text{Im} \, \ln(z_i - z_j)\right] \Psi_{old} \, . \tag{16}$$

If we apply this to Laughlin's wave function from eq.(12) we obtain the transformed state

$$\tilde{\Psi}(z_1, \ldots, z_N) = \prod_{i < j} |z_i - z_j|^m \tag{17}$$

which is purely real and is symmetric under particle exchange for both even and odd m. Hence we have the remarkable result that both fermion and boson systems map into bosons in this singular gauge. It is clear from eq.(17) that at least for the special case of Laughlin's wave function, the troublesome phase factors associated with all the zeros of the wave function have been eliminated. As we shall see below, replacing Ψ by $\tilde{\Psi}$ in eq.(1) will lead to ODLRO for Laughlin's state. Before doing this it is useful to examine the equivalent operation of evaluating the transformed density matrix:

$$\tilde{\rho}(z, z') = \frac{N}{Z} \int d^2 z_2 \ldots d^2 z_N$$
$$\times e^{-i\frac{e}{\hbar c} \int_z^{z'} dr \cdot \mathcal{A}_1} \Psi^*(z, z_2, \ldots, z_N) \, \Psi(z', z_2, \ldots, z_N), \tag{18}$$

where z and z' are vector representations of z and z'. The line integral in eq.(18) is multiple-valued but its exponential is single-valued because the flux tubes are quantized (i.e. the phase change around a flux tube is a multiple of 2π). The additional phases introduced by the singular gauge transformation will cancel the phases in Ψ nearly everywhere, and produce ODLRO in $\tilde{\rho}$, if and only if, the zeros of Ψ are bound to the particles. Thus ODLRO in $\tilde{\rho}$ always signals a "condensation"[29] of the zeros onto the particles. Thus $\tilde{\rho}$ is just what we have been seeking to measure the ordering in the FQHE state. This multi-particle object, which explicitly exhibits ODLRO, is very reminiscent of the topological order parameter in the XY model[18,19] and related gauge models[30,31] and is intimately connected with the frustrated XY model which arises in the correlated ring exchange theory[10,13].

Let us now actually compute the asymptotic behavior of the density matrix for Laughlin's wave function in the singular gauge. Using eq.(17) and the 2DOCP plasma analogy[14] we may write[5]:

$$\tilde{\rho}(z,z') = |z-z'|^{-m/2} \frac{N}{Z} \int d^2 z_2 \ldots d^2 z_N \; e^{-\beta(V+V')} \tag{19}$$

where $\beta = 2/m$ and

$$V = \sum_{1<i<j} -m^2 \ln|z_i - z_j| + \frac{m}{4} \sum_{1<k} |z_k|^2 \tag{20}$$

$$V' = \frac{m}{8}(|z|^2 + |z'|^2) - (\frac{m}{2})^2 \ln|z-z'| - \sum_{1<i} \frac{m^2}{2}\Big[\ln|z_i - z| + \ln|z_i - z'|\Big] . \tag{21}$$

The quantity V is the potential energy for a plasma of $N-1$ charge-m particles and V' is a perturbation consisting of two impurities of charge $m/2$ located at z and z'. Eq.(19) may be reexpressed as:

$$\rho(z,z') = \frac{v}{2\pi} e^{-\beta \Delta f(z,z')} |z-z'|^{-m/2} . \tag{22}$$

where $\Delta f(z,z')$ is the difference in free energy between two impurities of charge $m/2$ (located at z and z') and a single impurity of charge m (with arbitrary location). Because of complete screening of the impurities by the plasma, the energy difference Δf rapidly approaches a constant as $|z-z'| \to \infty$. Thus eq.(22) proves the existence of (algebraic) ODLRO in the Laughlin state. The decay is characterized by an exponent $\beta^{-1} = 2/m$ which significantly, is precisely equal to the plasma "temperature". The asymptotic value of Δf can be computed using thermodynamic integration of the screening charge density. For $m = 1$ this can be done exactly and one obtains $\beta \Delta f = -0.3942$. For general values of m the screening charge distribution can be found using the ion-disk approximation[14] or linear response based on the known static structure factor of the plasma[9]. The results for $m = 1,3$ and 5 are shown in Fig. 1. Note that the asymptotic form is achieved rather quickly due to the fact that perfect screening occurs within the plasma on the scale of the average particle spacing.

Let us now pause and summarize what we have so far. First, we have recognized that the fundamental ordering in the FQHE is the binding of the zeros of the wave function to the particles. Next we have found an object, namely the singular gauge density matrix whose infrared behavior measures this binding of the zeros. For the case of short-range repulsive interactions, Laughlin's wave function is the *exact* ground state[8,32]. We have found for the specific case of Laughlin's wave function that the zeros are bound directly on the particles. The phase factors associated with the zeros are precisely cancelled by the gauge field phase factors associated with the flux tubes that are attached to the particles. From this we

demonstrated that $\tilde{\rho}(z,z')$ exhibits algebraic ODLRO. We must now ask what happens for other interactions for which the ground state is not precisely Laughlin's state.

It has been found numerically that as the range of the interaction increases, $m-1$ of the zeros move away from each particle but remain bound in the vicinity[8,33]. The combination of the gauge and wave-function phase factors now appears as a bound vortex-antivortex pair. The zeros represent vortices near the particle position and the flux tube represents an antivortex directly on the particle. This is reminiscent of the situation in a superfluid at finite but low temperatures. In some sense the range of the particle interaction plays the role of temperature. As the interaction

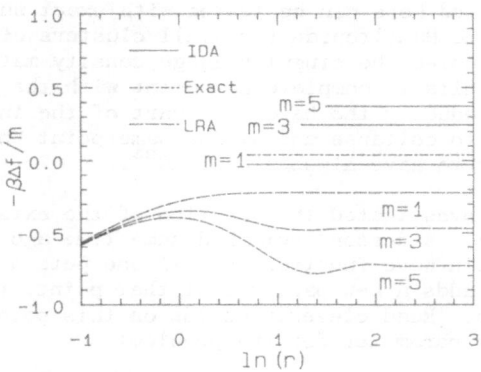

Fig. 1.   Plot of $-\beta\Delta f(z,z')/m$ vs. $r \equiv |z-z'|$ for filling factor $\upsilon = 1/m$. LRA is linear response approximation, IDA is ion-disk approximation (shown only for radii exceeding the sum of the ion-disk radii). Because the plasma is strongly coupled, the IDA is quite accurate at $m = 1$ (cf. the exact result) and improves further with increasing m. The LRA is less accurate at $m = 1$ and worsens with increasing m.

is softened a critical point is reached at which the gap collapses and the ground state probably becomes a Wigner crystal[8]. This gap collapse corresponds to the vortex unbinding transition in which the zeros unbind from the particles. While roughly analogous in spirit to the Kosterlitz-Thouless transition[18,19], this unbinding is probably first order since there is reason to believe that the new ground state is gapless only because it breaks translation symmetry[9].

As we have seen, the ordering in the FQHE is topological in nature rather like the ordering in the XY model. Kivelson has given[34] an interesting description of this ordering which I reproduce here in slightly modified form using the singular gauge transformation. Consider the following loop integral:

$$\langle \Sigma \rangle = \langle \exp\left[i \oint_\Gamma dr_1 \cdot [-\mathscr{A}_1(r_1) + \nabla\gamma(r_1)]\right] \rangle \ . \tag{23}$$

where Γ is a large closed contour around which we drag particle one. The quantity γ is the phase of the wave function and $\mathscr{A}_l$ is the singular gauge field. The brackets indicate the ensemble average over the positions of all the remaining particles. From the preceding discussion we recognize that the loop integral of the gauge field counts the number of flux tubes inside the contour and the loop integral of the wave function phase counts the number of zeros inside. For Laughlin's wave function these exactly cancel and $\langle \Sigma \rangle$ is exactly unity. If the zeros are not on the particles but are still bound to them then the mean square fluctuations in the exponent will be the proportional to the length of the contour and $\langle \Sigma \rangle$ will obey a *perimeter* law[35]; i.e. $\langle \Sigma \rangle \sim \exp(-$ perimeter of $\Gamma)$. If the zeros are unbound from the particles then $\langle \Sigma \rangle$ will obey an *area* law. This is precisely analogous to the way Wilson loops measure topological order in XY and gauge models[30].

The ideas discussed here can be tested with exact numerical diagonalizations of the Hamiltonian for small clusters of particles. Rezayi and Haldane[36] have studied the singular-gauge density matrix for small clusters and find results in complete agreement with the picture developed here. Artificially reducing the hard-core part of the interaction does indeed cause the gap to collapse and at the same point the singular-gauge density matrix ceases to have algebraic ODLRO[36].

Read[6] has also investigated the question of the existence of an order parameter in the FQHE. Anderson[1] remarked some time ago on the interesting feature of the Laughlin wave function that if one puts m quasiholes at some point, and then adds a new particle at that point, the state is essentially unchanged. Read cleverly builds on this point by proposing the following as an order parameter for the problem:

$$\phi(Z) = \int d^2z_1 \ldots d^2z_N \; \psi_m^{N+1}(z_1, \ldots, z_N, Z) \prod_{j=1}^{N} (z_j - Z)^m \; \psi_m^N(z_1, \ldots, z_N) \; . \quad (24)$$

This is merely the statement that if we start with the N-particle ground state for filling factor $\upsilon = 1/m$ and inject m quasiholes at position Z, the resulting state has a finite overlap with the $N+1$-particle ground state with one of its particles fixed at Z. In some sense this is like the idea behind the BCS picture of superconductivity where there is a coherence between states of N and $N+2$ particles. Read[6] has used this idea to develop a microscopic basis for the phenomenological Landau-Ginsburg picture[4] presented in the next section.

LANDAU-GINSBURG PICTURE OF THE FQHE

With all the analogies we have seen between the FQHE and superfluidity and superconductivity it is natural to ask whether there might not exist a phenomenological Landau-Ginsburg theory of the effect. The author has taken some tentative steps towards this goal[4] and the recent work of Read[6] appears to have taken us considerably closer. The central idea is based on the following facts:

1. There exists a type of topological order in the system similar to that in superfluids.

2. The phase of the order parameter must fluctuate severely if the particle density is to be well defined.

3. The collective mode has a gap ("mass").

4. Vortices have finite energy.

5. Vortices carry quantized (fractional) charge.

6. The system is incompressible, (i.e., it exhibits "charge exclusion". There is a cusp in the free energy at special charge densities).

Items 3-6 are reminiscent of the Anderson-Higgs[37-39] mechanism in superconductors. This is a set of effects resulting from the coupling of the order parameter to the vector potential:

1. The gapless Goldstone mode develops a mass. It becomes the plasmon and has finite energy (in three dimensions) because of the long range of the Coulomb interaction.

2. Quantized vortices have a finite energy.

3. Vortices carry a quantized magnetic flux.

4. The system exhibits "magnetic" incompressibility (i.e., flux exclusion, or in the case of Type II superconductivity which is the appropriate analog here, the system exhibits a cusp in the free energy at zero flux).

It would appear then from this analogy that we need a Landau-Ginsburg functional with a complex order parameter coupled to a gauge field. However we see from comparing the two lists of properties above that we must somehow interchange charge and flux. To make this more explicit consider the following action:

$$S = \int d^2r \, |(-i\boldsymbol{\nabla} + a)\psi(r)|^2 \,. \tag{25}$$

where ψ is a complex order parameter field and a is a gauge field. Now in a superconductor a is the ordinary vector potential due to the applied B field plus the B field which is generated by the currents flowing in the system. The currents in turn are determined by (the gradient of) the order parameter ψ. Hence we have a nonlinear self-coupled system. In the FQHE the Anderson-Higgs mechanism cannot be *literally* occurring because the flux generated by the small currents flowing in the inversion layer is negligibly small[40]. Nevertheless there is, as we shall see below, an effective gauge field in the problem. Before proceeding to this let us investigate the general properties of the action in eq.(25).

A superconductor is characterized at zero temperature by a spontaneously broken gauge symmetry in which the order parameter has a non-zero value everywhere in space[41]:

$$\psi(r) = e^{i\theta} \,. \tag{26}$$

The broken symmetry has the very important consequence that the energy of an isolated vortex diverges. To see this put the vortex at the origin so that eq.(26) is replaced by:

$$\psi(r) = f(r)e^{\pm i\varphi(r)} \tag{27}$$

where φ is the azimuthal angle and f goes to unity outside the core of the vortex. Outside the core, the gradient of ψ is

$$-i\nabla\psi = \pm\frac{\hat{\varphi}}{r} \qquad (28)$$

so that, in the absence of the gauge field, the action is

$$S = \int_{r_c}^{L} dr \; 2\pi r \; r^{-2} \quad \sim \quad \ln L/r_c \; , \qquad (29)$$

which diverges with the ratio of the system size L to the core radius r_c. It is clear from eqs.(25) and (28) that the action would be finite if at large distances we had:

$$\pm a = -\frac{\hat{\varphi}}{r} \; . \qquad (30)$$

But this implies that the flux Φ trapped in the vortex is quantized:

$$\oint_{\partial\Gamma} dr\cdot a = \int_{\Gamma} d^2r \; \nabla\times a\cdot\hat{z} = \Phi = \pm \; 2\pi \; . \qquad (31)$$

where Γ is a large region containing the vortex and $\partial\Gamma$ is its boundary. Independent of the physical origin of this flux, the system will gain an infinite amount of energy if the flux has the correct value. Hence the system will pay any finite energy cost to achieve this value of the flux. The superconductor does this by adjusting the radial distribution of the current near the vortex core. Once the flux is quantized, the currents far away vanish exponentially rather than as $1/r$ and the energy is finite. It is extremely important to note that all of this hinges on the broken symmetry. If the order parameter were non-zero only in a finite regime, there would be no divergence to enforce the flux quantization.

Now that we understand how flux quantization arises in the action in eq.(25) we are ready to think about how this can be applied to the FQHE. As we saw earlier, the finite order parameter is supposed to represent the condensation occurring in the system. We saw however from our study of the singular-gauge density matrix, that the FQHE does not represent the condensation of ordinary particles, but rather the condensation of composite objects consisting of a particle and a flux tube which generates a fake gauge field. It is this fake gauge field which plays the role of the vector potential in eq.(25). If $|\psi|^2$ represents the particle density[42] and if there are m flux quanta attached to each particle then we expect

$$\nabla\times a\cdot\hat{z} = 2\pi \; m \; (\; |\psi|^2 - 1) \qquad (32)$$

where the -1 term represents the true external B field. Notice the parity of the terms in eq.(32). The left hand side is a pseudoscalar and the right hand side is a scalar. This parity violation builds a handedness into the problem which represents the handedness enforced by the external B field[4]. Eq.(32) is the analog of the Maxwell equation

$$\nabla\times\nabla\times a = \frac{4\pi}{c} \; J[\psi] \qquad (33)$$

applicable to ordinary superconductors. We see however that eq.(32) implies that charge, not current generates "flux" in the FQHE case.

Before going on to study the implications of eq.(32), let us try to give it a new interpretation. This new interpretation has to do with the statistical mechanical concept of *frustration*. Frustration occurs for example, in an antiferromagnetically coupled set of spins on a triangular lattice. It is impossible to rotate the spins locally so that all the bonds are favorable. In a continuum model with action given by eq.(25), frustration is represented by the curl of a. If a has no curl, one can make a gauge transformation ("rotate the spins locally") to eliminate the gradient energy by choosing

$$\psi(r) = \psi(0) \exp\{-i \int_0^r dR \cdot a(R)\} \qquad (34)$$

to give $S = 0$. If however, a has a curl, then it is impossible to do this. The line integral in eq.(34) becomes ill-defined (path dependent). We cannot adjust the phase to achieve $S = 0$, so the energy rises and the system is frustrated[24].

Here is how frustration enters the FQHE. We know that if the density of particles obeys $\rho = \rho_m = 1/2\pi m$, the system can condense into the mth Laughlin state and gain considerable condensation energy[43]. If the density deviates from ρ_m the system is *frustrated* because it can *not* be in the Laughlin state (since the Laughlin state has quantized density). Let us make the ansatz that the case of constant order parameter

$$\psi = e^{i\theta} \qquad (35)$$

is a macroscopic representation of perfect microscopic condensation into the mth Laughlin state. Then if we further assume that the frustration is linearly proportional to the local density deviation, eq.(32) follows[44].

Thus we have two rather different interpretations of the problem which lead to the same form for the Landau–Ginsburg theory. What are the consequences of this theory? First consider the constant solution in eq.(35). From eq.(32), $\nabla \times a = 0$ and $\nabla \psi = 0$, so from eq.(25), $S = 0$. The fact that this is the minimum energy state is consistent with our ansatz that it represents the Laughlin state. Now consider the single–vortex state given by eq.(27). Applying eq.(32) to the flux–quantization condition of eq.(31) yields

$$2\pi m \int d^2r \ (|\psi|^2 - 1) = \pm 2\pi \ . \qquad (36)$$

The integrand is just the charge density, so this proves that the vortex carries fractional charge

$$q^* = \pm 1/m \qquad (37)$$

proportional to the winding number[45]. As we saw in the previous lecture[21], this is just the property required to represent Laughlin's quasiparticles.

Now consider what happens when the average density deviates from the quantized value so that $\langle |\psi|^2 \rangle \neq 1$. Eq.(32) shows that this is equivalent to a net flux $\int d^2r \ \nabla \times a$ applied to the system. Just as in a superconductor this will be accommodated in the form of quantized fluxoids (charged quasiparticles). Hence there is a cusp in the action at the density corresponding to the Laughlin state and the system is incompressible[4,14,22]. Another way to say this is that density fluctuations induce such strong phase fluctuations (due to frustration) that the density is pinned at the quantized value (see Item 2 in the first list at the beginning of this section).

The remarkably simple action represented by eqs.(25) and (32) correctly reproduces the phenomenology of the FQHE. We can put things on a fancier mathematical footing by writing a unified action which includes eq.(25) and yields eq.(32) as an equation of motion[4,5]:

$$S(\psi,\phi,a) = \int d^2r \ |(-i\mathbf{\nabla} + a)\psi(r)|^2 + i\int d^2r \ \phi(\psi^*\psi - 1)$$

$$- i\frac{\theta}{2\pi}\int d^2r \ \frac{1}{4\pi}(\phi\mathbf{\nabla}\times a + a\times\mathbf{\nabla}\phi) \tag{38}$$

where the 'vacuum angle' θ is given by $\theta = 2\pi/m$ and the last term on the right is a 'Chern-Simons' topological contribution to the action. We can interpret the Lagrange multiplier field ϕ as a scalar potential and note that both the charge density $\psi^*\psi$ and the 'flux density' $\mathbf{\nabla}\times a$ couple to the potential. This is what causes flux tubes to carry charge in this model (similar models have been discussed by field theorists, see Niemi and Semenoff 1983, de Vega and Schaposnik 1986, Schonfeld 1981, Jackiw and Templeton 1981, Deser et al. 1982a,b).

Other papers of particular interest by Cardy and Rabinovici (1982) and Cardy (1982) discuss the phenomenon of *oblique confinement*. The system they study contains an electric and a magnetic condensate and is characterized by a vacuum angle θ. The system exhibits phase transitions to special vacuum states at rational fractional values of $\theta/2\pi$ and the elementary excitations consist of objects with the same ratio of electric to magnetic charge as the condensate. The similarities to the hierarchy of states in the FQHE are quite striking and should be pursued.

SUMMARY

As I have tried to show in these two lectures, the fractional quantum Hall effect exhibits deep analogies with superfluidity and superconductivity. In this second lecture I have examined the topological nature of the ordering. It is the binding of the zeros of the wave function to the particles which produces the order in the FQHE. The singular-gauge density matrix is a measure of this order and shows that in a sense, the objects which are condensing are not ordinary particles, but rather composite objects consisting of a particle and a gauge flux tube. A phenomenological Landau-Ginsburg theory for this condensation was presented and shown to correctly reproduce the observed features of the effect.

The purpose of this approach to the problem has been to unify the differing pictures of the FQHE now in existence and deepen our understanding of the meaning of Laughlin's remarkable wave function. The Landau-Ginsburg theory presented here is primarily of conceptual value and makes no experimental predictions. If however further work such as that of Read[6] puts these ideas on a firmer microscopic basis, then a Landau-Ginsburg theory will prove very useful just as it has in superconductivity. One could for instance add terms to the action representing a coupling of the density to a disorder potential and investigate the possible quasiparticle localization, gap collapse and so forth[4]. Just as for the case of superconductivity, this would be much easier to do in a macroscopic Landau-Ginsburg theory than in full microscopic theory with all its attendant complications.

ACKNOWLEDGMENTS

The author is grateful to S. B. Libby for bringing to his attention the phenomenon of oblique confinement and for emphasizing its possible relevance to the FQHE. The author would also like to express his thanks for useful conversations to numerous others including G. Semenoff, Z. Tesanovic, P. A. Lee, D. R. Nelson, P. W. Anderson, N. Read, C. Kallin, D. Arovas and S. Kivelson. This lecture is based on work done in collaboration with my colleague A. H. MacDonald.

REFERENCES

1. P. W. Anderson, Phys. Rev. B 28:2264 (1983).
2. R. Tao and Yong-Shi Wu, Phys. Rev. B 30:1097 (1984).
3. D. J. Thouless, Phys. Rev. B 31:8305 (1985).
4. S. M. Girvin, Chapter 10, in: "The Quantum Hall Effect", R. E. Prange and S. M. Girvin, ed., Springer-Verlag, New York (1987).
5. S. M. Girvin and A. H. MacDonald, Phys. Rev. Lett. 58:1252 (1987).
6. N. Read, Bull. Am. Phys. Soc. 32:923 (1987) and (unpublished).
7. The apparent symmetry breaking associated with the discrete degeneracy of the ground state in the Landau gauge[2] is an artifact of the toroidal geometry[3,8] and is not at issue here.
8. F. D. M. Haldane, Chapter 8, in: "The Quantum Hall Effect", R. E. Prange and S. M. Girvin, ed., Springer-Verlag, New York (1987).
9. S. M. Girvin, A. H. MacDonald and P. M. Platzman, Phys. Rev. Lett. 54:581 (1985); Phys. Rev. B 33:2481 (1986); S. M. Girvin, Chapter 9, in: "The Quantum Hall Effect", R. E. Prange and S. M. Girvin, ed., Springer-Verlag, New York (1987).
10. S. Kivelson, C. Kallin, D. P. Arovas and J. R. Schrieffer, Phys. Rev. Lett. 56:873 (1986) and (unpublished).
11. S. T. Chui, T. M. Hakim and K. B. Ma, Phys. Rev. B 33:7110 (1986); S. T. Chui (unpublished).
12. G. Baskaran, Phys. Rev. Lett. 56:2716 (1986) and (unpublished).
13. D. H. Lee, G. Baskaran and S. Kivelson, Bull. Am. Phys. Soc. 32:923 (1987) and (unpublished).
14. R. B. Laughlin, Phys. Rev. Lett. 50:1395 (1983); Chapter 7, in: "The Quantum Hall Effect", R. E. Prange and S. M. Girvin, ed., Springer-Verlag, New York (1987).
15. R. P. Feynman, Phys. Rev. 91:1291 (1953).
16. C. N. Yang, Rev. Mod. Phys. 34:694 (1962).
17. P. W. Anderson, "Basic Notions of Condensed Matter Physics," Benjamin, Menlo Park (1984); Rev. Mod. Phys. 38:298 (1966).
18. J. V. José, L. P. Kadanoff, S. Kirkpatrick and D. R. Nelson, Phys. Rev. B 16:1217 (1977).
19. S. K. Ma, "Statistical Mechanics," World Scientific, Singapore (1985).
20. This is true except in the vortex core where the order parameter must vanish.
21. S. M. Girvin, Phonons, Rotons and Fractionally-Charged Vortices in the Quantum Hall Effect, in: "Interfaces, Quantum Wells and Superlattices," Chapter 17, C.Richard Leavens and Roger Taylor eds., Plenum, New York (1988).
22. B. I. Halperin, Helv. Phys. Acta 56:75 (1983).
23. The parameter m is an irrelevant constant in this context but will prove useful when we look at the plasma corresponding specifically to the mth Laughlin state.
24. E. Brézin, D. R. Nelson and A. Thiaville, Phys. Rev. B 31:7124 (1985).
25. F. Wilczek, Phys. Rev. Lett. 49:957 (1982).
26. D. P. Arovas, J. R. Schrieffer, F. Wilczek and A. Zee, Nucl. Phys. B 251:117 (1985).
27. D. Arovas, J. R. Schrieffer and F. Wilczek, Phys. Rev. Lett. 53:722 (1984).

28. S. Kivelson (private communication) has pointed out that the statement in refs. 5 and 26 that this gauge field corresponds to λ (instead of $\lambda/2$) flux quanta per particle is incorrect. One must be careful to count the phase change associated with the motion of the flux tube in the presence of the other charge as well as the motion of the charge in the presence of the other flux tube. This minor error has no effect on the conclusions of either ref. 5 or 26.

29. We loosely (and not strictly correctly) refer to this as condensation because of the slow power-law decay even though the largest eigenvalue $\lambda \equiv \int d^2 z \, \tilde{\rho}(z,z')$ of the density matrix diverges only for $m \leq 4$. See Yang's discussion[16] of this point.

30. J. B. Kogut, Rev. Mod. Phys. 51:659 (1979).

31. J. L. Cardy and E. Rabinovici, Nucl. Phys. B 205:1 (1982); J. L. Cardy, ibid. 205:17 (1982).

32. S. A. Trugman and S. Kivelson, Phys. Rev. B 31:5280 (1985).

33. D. J. Yoshioka, Phys. Rev. B 29:6833 (1984).

34. S. Kivelson (private communication).

35. Only particles and their zeros near the boundary can contribute (because the particle must be inside and one of its zeros outside, or vice versa). Kivelson assumes that the value of the contour integral is gaussian distributed with mean square value proportional to the perimeter. The expectation value of the exponential then follows from its second cumulant.

36. E. A. Rezayi, (private communication).

37. P. W. Anderson, Phys. Rev. 112:1900 (1958); 130:439 (1963).

38. P. W. Higgs, Phys. Lett. 12:132 (1964); Phys. Rev. Lett. 13:508 (1964); Phys. Rev. 145:1156 (1966).

39. K. Moriyasu, "An Elementary Primer for Gauge Theory," World Scientific, Singapore (1983).

40. F. D. M. Haldane and L. Chen, Phys. Rev. Lett. 53:2591 (1984).

41. Polynomial terms of the form $(|\psi|^2-1)^2$ which enforce the non-zero value of ψ have not been explicitly included in the action in eq.(25). These will not be necessary in the FQHE where $|\psi|^2$ is defined by the total particle density[42].

42. In a superconductor we do not envision that $|\psi|^2$ represents the total particle density since only a tiny fraction of the particles (those near the Fermi energy) actually take part in the pairing. We see from Laughlin's wave function however that the FQHE ground state is very much different. Every electron is coupled to every other on an equal footing. Every electron is a source of m zeros seen by all the others. This is at least a hand-waving justification for the interpretation of $|\psi|^2$ as the total charge density.

43. For simplicity I am only considering the parent Laughlin states at filling factor $v = 1/m$ and ignore the hierarchy of rational fractional states[8].

44. The value of the coefficient in the linear term can be obtained from the previous argument using the singular gauge transformation. There is a unique value of the gauge flux per particle which leads to ODLRO (namely the value which allows the gauge field a to cancel on the average the vector potential of the applied field).

45. This is where the handedness described earlier appears.

FIBONACCI SUPERLATTICES

A.H. MacDonald*

National Research Council of Canada
Ottawa K1A 0R6 Canada

INTRODUCTION

Many of the interesting properties of semiconductor superlattices are consequences of the artificially imposed one-dimensional (1D) periodicity along the growth direction. Fibonacci semiconductor superlattices have a 1D structure along the growth direction which is quasiperiodic, i.e. it is characterized by two different fundamental periods whose ratio is irrational. Both the structural properties and the various spectral properties (e.g. plasmon, phonon and electron) are very different for quasiperiodic structures when compared to those of periodic structures. These materials provide yet another example of the enormous variety of physical phenomena which can occur in semiconductor multi-layer systems.

Interest in the spectral properties of quasiperiodic Schrodinger operators predates[1] the discovery of quasicrystals[2], but was reinvigorated by the event. In particular, Merlin et al.[3] realized that a 1D version of the model for quasicrystals proposed by Levine and Steinhardt[4] could be fabricated in semiconductor multi-layer systems. More recently quasiperiodic superconducting multi-layer systems[5] and two-dimensional (2D) quasiperiodic superconducting networks[6] have been fabricated and their properties have been studied. In view of the complicated structures[7] of naturally occurring quasicrystals, the comparisons between theory and experiment for these artificially fabricated quasiperiodic structures have a useful role to play in this general subject. In this article we discuss the properties of the

*Present and permanent address: Department of Physics, Swain Hall West, Indiana University, Bloomington, IN 47405, USA

layered semiconductor structures of Merlin et al. and, for reasons to be explained shortly, we refer to them as Fibonacci superlattices.

FIBONACCI NUMBERS, FIBONACCI SEQUENCES AND FIBONACCI LATTICES

The Fibonacci numbers are generated by the recursion relation

$$F_{i+1} = F_i + F_{i-1} \tag{1}$$

with the initial condition $F_0 = 0, F_1 = 1$. The first several Fibonacci numbers are listed in Table 1. Note that the recursion relation can be reversed and gives $F_{-p} = (-)^{p+1} F_p$. The Fibonacci numbers are related to a certain irrational number, known as the golden mean, which is the positive solution of the quadratic equation,

$$\tau^2 - \tau - 1 = 0. \tag{2}$$

Eq. (2) arises in considering the construction of a decagon. The positive solution to Eq. (2) is

$$\tau = \frac{1 + \sqrt{5}}{2} = 1.618034\ldots \tag{3}$$

and $\cot^{-1}(\tau) = \pi/5$. The ratio of successive Fibonacci numbers converges rapidly to the golden mean. Several useful identities, related to this fact, can be established by induction starting with Eqs. (1) and (2):

$$\tau^{p-1} + \tau^p = \tau^{p+1} \tag{4a}$$

$$F_p + \tau F_{p+1} = \tau^{p+1} \tag{4b}$$

$$F_{p+1} - \tau F_p = (-1/\tau)^p . \tag{4c}$$

In Eq. (4) p is any integer, whether positive, negative or zero.

Two symbols, A and B, can be arranged in a Fibonacci sequence by the following recursive procedure. At the first generation the sequence consists of the symbol B ($S_1 = B$), at the second generation the sequence consists of the symbol A ($S_2 = A$) and for subsequent generations the next sequence is constructed by following the current sequence with the previous, i.e.

$$S_{n+1} = S_n S_{n-1} . \tag{5}$$

The first several Fibonacci sequences are listed in Table 2. It follows from this definition that the sequence S_n has F_n symbols, F_{n-1} of which are A symbols and F_{n-2} of which are B symbols. It also follows that the symbols may be grouped into "words" of any size and the "words" will occur in precisely the same sequence. To be precise, if we define the "words",

$$W_B^k = S_k \tag{6a}$$

and

$$W_A^k = S_{k+1} \qquad\qquad (6b)$$

Table 1. Fibonacci Numbers

i	F_i	i	F_i	i	F_i
0	0	6	8	12	144
1	1	7	13	13	233
2	1	8	21	14	377
3	2	9	34	15	610
4	3	10	55	16	987
5	5	11	89	17	1597

Table 2. Fibonacci Sequences Written in Terms of Elementary
Symbols and Words (W_B^k, W_A^k) of Level k

$S_1 = B$ $\qquad$ $S_5 = ABAAB$

$S_2 = A$ $\qquad$ $S_6 = ABAABABA$

$S_3 = AB$ $\qquad$ $S_7 = ABAABABAABAAB$

$S_4 = ABA$ $\qquad$ $S_8 = ABAABABAABAABABAABABA$

$W_B^2 = A$ $\qquad$ $W_A^2 = AB$

$S_2 = W_B^2$ $\qquad$ $S_5 = W_A^2 W_B^2 W_A^2$

$S_3 = W_A^2$ $\qquad$ $S_6 = W_A^2 W_B^2 W_A^2 W_A^2 W_B^2$

$S_4 = W_A^2 W_B^2$ $\qquad$ $S_7 = W_A^2 W_B^2 W_A^2 W_A^2 W_B^2 W_A^2 W_B^2 W_A^2$

$W_B^3 = AB$ $\qquad$ $W_B^3 = ABA$

$S_3 = W_B^3$ $\qquad$ $S_6 = W_A^3 W_B^3 W_A^3$

$S_4 = W_A^3$ $\qquad$ $S_7 = W_A^3 W_B^3 W_A^3 W_A^3 W_B^3$

$S_5 = W_A^3 W_B^3$ $\qquad$ $S_8 = W_A^3 W_B^3 W_A^3 W_A^3 W_B^3 W_A^3 W_B^3 W_A^3$

then the Fibonacci sequence at the nth generation ($n \geq k$) can be considered as being composed of words W_B^k and W_A^k arranged in a generation $n+1-k$ Fibonacci sequence (see Table 2). The simple behaviour of the Fibonacci sequence under such an inflation transformation is responsible for many of the properties which we discuss below.

We define a Fibonacci lattice as a set of points separated by intervals of lengths d_A and d_B which are arranged in a Fibonacci sequence. The semi-conductor Fibonacci superlattice is fabricated by growing arbitrary thin films of thicknesses d_B and d_A and alternating them in a Fibonacci sequence.

STRUCTURE FACTOR OF THE FIBONACCI LATTICE

We define the structure factor for a set of points $\{\xi_j\}$ in 1D as

$$S(k) = \frac{1}{L} \sum_j e^{-ik\xi_j} \tag{7}$$

where L is the length of the 1D system. For a periodic system with lattice constant d

$$S(k) = S_p(k) \equiv d^{-1} \sum_n \delta(k, k_n) \tag{8}$$

where $k_n = 2\pi n/d$. Many quasiperiodic structures, and we demonstrate below that the Fibonacci crystal belongs to this class, may be obtained by projecting from a higher dimensional periodic lattice[7-11]. We consider a 2D rectangular lattice[12] with lattice constants a_x and a_y and introduce a coordinate system rotated by θ with respect to the principal axes of the lattice,

$$\xi = x\cos\theta + y\sin\theta \tag{9}$$

$$\eta = -x\sin\theta + y\cos\theta. \tag{10}$$

(See Fig. 1) We want the set of points obtained by projecting the 2D lattice points satisfying $|\eta-\eta_0|<w$ onto the $\eta = 0$ line (the ξ axis) to be the points of the Fibonacci lattice. A path can be defined in the 2D lattice which connects points satisfying $|\eta-\eta_0|<w$ in the order of increasing ξ. Steps made along this path by moving along the x-axis by one lattice constant will produce an interval of width

$$d_A = a_x\cos\theta \tag{10a}$$

between points along the ξ axis, while steps made by moving along the y-axis by one lattice constant will produce an interval of width

$$d_B = a_y\sin\theta . \tag{10b}$$

In the Fibonacci sequence only two interval lengths occur and this property is guaranteed if we choose w so that it is always possible to step either in

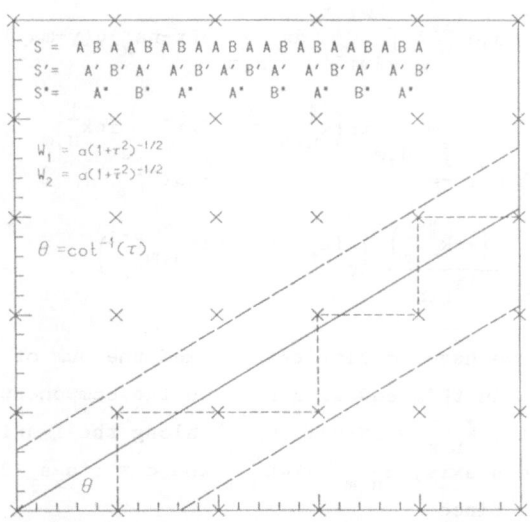

Fig. 1. Fibonacci lattice obtained via a projection from 2D.

the x-direction or the y-direction but never possible to step in both directions. Since a step in the x-direction changes η by

$$\Delta\eta_x = - a_x\sin\theta \qquad (11a)$$

while a step in the y-direction changes η by

$$\Delta\eta_y = a_y\cos\theta \qquad (11b)$$

we must choose the interval width of accepted η values, $2w$, to obey

$$2w = (\Delta\eta_y - \Delta\eta_x) = (a_y\cos\theta + a_x\sin\theta). \qquad (12)$$

Next, we know that A intervals occur τ times more often than B intervals in a long Fibonacci sequence. To obtain this property we require that

$$\cot\theta = \frac{\tau a_x}{a_y} \qquad (13)$$

which in turn implies, by comparison with Eqs. (10) that

$$\frac{d_A}{d_B} = \tau\left(\frac{a_x}{a_y}\right)^2 . \qquad (14)$$

Finally, we must choose η_0 so that the intervals occur in a Fibonacci sequence. In Appendix A we give an algebraic proof that $\eta_0 = w\tau^{-3}$ produces the Fibonacci sequence so that the condition on η is $-w_1 < \eta < w_2$ where $w_1 = w(1+\tau^3) = 2w\tau^{-1}$ and $w_2 = w(1-\tau^{-3}) = 2w\tau^{-2}$.

With the projection from 2D established the evaluation of the structure factor for the Fibonacci lattice, S_F, is straight-forward:

$$S_F(k) = \frac{1}{L_\xi} \int_{-\infty}^{\infty} d\xi e^{-ik\xi} \int_{-2w\tau^{-1}}^{2w\tau^{-2}} d\eta \sum_{n,m} \delta(x-na_x)\delta(y-ma_y)$$

$$= \frac{1}{a_x a_y L_\xi} \sum_{n,m} \int_{-\infty}^{\infty} d\xi e^{i\xi\left(k_{n,m}^{\parallel}-k\right)} \int_{-2w\tau^{-1}}^{2w\tau^{-2}} e^{i\eta k_{n,m}^{\perp}}$$

$$= \frac{\tau^2}{d} \sum_{n,m} \frac{\delta\left(k-k_{n,m}^{\parallel}\right)}{iz_{n,m}} \left(e^{iz_{n,m}\tau^{-2}} - e^{-iz_{n,m}\tau^{-1}} \right) . \tag{15}$$

To obtain Eq. (15) we have Fourier transformed the sum of delta functions on 2D lattice points. In this equation $k_{n,m}^{\parallel}$ is the component of the 2D reciprocal lattice vector $\vec{k}_{n,m} = 2\pi(n/a_x, m/a_y)$ along the ξ axis, $k_{n,m}^{\perp}$ is its component along the η axis, $z_{n,m} = 2w k_{n,m}^{\perp}$ and $d = \tau^2 a_x a_y/2w = a_y/\sin\theta = \tau a_x/\cos\theta = d_B + \tau d_A$. Since

$$k_{n,m}^{\parallel} = \frac{2\pi n\cos\theta}{a_x} + \frac{2\pi m\sin\theta}{a_y} = \frac{2\pi}{d}(m+n\tau) \tag{16}$$

we see that $S_F(k)$, like $S_p(k)$, has a set of δ-function peaks but, unlike $S_p(k)$, the set of k points is labelled by two indices associated with two fundamental periods whose ratio is τ. The weight of the delta function peaks depends on $k_{n,m}^{\perp}$, approaching τ^2/d for $k_{n,m}^{\perp} \to 0$ and decreasing as $|k_{n,m}^{\perp}|^{-1}$ for $z_{n,m} \gg 1$:

$$z_{n,m} = \frac{4\pi wm\cos\theta}{a_y} - \frac{4\pi ns\sin\theta}{a_x} = \frac{2\pi(md_A - nd_B)}{(\tau^{-2}d_B + \tau^{-1}d_A)} . \tag{17}$$

Note that for $d_A = d_B$, $z_{n,m} = 2\pi(m-n)$ so that $S_F\left(k_{n,m}^{\parallel}\right) = 0$ unless $n = m$ and the results for a periodic system are recovered. Also note that τ^2/d, the $k_{n,m}^{\perp} \to 0$ limit of $S_F(k_{n,m}^{\parallel})$, is the reciprocal of the mean distance between points on the Fibonacci lattice.

FORM FACTORS AND FOURIER TRANSFORMS

Many physical probes measure Fourier transforms of some function $F(\vec{r})$ of position in a system. For example elastic X-ray scattering measures the Fourier transform of the electron density while elastic neutron scattering measures the Fourier transform of the nuclear density. For a periodic superlattice[13], with period d,

$$F(\vec{q}) = \frac{1}{V} \int_{-\infty}^{\infty} d^3r e^{-i\vec{q}\cdot\vec{r}} F(\vec{r})$$

$$= \frac{1}{L_z} \sum_{n=-\infty}^{\infty} e^{-iq_z dn} f(\vec{q}) = S_p(q_z) f(\vec{q}) \tag{18}$$

where, the form factor

$$F(\vec{q}) = \int_0^d dz\, e^{-iq_z z} \int \frac{d^2 r}{L_x L_y} e^{i\vec{q}_\perp \cdot \vec{r}_\perp} F(\vec{r}) \ . \tag{19}$$

The expression of $F(\vec{q})$ as the product of the static structure factor, discussed above, and a form factor is often useful in periodic systems. We show below that a similar separation is possible for quasiperiodic systems.

For a Fibonacci superlattice, $F(\vec{r})$ has the same dependence on position within each A interval and within each B interval. It follows that for a Fibonacci superlattice

$$F(\vec{q}) = S_A(\vec{q}) f_A(\vec{q}) + S_B(\vec{q}) f_B(\vec{q}) \tag{20}$$

where $f_A(\vec{q})$ and $f_B(\vec{q})$ are form factors calculated as in Eq. (19) by integrating over A intervals and B intervals respectively, and the structure factors $S_A(\vec{q})$ and $S_B(\vec{q})$ are defined by

$$S_X(\vec{q}) = \frac{1}{L_\xi} \sum_{j \in \{j\}_X} e^{-iq\xi_j} \tag{21}$$

where $\{j\}_X$ is the set of labels of points on the Fibonacci lattice which are followed by X intervals. For $S_A(\vec{q})$ we note that each point on the Fibonacci lattice corresponding to the "words" $W_A^2 = AB$ and $W_B^2 = A$ (see Table 2 and the associated discussion) is followed by an A interval. With these words as elementary symbols $d_A \to d_A^{(2)} = d_A + d_B$, $d_B \to d_B^{(2)} = d_A$ so that $d_B^{(2)} + \tau d_A^{(2)} = \tau d$, $k_{n+m,n}^{\parallel (2)} = 2\pi(n+(n+m)\tau)/d^{(2)} = k_{n,m}^\parallel$ and $z_{n+m,n}^{(2)} = -z_{n,m}\tau^{-1}$. Inserting these results in Eq. (15) gives

$$S_A(\vec{q}) = \frac{\tau^2}{d} \sum_{n,m} \frac{\delta\left(q, k_{n,m}^\parallel\right)}{iz_{n,m}} \left(e^{iz_{n,m}\tau^{-2}} - e^{-iz_{n,m}\tau^{-3}} \right) \ . \tag{22}$$

Similarly, to evaluate $S_B(\vec{q})$ we note that with $W_A^3 = ABA$ and $W_B^3 = AB$ as elementary symbols, there is a B interval d_A to the right of each lattice point. Using $d^{(3)} = \tau d^{(2)} = \tau^2 d$, $k_{2n+m,n+m}^{\parallel (3)} = k_{n+m,n}^{\parallel (2)} = k_{n,m}^\parallel$, and $z_{2n+m,n+m}^{(3)} = -z_{n+m,n}^{(2)}\tau^{-1} = \tau^{-2} z_{n,m}$ in Eq. (15) then gives

$$S_B(q) = \frac{e^{-iqd_A}}{d}\tau^2 \sum_{n,m} \frac{\delta\left(q, k_{n,m}^\parallel\right)}{iz_{n,m}} \left(e^{iz_{n,m}\tau^{-4}} - e^{-iz_{n,m}\tau^{-3}} \right) \ . \tag{23}$$

Eqs. (20), (22) and (23) can be used to describe experiments which probe the structural properties of Fibonacci superlattices.

Elastic X-ray scattering measures the Fourier transform of the electronic density. For the sake of definiteness we consider here a model in which the X intervals consist of N_X layers separated by a lattice constant a_X and consisting entirely of atom X (X=A or B). We thus avoid the problem of alloy disorder, which would occur, for example, in the Ge_xSi_{1-x} layers o a Si/Ge_xSi_{1-x} superlattice and which is irrelevant to our present interest. We also avoid keeping track of the algebraic details associated with the presence of two types of layers within each interval in superlattices composed of III-V semiconductors, e.g. GaAs/AlAs. We assume identical 2D lattices in each layer and take the electron density to be the sum of overlapping charge densities centered on each atomic site. After expressing th atomic charge density in terms of its Fourier transform, $n_0(\vec{q})$ and insertin into Eq. (19) an elementary calculation yields[14] the familiar result

$$f_X(\vec{G}_\perp, q_z) = A_0^{-1} n_0(\vec{G}_\perp, q_z) \left(\frac{1-e^{-iq_z N_X a_X}}{1-e^{-iq_z a_X}} \right) \qquad q_z \neq \frac{2\pi k}{a_X} \qquad (24)$$
$$= N_X \qquad q_z = \frac{2\pi k}{a_X} \quad .$$

A_0 is the unit cell area for the 2D lattice and we have noted that f_X is zero unless $\vec{q}_\perp$ equals a reciprocal lattice vector of the 2D lattice ($\vec{G}_\perp$). Note that $f_X(\vec{G}_\perp, q_z)$ has a strong peak, for $N_X \gg 1$, when q_z is near a recipro cal lattice vector, $G_z^X = 2\pi k/a_X$, of the bulk material X. For that reason, it is instructive to write $q_z = G_z^X + \tilde{q}_z$ and make the, usually appropriate, continuum approximation $\tilde{q}_z a_X \ll 1$ which, for $\tilde{q}_z \neq 0$ gives

$$f_X(\vec{G}_\perp, q_z) \simeq (a_X A_0)^{-1} n_0^X(\vec{G}_\perp, G_z^X)(1-e^{-i\tilde{q}_z d_X})/(i\tilde{q}_z) \qquad (25)$$

where $d_X = N_X a_X$ and we have noted that $n_0(|\vec{G}|)$ varies with $\vec{G}$ on a scale $\sim a_X^-$

We first consider the case of a lattice-matched Fibonacci superlattice (e.g. GaAs/AlAs) where, to sufficient accuracy, we may take $a_A = a_B = a$. Then adopting the continuum approximation and comparing Eqs. (25), (20), (22) and (23) we have, after some algebra, that

$$n(\vec{G}_\perp, G_z + \tilde{q}_z) = 4\tau^2 (n_0^A(\vec{G}) - n_0^B(\vec{G}))$$

$$\sum_{n,m} \frac{\delta(\tilde{q}_z, k_{n,m}^\parallel) e^{-i\left(z_{n,m}\tau^{-3} + k_{n,m}^\parallel d_A\right)/2}}{A_0 a z_{n,m} k_{n,m}^\parallel d} \sin\left(\frac{z_{n,m}}{2\tau}\right) \sin\left(\frac{k_{n,m}^\parallel d_A}{2}\right) \quad . \qquad (26)$$

In Eq. (26) we have explicitly separated the phase factor since the experiment measures only $|n(\vec{G})|^2$. The intensity of the satellites in the X-ray diffraction pattern around reciprocal lattice vector $\vec{G}$ is therefore proportional to $|n_0^A(\vec{G}) - n_0^B(\vec{G})|^2$, decreases as $z_{n,m}^{-2}$ and as $k_{n,m}^{\|^2}$ if either $z_{n,m}$ or $k_{n,m}^{\|}$ becomes large, and is largest if both $z_{n,m}$ and $k_{n,m}^{\|} d_A$ are as small as possible. The largest satellites generally occur for $k_{n,m}^{\|} = \pm k_p^{\|}$ where,

$$k_p = k_{F_p,F_{p-1}}^{\|} = \frac{2\pi}{d}(F_p + F_p \tau) = \frac{2\pi\tau^p}{d} . \tag{27}$$

and we have used Eq. (4b).

This is especially true for the case where d_A and d_B are also in the ratio of the golden mean. Then using Eq. (4c)

$$z_p = z_{F_p,F_{p-1}} = \frac{(-)^p 2\pi\tau^{-p+2}}{(\tau^{-1}+\tau)} \tag{28}$$

$$|n(\vec{G}_\perp, G_z + k_p)|^2 =$$
$$\tau^2 \frac{|n_0^A(\vec{G}) - n_0^B(\vec{G})|^2 (\tau^{-1}+\tau)^2 \sin^2(\pi\tau^p/(\tau^{-1}+\tau)) \sin^2(\pi\tau^{1-p}/(\tau^{-1}+\tau))}{(2\pi a A_0)^2} \tag{29}$$

which is invariant under $p \to 1-p$ and is largest for $p = -1, 0, 1, 2$. For large negative p ($k_{n,m}^{\|}$ very small) $|n(\vec{G}_\perp, G_z + k_p)|^2$ decreases because $z_{n,m} \propto k_{n,m}^{\perp}$ becomes large, while for large positive p ($k_{n,m}^{\|}$ very large) $|n(\vec{G}_\perp, G_z + k_p)|^2$ vanishes because the form factor vanishes. Unless $|p|$ becomes quite large, these conclusions are altered only by substantial deviations of d_A/d_B from τ. In Fig. 2 we plot $|n(\vec{G}_\perp, G_z)|$ versus G_z for a Fibonacci superlattice with A intervals composed of $F_9 = 34$ layers of material A and B intervals composed of $F_8 = 21$ layers of material B. These curves were obtained from Eq. (24) without using the continuum approximation which we expect to break down for $k^{\|} d \backsim d/a = (N_B + \tau N_A) = \tau^9$. The X-ray diffraction pattern for GaAs/AlAs superlattices has been measured by Merlin and co-workers[3,15] who have observed satellites at $G_z + k_p^{\|}$ for p as large as 5. The occurrence of the main-diffraction peaks in a geometric progression rather than the arithmetic progression which occurs in periodic systems, is a consequence of the simple behaviour of the system when the length scale is changed by a factor of τ by an inflation transformation.

We now turn to the case of strained-layer superlattices ($a_A \neq a_B$) for which, to our knowledge, no X-ray diffraction studies have been reported. Here the form factors of the A and B intervals peak at different wavevectors ($2\pi k/a_A$ and $2\pi k/a_B$ respectively). In addition we note that

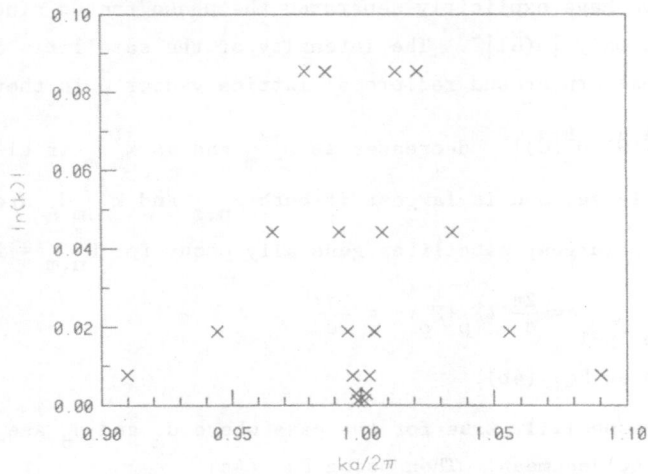

Fig. 2. Bragg diffraction satellites for a lattice-matched Fibonacci superlattice (see text).

$$k^{\parallel}_{n+kN_A,\,m+kN_B} = k^{\parallel}_{n,m} + \frac{2\pi k}{\bar{a}} \qquad (30a)$$

where

$$\bar{a} = (a_B N_B + \tau a_A N_A) \,/\, (N_B + \tau N_A) \qquad (30b)$$

is the average lattice constant, while

$$z^{\parallel}_{n+kN_A,\,m+kN_B} = z_{n,m} + 2\pi k \left[\frac{a_A - a_B}{\bar{a}}\right] \frac{N_B N_A}{N_B + \tau N_A} \;. \qquad (31)$$

Thus, provided $(a_A - a_B)/\bar{a}$ is small enough, the structure factors for the A and B sublattices have their strongest δ-function peaks centered about $2\pi k/\bar{a}$.

In Fig. 3 we plot $|n(\vec{G}_\perp, G_Z)|$ versus G_Z for a superlattice identical to that used for Fig. 2 except for the introduction of a ∿6% mismatch between a_A and a_B ($a_A < a_B$). For $G_Z < 2\pi/\bar{a}$ the diffraction satellites come mainly from the B intervals and reflect the structure of the B sublattice while for $G_Z > 2\pi/\bar{a}$ the diffraction satellites come mainly from the A intervals and reflect the structure of the A sublattice.

RAMAN SCATTERING FROM ACOUSTIC PHONONS

Raman scattering has been used extensively and successfully to study acoustic phonons in periodic semiconductor superlattices. (See, for example, Refs. 16 and 17 and work quoted therein.) Actually, as we see below, these experiments provide mainly structural information. In the photoelastic

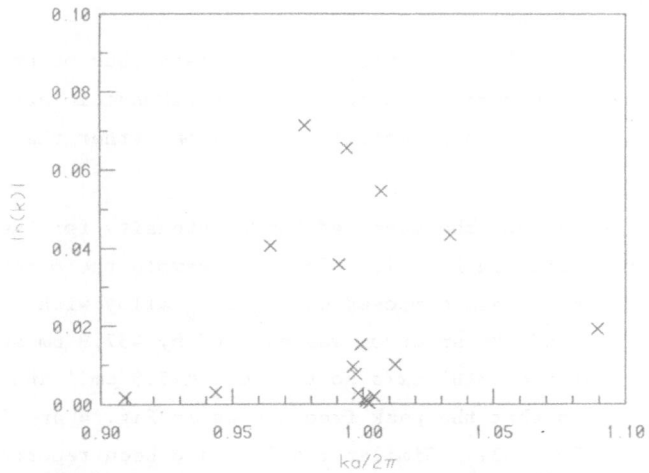

Fig. 3. Bragg diffraction satellites for a strained layer Fibonacci superlattice (see text).

continuum model the intensity for Raman scattering by acoustic phonons is given by:[11]

$$I(\omega) \propto \sum_j \delta(\omega-\omega_j)\left(\frac{n(\omega_j)+1}{\omega_j}\right)\left|\int_{-\infty}^{\infty} dz e^{-iqz}P(z)\frac{\partial U_j(z)}{\partial z}\right|^2 \qquad (32)$$

where q is the light-wavevector, ω_j is the normal-mode frequency, P(z) is the local photoelastic coefficient and $U_j(z)$ is normalized and gives the local displacement from equilibrium of the j-th normal mode[18]. The process described by the photoelastic coefficient is one in which the photon excites a particle-hole pair in the electron system, which subsequently decays to its ground state by emitting a phonon. P(z) is given locally by the bulk photoelastic coefficient of the immediate environment.

Typically the sound velocities of a superlattice's constituents differ far less than do the photoelastic coefficients. It is therefore a good approximation to replace the sound velocity by its root mean square average, $\bar{v}$, so that the phonon modes may be labelled by wavevector, $\omega_k = \bar{v}k$ and $U_k(z) = L_z^{-\frac{1}{2}}e^{ikz}$. With these simplifications, Eq. (32) reduces, in the low temperature limit, to

$$I(\omega) \propto \sum_k \delta(\omega-\bar{v}|k+q|)|P(k)|^2 \qquad (33)$$

where P(k) is the Fourier transform of P(z). As discussed in the previous section the most intense δ-function peaks of P(k) in the Fibonacci superlattice will occur for $k = \pm k_p = \pm 2\pi\tau^P/d$ and at k = 0, so that the main peaks in the Raman intensity should occur for $\omega = \omega_{BR} = \bar{v}q$ and at

$$\omega_p^{\pm} = \left| \omega_0 \tau^P \pm \omega_{BR} \right| \tag{33}$$

where $\omega_0 = \bar{v} 2\pi/d$. The most intense Raman peaks thus occur in a set of doublets, as in the periodic case. For the Fibonacci lattice, however, the mid-points increase in a geometric progression rather than in an arithmetic progression.

In Fig. 4 we show the measured Raman intensity for the Si/Ge_xSi_{1-x} superlattice studied in Ref. 12. For this sample the A regions are composed of Si, the B regions are composed of Ge_xSi_{1-x} alloy with $x \approx 0.48$, $d_A = 5.4$ nm, $d_B = 4.2$ nm and the spectrum was excited by 457.9 nm argon laser light. We estimate that $\bar{v} \approx 8.0 \times 10^5$ cm/s so that $\omega_{BR} = 5.3$ cm^{-1} and $\omega_0 = 20.7$ cm^{-1}. It is easily seen that the peak frequencies in Fig. 4 are in quantitative agreement with Eq. (33). Similar results have been reported for[19,20,21] GaAs/AlAs and for[22] GaAs/Ga$_{1-x}$Al$_x$As Fibonacci superlattices with a slightly different geometry. In the latter case both A and B intervals have an Al$_x$Ga$_{1-x}$As (x = 0.25 in Ref. 21 and x = 1.0) layer of thickness d_2 and a GaAs layer whose thickness differs in A and B layers. Writing

$$P(k) = \sum_{n,m} \left(\delta\left(k, k_{n,m}^{\parallel}\right)\left(P_1 - P_2\right) S\left(k_{n,m}^{\parallel}\right) + \bar{P}\delta(k) \right), \tag{34}$$

eqs. (19), (22) and (23) give

$$\left| S\left(k_{n,m}^{\parallel}\right) \right| = \left| \frac{4\tau^2 \sin(k_{n,m} d_A/2)\sin(z_{n,m}/2\tau)}{k_{n,m} z_{n,m} d} \right| \tag{35}$$

for the geometry of the Si/Ge_xSi_{1-x} sample and

$$\left| S\left(k_{n,m}^{\parallel}\right) \right| = \left| \frac{4\tau^2 \sin(k_{n,m}^{\parallel} d_2/2)\sin(z_{n,m}/2\tau)}{k_{n,m} z_{n,m} d} \right| \tag{36}$$

for the geometry of the GaAs/Al$_x$Ga$_{1-x}$As samples. Eqs. (35) and (36) can be understood by realizing that, for $k \neq 0$, any constant shift in P(z) does not change P(k). In Eq. (35), P(k) equals the structure factor of the A-sublattice times the form factor, $(P_{Si} - P_{Ge})(1 - e^{-ik_{n,m} d_A})/ik_{n,m}$, obtained by integrating $P(z) - P_{Ge}$ over the Si intervals. In Eq. (36) P(k) equals the structure factor of the full Fibonacci lattice times the form factor $(P_{Al_xGa_{1-x}As} - P_{GaAs})(1 - e^{-ik_{n,m} d_2})/ik_{n,m}$, obtained by integrating $P(z) - P_{GaAs}$ over the Al$_x$Ga$_{1-x}$As intervals. For the same d_A and d_B, $d_2 < d_A$ so that $S(k_{n,m}^{\parallel})$ decreases at larger values of $k_{n,m}^{\parallel}$ and smaller values of $k_{n,m}^{\perp}$ in the latter case. Correspondingly, the four values of p which produce the most intense peaks, are p = 0,1,2, and 3 rather than p = -1,0,1 and 2.

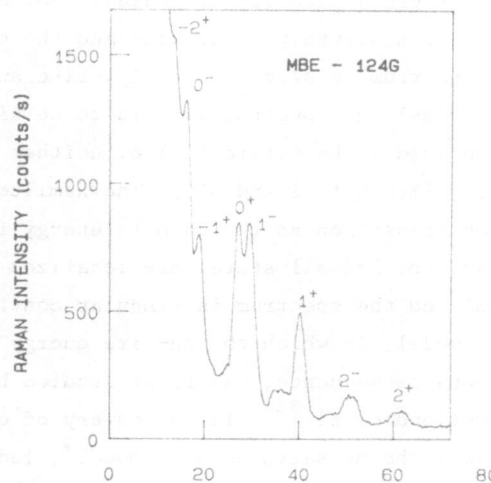

Fig. 4. Raman spectrum for a Fibonacci superlattice.

While the photoelastic continuum theory does predict the peak frequencies for the Raman spectrum accurately, it does considerably more poorly for the peak intensities. Part of the discrepancy is due to the broadening of the interfaces between layers which occurs in real systems and, in fact, the peak intensities give some information on this important materials parameter[22,23]. Some of the discrepancy is due to our neglect of the sound velocity difference between the two materials and we will return to this later. In all probability, the local approximation for the photoelastic coupling also introduces some inaccuracy, since the thin films may consist of as few as ~10 atomic layers. Finally, we mention that the photoelastic continuum model becomes quite poor when the exciting frequency is near an electronic resonance. Merlin and co-workers[3,20] have attempted to associate features in the Raman spectrum in this regime with gaps in the phonon density of states.

SPECTRAL PROPERTIES OF QUASIPERIODIC SCHRODINGER OPERATORS

The spectra of 1D quasiperiodic Schrodinger operators can show a variety of behaviours. A frequently studied example[25-30] is the discrete Schrodinger equation

$$\psi_{n+1} + \psi_{n-1} + \lambda \cos(2\pi\sigma n + \theta)\psi_n = E\psi_n \qquad (37)$$

where σ is an irrational number. The spectra can be classified[30], at least qualitatively, according to the behaviour of

$$\lim_{\Delta E \to 0} \left(N(E + \Delta E) - N(E) \right) \sim (\Delta E)^{\alpha} \qquad (38)$$

where N(E) is the integrated density of states. For $\alpha = 1$ in Eq. (38), the spectrum is said to be absolutely continuous and the eigenstates are extended; for $\alpha = 0$ the spectrum is said to be point-like and the eigenstates are localized; and for $0 < \alpha < 1$ the spectrum is said to be singular continuous and the eigenstates are said to be critical, i.e. neither localized nor extended in the usual sense. (See Ref. 1 and 31.) The Hamiltonian of Eq. (37) exhibits a localization transition as the on-site energy increases. For $\lambda < 2$ all states are extended, for $\lambda > 2$ all states are localized and for $\lambda = 2$ all states are critical and the spectrum is singular continuous. The corresponding Fibonacci model, in which the on-site energy has two values which alternate in a Fibonacci sequence, was first studied by Kohmoto et al.[32] and independently by Ostlund et al.[33]. The discovery of quasicrystals[2], for which these Fibonacci chains serve as a 1D model[4], led to many new investigations[34-42]. For the Fibonacci chains, the spectrum is singular continuous and the eigenstates are critical for any strength of on-site interaction. We therefore expect that plasmons, phonons, electrons or any other elementary excitations of the semiconductor Fibonacci superlattices will have singular continuous spectra.

For Fibonacci systems the limit $\Delta E \to 0$ in Eq. (38) may be taken by considering a sequence of periodic approximations in which the Fibonacci sequence is stopped at the kth generation and repeated periodically. For the Fibonacci chains mentioned above, for example, the tight-binding band would break up into F_k subbands so that

$$N(E_i^{(k)} + W_i^{(k)}) - N(E_i^{(k)}) = (F_k)^{-1} \tag{39}$$

where $E_i^{(k)}$ is the bottom of the ith subband, $W_i^{(k)}$ is its width and the integrated density of states is expressed in units of electrons per site. Comparing with Eq. (38) we see that

$$\alpha = - \lim_{k \to \infty} \frac{\ln(F_k)}{\ln(W_i^{(k)})} . \tag{40}$$

However, as we discuss below, when k is increased to $k+k'$, each subband of the generation-k periodic approximation breaks up into smaller subbands in the same way as the original full band is broken up into subbands. This self-similarity leads to Cantor-set spectra where α has fractional values. As we see from Eq. (40), α depends only on the "tail" of the sequence of divisions into subbands and, unless it is a constant, it will be a strongly singular function of energy. The set of energies in a Fibonacci spectrum is an example of a multifractal object, i.e. many different values of the scaling index α occur.

Multifractal, or "strange", sets have arisen in a number of contexts, (e.g. Ref. 43-47) and Halsey et al.[48] have proposed a very powerful way of characterizing them.[49] Before briefly describing their method we mention the simplest example of a fractal set, the classical Cantor set. Consider a single band of width W which we divide into subbands by the following iterative procedure. At each generation we obtain the next generation by dividing each subband into m smaller subbands and dividing the subband width by n>m. Then

$$\alpha = - \lim_{k \to \infty} \frac{\ln(m^k)}{\ln(Wn^{-k})} = \frac{\ln m}{\ln n} . \tag{41}$$

For this case, then, the set of energies in the spectrum has fractal dimension[50], f, given by f = α, and every energy in the spectrum is characterized by a scaling index α.

The Fibonacci system spectrum is more complicated: a range of scaling indices, α, occur. We want to characterize the spectrum by calculating the fractal dimension of the set of energies where the scaling index is α, f(α). To evaluate f(α) we consider the quantity

$$\Gamma(q,\tau) \equiv \lim_{k \to \infty} \Gamma^{(k)}(q,\tau) \tag{42}$$

where

$$\Gamma^{(k)}(q,\tau) = F_k^{-q} \sum_{i=1}^{F_k} (W_i^{(k)})^{-\tau} . \tag{43}$$

By the definition of the scaling index, if subband i occurs in a sequence characterized by scaling index α

$$\lim_{k \to \infty} W_i^{(k)} \equiv W(\alpha) = W_0(\alpha)(F_k)^{-1/\alpha} \tag{44}$$

and, by the definition of the fractal dimension, the number of subbands with scaling index between α and α+dα as k→∞ is

$$M(\alpha)d\alpha = d\alpha M_0(\alpha) \ (W(d))^{-f(\alpha)} = d\alpha M_0(\alpha) W_0(\alpha)^{-f(\alpha)} F_k^{f(\alpha)/\alpha} . \tag{45}$$

In Eqs. (44) and (45) $W_0(\alpha)$ and $M_0(\alpha)$ are independent of k and we will not be able to determine their values. Substituting Eqs. (44) and (45) into Eqs. (43) and (42) gives

$$\Gamma(q,\tau) = \lim_{k \to \infty} \int d\alpha' \ M_0(\alpha') [W_0(\alpha')]^{-f(\alpha')-\tau} F_k^{(f(\alpha')+\tau)/\alpha'-q}$$
$$= C \lim_{k \to \infty} F_k^{f(\alpha(\tau))-q} \tag{46}$$

where we have expanded $(f(\alpha')+\tau)/\alpha'$ about its maximum with respect to α, α(τ), so that

$$f'(\alpha(\tau)) = (f(\alpha(\tau)) + \tau)/\alpha(\tau) \tag{47}$$

and C has only a logarithmic dependence on k, which we ignore. The second derivative of $(f(\alpha') + \tau)/\alpha'$ with respect to α' at the maximum is $f''(\alpha(\tau)) < 0$. From Eq. (46) we see that $\Gamma(q,\tau)$ remains finite as $k \to \infty$ only if $\tau = \tau^*(q)$ where

$$f'(\alpha)\big|_{\alpha=\alpha(\tau^*(q))} = q \ . \tag{48}$$

Differentiating Eq. (48) with respect to q gives

$$\frac{d\tau^*(q)}{dq} = \left[f''(\alpha) \frac{d\alpha(\tau)}{d\tau} \big|_{\tau=\tau^*} \right]^{-1} = \alpha(\tau^*(q)) \tag{49}$$

where the second form follows from differentiating Eq. (47) with respect to τ. On the other hand, from Eqs. (47) and (48)

$$f(\alpha)\big|_{\alpha=\alpha(\tau^*(q))} = \alpha q - \tau^*(q) \ . \tag{50}$$

Equations (49) and (50) can be used to calculate the distribution of scaling indices and the fractal dimension associated with each scaling index, $f(\alpha)$, for any multifractal set. For large negative q, $\tau^*(q)$ will be negative[51], $\Gamma^{(k)}(q,\tau)$ will reflect only the widest bands, i.e. those with the maximum scaling index $\alpha=\alpha_{MAX}$. Normally α achieves its maximum at a countable set of energies so that $f_{MAX} = 0$ and from Eq. (50)

$$\lim_{q\to-\infty} \frac{\tau^*(q)}{q} = \alpha_{MAX} \ . \tag{46}$$

For q = 0 $f'(\alpha) = 0$ so that $f(\alpha)\big|_{\alpha=\alpha(\tau^*(0))}$ is the maximum fractal dimension. In fact the maximum of $f(\alpha)$ is the (Hausdorf) fractal dimension of the spectrum since the fractal dimension of the union of several sets is the maximum of the fractal dimensions of the sets individually. Note from Eq. (50) that

$$\tau^*(q=0) = -f\big[\alpha(\tau^*(q=0))\big] \ . \tag{51}$$

For q=1, we note that

$$\Gamma^{(k)}(q=1, \tau=0) = 1 \tag{52}$$

so that $\tau^*(q=1) = 0$ and from Eq. (50)

$$f\big[\alpha(\tau^*(q=1))\big] = \alpha\big[\tau^*(q=1)\big] \ . \tag{53}$$

For q large and positive, $\tau^*(q)$ becomes positive and $\Gamma^{(k)}(q,\tau)$ will reflect the narrowest bands, i.e. those with $\alpha=\alpha_{MIN}$. Assuming again that $f(\alpha_{MIN}) = 0$ we have

$$\lim_{q\to\infty} \frac{\tau^*(q)}{q} = \alpha_{MIN} \ . \tag{54}$$

The $f(\alpha)$ curves thus provide a lot of revealing information about the multifractal spectra of Fibonacci systems. Note that we are not guaranteed, in general, that we will be able to isolate the sets of each scaling index which occurs, by this procedure. In particular, if $f''(\alpha)>0$, the large k behaviour of $\Gamma^{(k)}(q,\tau)$ will not be dominated by the set with scaling index α, for any value of τ.

THE METHOD OF KOHMOTO, KANDANOFF AND TANG

The model originally considered by Kohmoto et al.[32] and Ostlund et al.[33] is a one-dimensional hopping Hamiltonian

$$t\psi_{i+1} + t\psi_{i-1} + V_i\psi_i = E\psi_i \tag{55}$$

where the on-site energies, $\{V_i\}$, alternate in a Fibonacci sequence. We discuss this case in some detail, following the methods of the former authors, since the same techniques can be used to treat models with more direct experimental relevance and, in fact, many properties of the spectra will prove to be generic. We use a transfer matrix technique defining $M(n)$ by

$$\begin{bmatrix} \psi_{n+1} \\ \psi_n \end{bmatrix} = M(n) \begin{bmatrix} \psi_n \\ \psi_{n-1} \end{bmatrix} \tag{56}$$

so that

$$M(n) = \begin{pmatrix} (E-V_n)/t & -1 \\ 1 & 0 \end{pmatrix} \tag{57}$$

and

$$\begin{bmatrix} \psi_{N+1} \\ \psi_N \end{bmatrix} = M(N)M(N-1)\dots M(1) \begin{bmatrix} \psi_1 \\ \psi_0 \end{bmatrix} . \tag{58}$$

Defining $M_j=M(F_j)M(F_j-1)\dots M(1)$ (i.e. M_j is the transfer matrix from the beginning to the end of the Fibonacci sequence at generation j), it follows from the definition of the Fibonacci sequence that

$$M_{j+1} = M_{j-1} M_j . \tag{59}$$

As emphasized previously the change in the Fibonacci sequence with increasing generation can be thought of either as lengthening the sequence or as maintaining the same sequence but with an increase in the length of the elementary building blocks (words). Thus the recursion relation, Eq. (59), can be thought of as a renormalization group transformation.

We approach the spectrum of an infinite Fibonacci lattice by periodically continuing after j generations and taking the limit that j goes to infinity. The Bloch condition after j generations is

$$\det|M_j-e^{ikd_j}| = (M_{11}-e^{ikd_j})(M_{22}-e^{ikd_j}) - M_{12}M_{21} = 0 . \tag{60}$$

Noting that $\det(M_1) = \det(M_2) = 1$, which implies that $\det(M_j) = 1$, Eq. (60) leads to

$$x_j \equiv \frac{1}{2} \, \text{tr}(M_j) = \cos(kd_j) \ . \tag{61}$$

(In Eqs. (60) and (61) $k \in [-\frac{\pi}{d_j}, \frac{\pi}{d_j}]$ and $d_j = F_j a$ where a is the lattice constant.) It follows from Eq. (61) that E is in the spectrum if $|x_j(E)| \leq 1$. A recursion relation for the x_j's can be obtained from eq. (59) by noting that $M_{j-2} = M_{j-1}^{-1} M_j$ so that

$$x_{j+1} + x_{j-2} = \frac{1}{2} \, \text{Tr}\left([M_{j-1}^{-1} + M_{j-1}] M_j \right) = x_{j-1} \text{Tr}(M_j) = 2x_{j-1} x_j . \tag{62}$$

Thus studying the eigenvalue spectrum reduces to studying the properties of a mapping of the traces of the eigenvalue spectrum. The same trace map will apply to all the Fibonacci models we consider. The spectrum for the diagonal Fibonacci model is illustrated in Fig. 5.

Numerical studies[32] show that the map of Eq. (62) has a strange repeller; i.e. all starting points of the map, except for points in a strange (multifractal) set will produce escaping trajectories, i.e. trajectories for which $|x_k| > 1$ for $k > k_0$. Thus almost all energies are not in the spectrum as $k \to \infty$. In fact numerical studies show that

$$\sum_i W_i^{(k)} \sim F_k^{-\delta} \ . \tag{63}$$

The map also has an important constant of the motion,

$$I \equiv x_{j+1}^2 + x_j^2 + x_{j-1}^2 - 2x_{j+1} \, x_j \, x_{j-1} - 1. \tag{64}$$

For the diagonal discrete Fibonacci model

$$x_1 = \frac{1}{2} \, \text{tr}(M_B) = (E - V_B)/2t \tag{65a}$$

$$x_2 = \frac{1}{2} \, \text{tr}(M_A) = (E - V_A)/2t \tag{65b}$$

and

$$x_3 = \frac{1}{2} \, \text{tr}(M_B M_A) = 2x_1 x_2 - 1 \tag{65c}$$

so that

$$I = \frac{1}{4} \left(\frac{V_B - V_A}{t} \right)^2 \geq 0 \ . \tag{66}$$

We will argue that to a large degree the scaling properties of the spectrum depend only on I. The first evidence of this comes from studies[52] which show that, with x_1 and x_2 chosen at random and x_3 chosen to give a fixed value of I, the probability of a point escaping to infinity after k iterations varies as $F_k^{-\delta'}$ and, to numerical accuracy, $\delta' = \delta$ with δ calculated from Eq. (63), for the same value of I. For fixed I, the starting point of the trace map for the diagonal discrete Fibonacci model, traces out a line in (x_1, x_2) space. We see later that other models trace out different lines

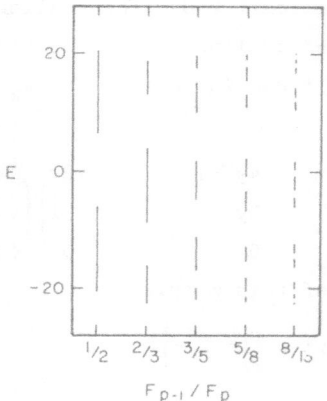

Fig. 5. Spectrum for the diagonal Fibonacci model.

in (x_1,x_2) space and the fact that $\delta'=\delta$ here probably indicates that δ depends only on I. Comparing Eqs. (63), (43) and (46) we see that

$$\delta = f'(\alpha(-1)) = \frac{1 - f(\alpha(-1))}{\alpha(-1)} \quad , \tag{67}$$

which can be evaluated from the $f(\alpha)$ curve.

On a qualitative level the self-similarity of the eigenvalue spectrum is already apparent from the trace map, since for $|E-E_j^{(k)}| \lesssim W_j^{(k)}/2$ we can approximate

$$x_k(E) \simeq 2(E-E_j^{(k)})/W_j^{(k)} . \tag{68}$$

(Here $E_j^{(k)}$ is the center of the j-th band at generation k.) Within this narrow energy range $x_k(E)$ has the same form as $x_1(E)$ has over its bandwidth $(W^{(1)}=4t)$ so that it breaks up into still smaller bands with increasing generation just as the original bands broke up to form the band of width $W_j^{(k)}$ at generation k. The non-escaping trajectories of the trace map are either chaotic or cyclic and, for the latter class, a quantitative treatment of the scaling behaviour is possible. For example, it follows from Eq. (62) that a 2-cycle of the trace map, $(y \rightarrow z \rightarrow y \rightarrow z...)$, can occur if $y = z/(2z-1)$. To analyze this case it is convenient to express y and z in terms of a single free parameter J so that

$$y = J + (J^2-J)^{\frac{1}{2}} \tag{69a}$$

$$z = J - (J^2-J)^{\frac{1}{2}} . \tag{69b}$$

Fixing the "constant of the motion" I fixes J by Eq. (64) from which it follows that

$$J = \frac{3 + \sqrt{25+16I}}{8} \tag{70}$$

which is larger than one for I > 0. The scaling of the spectrum near a 2-cycle may be determined by linearizing the trace map around this fixed point. Defining $x_k = y + \delta x_k$ for k even and $x_k = z + \delta x_k$ for k odd we find that, for k even,

$$\begin{pmatrix} \delta x_{k+2} \\ \delta x_{k+1} \\ \delta x_k \end{pmatrix} = \begin{pmatrix} 4yz + 2z & 4y^2-1 & -2y \\ 2z & 2y & -1 \\ 1 & 0 & 0 \end{pmatrix} \begin{pmatrix} \delta x_k \\ \delta x_{k-1} \\ \delta x_{k-2} \end{pmatrix} . \tag{71}$$

The largest eigenvalue of this matrix is

$$\epsilon_2 \equiv \left(4J - \tfrac{1}{2}\right) + \left(\left(4J - \tfrac{1}{2}\right)^2 - 1\right)^{\frac{1}{2}} > 1 \tag{72}$$

so that for large p, $\delta x_{k+2p} \sim \epsilon_2^p \, \delta x_k$ and from Eq. (68)

$$W^{(k+2p)} \sim W^{(k)}/\epsilon_2^p . \tag{73}$$

Using Eq. (73) and noting that $\ell n(F_k) \sim \ell n(\tau^k) \sim k\ell n(\tau)$ it follows that for a 2-cycle of the trace map $\alpha = \alpha_2$ where

$$\alpha_2 = \frac{2\ell n(\tau)}{\ell n\,\epsilon_2} . \tag{74}$$

The two cycle governs the spectrum near its edges, or the edges of its subclusters, where the bands are typically most dense so that α_2 is usually expected to be the minimum scaling index. This type of fixed point analysis was first done by Kohmoto and Oono[32] who also identified a six-cycle of the map (y → 0 → 0 → y → 0 → 0), which dominates the scaling near band centers and is typically found to be the maximum scaling index. They find that

$$\alpha_6 = \frac{6\ell n\tau}{\ell n\,\epsilon_6} \tag{75}$$

where

$$\epsilon_6 = \{(1+4(1+I)^2)^{\frac{1}{2}} + 2(1+I)\}^2 . \tag{76}$$

In Fig. 6 α_2 and α_6 are plotted versus I. (Note that for I>>1 $\alpha_2 = 4/\ell n(I/\tau) > \alpha_6 = 3/\ell n(I/\tau)$.) For I → 0, α_6 approaches 1 while α_2 approaches the scaling index of the band edges of the periodic system, $\tfrac{1}{2}$. Note that both α_2 and α_6 depend only on I and not on the initial conditions for the trace map appropriate to the diagonal Fibonacci model. Also note that if $(x_j, x_{j+1}, x_{j+2}, \ldots)$ is a trajectory of the trace map, so too is $(x_{j+2}, x_{j+1}, x_j, \ldots)$. Thus if λ is an eigenvalue of the linearized trace map, 1/λ is also an eigenvalue and we are guaranteed that the spectrum is not point-like. We see below that many spectral properties can be calculated from the same trace map. The different situations are distinguished entirely by the initial conditions of the map.

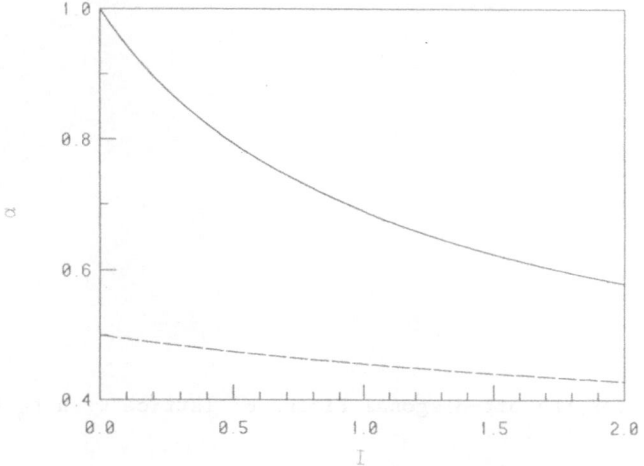

Fig. 6.   Band-edge (α_2:dashed line) and band-center (α_6:solid-line) scaling
indices versus I.

OTHER MODELS FOR DISCRETE SYSTEMS

Another natural model for the electronic properties of a 1D quasiperiodic system is the off-diagonal discrete Fibonacci model where the on-site
energy is constant and the hopping integrals between sites, t_A and t_B, alternate in a Fibonacci sequence. This model has been studied numerically by
several workers[35-37] and via the trace map by Kohmoto and co-workers.[34,41]
In this case it is convenient to choose as the elementary building blocks of
the Fibonacci sequence W_B^2 = A and W_A^2 = AB so that, choosing the on-site
energy to be zero,

$$x_1 = E/2t_A \tag{77a}$$

$$x_2 = E^2/2t_A t_B - (t_B/t_A + t_A/t_B)/2 \tag{77b}$$

and

$$x_3 = \frac{E^3}{2t_A^2 t_B} - \frac{E}{2}\left(\frac{t_B}{t_A^2} + \frac{1}{t_B} + \frac{1}{t_A}\right) \tag{77c}$$

and

$$I = \frac{1}{4}\left(\frac{t_B}{t_A} - \frac{t_A}{t_B}\right)^2 . \tag{78}$$

Despite the absence of any simple transformation between the diagonal and
off-diagonal models, for the same value of I the scaling behavior of the
spectra is nearly the same in the two cases. In Fig. 7 we show the $f(\alpha)$
curve for the off-diagonal Fibonacci model for I = 9/16 (t_B/t_A = 2). Note
that the minimum and maximum values of α obtained from the numerical calculation agree quite well with the expressions for α_2 and α_6 respectively.

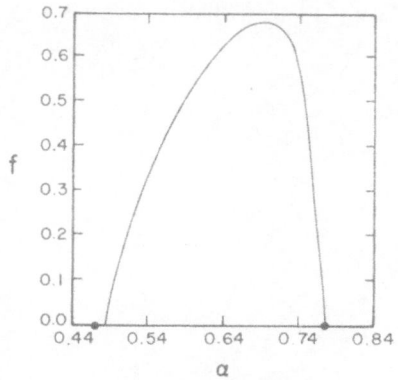

Fig. 7. f(α) for the off-diagonal Fibonacci lattice with $t_B/t_A = 2$. (From Kohmoto et al.).

Another spectrum which has been studied is that of the lattice vibrations of a 1D model with force constants alternating in a Fibonacci sequence.[34,36,40] For this case the equation of motion,

$$-\omega^2 U_n = K_{n+1}(U_{n+1} - U_n) - K_n(U_n - U_{n-1}),$$ (79)

where U_n is the displacement on the n-th site and we have set the mass equal to one, can be written as a transfer matrix

$$\begin{pmatrix} U_{n+1} \\ U_n \end{pmatrix} = \begin{pmatrix} (K_{n+1} + K_n - \omega^2)/K_{n+1} & -K_n/K_{n+1} \\ 1 & 0 \end{pmatrix} \begin{pmatrix} U_n \\ U_{n-1} \end{pmatrix}.$$ (80)

This can be mapped onto a version of the off-diagonal Fibonacci model where the on-site energy has the value E_{AA} for sites surrounded by two A links and the value E_{AB} for sites surrounded by an A link and a B link. ($E \to \omega^2$, $t_A \to K_A$, $t_B \to K_B$, $E_{AA} \to 2K_A$ and $E_{AB} \to K_A + K_B$.) It follows that Eqs. (77) apply with E replaced by $E - E_{AA}$ in Eq. (77a), by $E - E_{AB}$ in Eq. (77b) and x_3 calculated from the recursion relation with $x_0 = (2E_{AB} - E_{AA})/2t_B$. For this electronic model

$$I = \frac{1}{4} [\frac{t_A}{t_B} - \frac{t_B}{t_A} + \frac{E_{AB}(E_{AB} - E_{AA})}{t_B t_A}]^2.$$ (81)

Thus for the corresponding phonon problem we obtain

$$I = \frac{\omega^4}{4} [\frac{1}{K_B} - \frac{1}{K_A}]^2$$ (82)

in agreement with Ref. 34. Note that $I \to 0$ as $\omega \to 0$ so that we can expect all of the scaling indices to approach 1 in this region of the spectrum. This result reflects the physics that the low energy excitations are characterized by length scales much larger than a lattice constant and therefore tend to reflect mainly the average force constant. We see later that a similar result is obtained for electromagnetic waves in a layered dielectric medium and for a continuum description of acoustic phonons in a Fibonacci semiconductor superlattice.

PLASMONS

Plasmons in semiconductor superlattices are now well understood.[52-54] The theoretical model which has been successfully applied to periodic systems assumes that the electrons in the quantum wells form independent 2DEG systems. In the plasmon mode the electrostatic potential energy in one 2DEG layer is self-consistently generated by the charge density it induces in all layers. Thus for a mode with in-plane wavevector q and frequency ω,

$$\phi_\ell = \sum_{\ell'} \frac{2\pi e^2}{\epsilon q} e^{-q|z_\ell - z_{\ell'}|} \pi(q,\omega) \, \phi_{\ell'} \tag{83}$$

where $\pi(q,\omega)$ is the 2DEG polarizability and $\{z_\ell\}$ are the 2DEG layer positions along the superlattice growth direction. Defining

$$X \equiv \frac{2\pi e^2}{\epsilon q} \pi(q,\omega) \tag{84}$$

and noting that for $q \to 0$ we can write

$$X \simeq \frac{\omega_q^2}{\omega^2} \tag{85}$$

we see that Eq. (83) can be rewritten in the form

$$\omega^2 \phi_\ell = \sum_{\ell' \neq \ell} \omega_q^2 e^{-q|z_\ell - z_{\ell'}|} \phi_{\ell'} + \omega_{\vec{q}}^2 \phi_\ell . \tag{86}$$

Here $\omega_{\vec{q}} = (2\pi\sigma e^2 q/\epsilon m^*)^{\frac{1}{2}}$ is the 2DEG plasma frequency. Eq. (86) led Das Sarma et al.[55-56] to suggest that studies of superlattice plasmons might provide experimental information on the properties of 1D quasiperiodic or disordered systems. Note that in Eq. (86) the effective on-site energy, ω_q^2, can be controlled by controlling the carrier density in the wells while the effective hopping terms $\omega_q^2 e^{-q|z_\ell - z_{\ell'}|}$ can be controlled by controlling the layer separations. However, the powerful transfer matrix methods discussed above cannot be used directly on Eq. (86) because, especially as $q \to 0$, distant hops become important. We therefore follow Hawvrylak et al.[57-58] in adopting an alternative formulation of the problem.

In the carrier free region between the ℓth and $(\ell+1)$th 2DEG layer the Poisson equation allows us to write the electrostatic potential energy as

$$\phi_\ell(z) = A_\ell e^{-q(z-z_\ell)} + B_\ell e^{q(z-z_\ell)} \tag{87}$$

where q is the in-plane wave vector. Continuity of the potential across the (ℓ+1)th 2DEG layer requires that

$$A_{\ell+1} + B_{\ell+1} = A_\ell e^{-qd_\ell} + B_\ell e^{qd_\ell} \tag{88a}$$

and integration of the Poisson equation across the layer relates the electric field discontinuity to the density induced <u>in the layer</u>, i.e.

$$\frac{\partial \phi_{\ell+1}}{\partial z}\bigg|_{z_{\ell+1}^+} = \frac{\partial \phi_\ell}{\partial z}\bigg|_{z_{\ell+1}^-} + \frac{4\pi e^2}{\epsilon}\, \Pi\, \phi(z_{\ell+1}). \tag{88b}$$

Combining Eqs. (88) defines the transfer matrix:

$$\begin{pmatrix} A_{\ell+1} \\ B_{\ell+1} \end{pmatrix} = \begin{pmatrix} (1+X)e^{-qd_\ell} & X\,e^{qd_\ell} \\ -X\,e^{-qd_\ell} & (1-X)\,e^{qd_\ell} \end{pmatrix} \begin{pmatrix} A_\ell \\ B_\ell \end{pmatrix}. \tag{89}$$

We consider explicitly the case in which the 2DEG layers are identical and the layer separations d_A and d_B alternate in a Fibonacci sequence. Using Eq. (89) the starting conditions for the trace map are found to be,

$$x_1 = \cosh(qd_B) - X \sinh(qd_B) \tag{90a}$$

$$x_2 = \cosh(qd_A) - X \sinh(qd_A) \tag{90b}$$

and

$$x_3 = 2x_1 x_2 - \cosh(q(d_A - d_B)), \tag{90c}$$

from which we find that

$$I = X^2 \sinh^2(q(d_A - d_B)) . \tag{91}$$

Note that for $qd_B \gg 1$, $qd_A \gg 1$ the starting point of the trace map has $|x_i| \gg 1$ unless X is near to one. In this limit energies will not be in the spectra unless $X \simeq 1$, i.e. unless $\omega \simeq \omega_q$. Also note that as $q(d_A - d_B)$ becomes small, I becomes small and the scaling indices should approach 1. For given q, d_A and d_B, $I \propto X^2$ so that the scaling indices will be lower (the bands will be narrower) in the low frequency part of the spectrum, as found in numerical studies.[56-58]

ACOUSTIC PHONONS AND ELECTROMAGNETIC WAVES

In a continuum model, the equation of motion for the acoustic phonons in a Fibonacci semiconductor superlattice is[59]

$$-v^2(\xi) \frac{d^2 U}{d\xi^2} = \omega^2 U \tag{92}$$

where $V(\xi)$ is a constant, characteristic of the constituent material within each layer. (The same equation, and subsequent developments in this section, can be applied to discuss propagating electromagnetic waves in a layered dielectric medium[60]. In that case $V(\xi)$ is the local light velocity.) Following Kohmoto[61] we introduce a transfer matrix for this continuum case by defining an interval of width 2ϵ, which can be subsequently set to zero, centered on each Fibonacci lattice point in which the sound velocity has some reference value, V_0. Then we can write

$$U(\xi) = \psi_n e^{ik(x-x_n)} + \phi_n e^{-ik(x-x_n)} \qquad |x-x_n| < \epsilon \qquad (93)$$

where $k = \omega/V_0$. The transfer matrix is defined by requiring $U(\xi)$ and the stress, which is proportional to $U'(\xi)$, to be continuous at each interface. The most general case we consider is one where an X (A or B) interval of the Fibonacci superlattice has layers composed of material 1 and of thickness d_{1r}^X and $d_{1\ell}^X$ on the left and right respectively, sandwiching a layer of material 2 of thickness d_2^X. In this case

$$\begin{pmatrix} \psi_{n+1} \\ \phi_{n+1} \end{pmatrix} = \begin{pmatrix} U_n^* & -V_n^* \\ -V_n & U_n \end{pmatrix} \begin{pmatrix} \psi_n \\ \phi_n \end{pmatrix} \qquad (94)$$

where

$$U_X = e^{-ik_1 d_1^X} \left[\cos(k_2 d_2^X) - i \sin(k_2 d_2^X) \, (k_1/2k_2 + k_2/2k_1) \right], \qquad (95a)$$

$$V_X = i(k_2/2k_1 - k_1/2k_2) \, e^{ik_1(d_{1\ell}^X - d_{1r}^X)} \sin(k_2 d_2^X) . \qquad (95b)$$

In the above U_X and V_X are related to the transmission and reflection coefficients, t_X and r_X, for a wave incident from the reference medium and passing through an X interval, by

$$U_X = t_X^{-1} \qquad (96a)$$

and

$$V_X = r_X/t_X \qquad (96b)$$

so that the determinant of the transfer matrix is

$$|U_X|^2 - |V_X|^2 = |r_X|^2 + |t_X|^2 = 1. \qquad (97)$$

The starting point of the trace map is

$$x_1 = \text{Re}(U_B) \qquad (98a)$$

$$x_2 = \text{Re}(U_A) \qquad (98b)$$

and

$$x_3 = \text{Re}(U_B U_A + V_B V_A^*) \qquad (98c)$$

so that

$$I = \left(\text{Re}(V_B V_A^*) - \text{Im}(U_B) \text{Im}(U_A) \right)^2 - \left(|V_B|^2 - \text{Im}^2(U_B) \right) \left(|V_A|^2 - \text{Im}^2(U_A) \right). \quad (99)$$

The integrated density of acoustic phonon modes for a superlattice with A intervals consisting of 6nm of GaAs and B intervals consisting of 4nm of AℓAs is shown in Fig. 8. This calculation was done with a 13 generation Fibonacci lattice, periodically continued and is based on a sound velocity of 4.72×10^5 cm/s for GaAs and 5.62×10^5 cm/s for AℓAs. Notice that the gaps are small compared to the bands, in this periodic approximation, so the scaling indices are expected to be close to 1. This is confirmed by the small values for I, whose ω dependence is also illustrated in Fig. 8. For the case where the A intervals are composed entirely of material 1 and the B intervals are composed of material 2, Eq. (99) reduces to

$$I = \left[\frac{k_2}{2k_1} - \frac{k_1}{2k_2}\right]^2 \sin^2(k_1 d_1^A) \, \sin^2(k_2 d_2^B). \tag{100}$$

Note that $I(\omega)$ vanishes like ω^4 for small ω, just as it did in the discrete phonon problem, but that for the continuum case it also vanishes when $k_1 d_1^A$ or $k_2 d_2^B$ are multiples of π. The scaling indices approach 1 for energies near any zero of $I(\omega)$, and this property is reflected in the integrated density of states calculation. At the zeroes of $I(\omega)$, the transmission probability approaches 1 (see Fig. 8). This is the analog for a

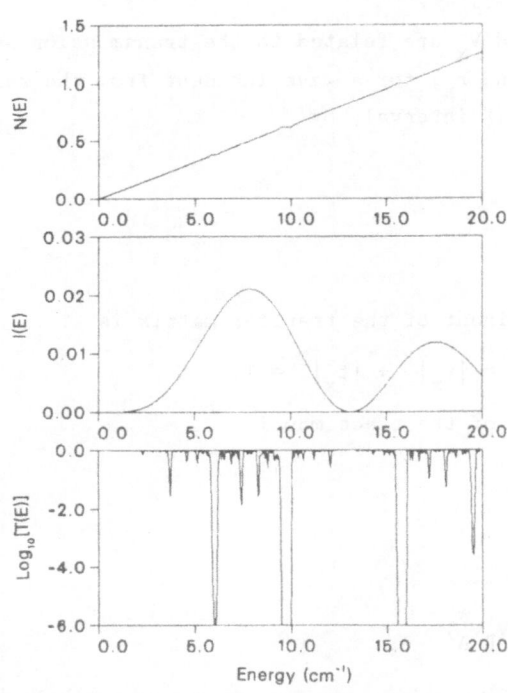

Fig. 8. Integrated density of states, I(E) and transmission coefficient for acoustic phonons for the Fibonacci superlattice discussed in the text.

quasiperiodic system of the perfect transmission at resonance in a certain class of disordered 1D systems discussed by Tong.[61,62]

For typical superlattices, the gaps in the phonon spectrum resulting from differences of the sound velocities of two semidconductors may be estimated by treating the sound velocity difference perturbatively. In Fig. 9 we plot the gap width versus the frequency at gap center for a Fibonacci superlattice with A and B intervals having 8 monolayers of $Ga_{1-x}Al_x As$ (material 2), A intervals having 18 monolayers of GaAs (material 1) and B intervals having 8 monolayers of GaAs. The Fourier component of $V^2(\xi)$ with wavevector $k_{n,m}^{\parallel}$ (see Eq. 15) couples the degenerate phonon modes at $q = \pm k_{n,m}^{\parallel}/2$ producing a gap centered on frequency $\bar{V}k_{n,m}^{\parallel}/2$ and of magnitude

$$\Delta_{n,m} = \left| \frac{V_2-V_1}{\bar{V}} \right| \frac{\omega_0\tau^2}{\pi} \left| \frac{\sin(k_{n,m}^{\parallel}d_2/2)\ \sin(z_{n,m})}{z_{n,m}} \right| \leq \left| \frac{V_2-V_1}{\bar{V}} \right| \frac{\omega_0\tau^2}{\pi}.$$

(101)

The gaps calculated from Eq. (101) agree quite precisely with those shown in Fig. 9. Note that the maximum gap is proportional to the sound velocity difference between the two materials and inversely proportional to the layer thicknesses.

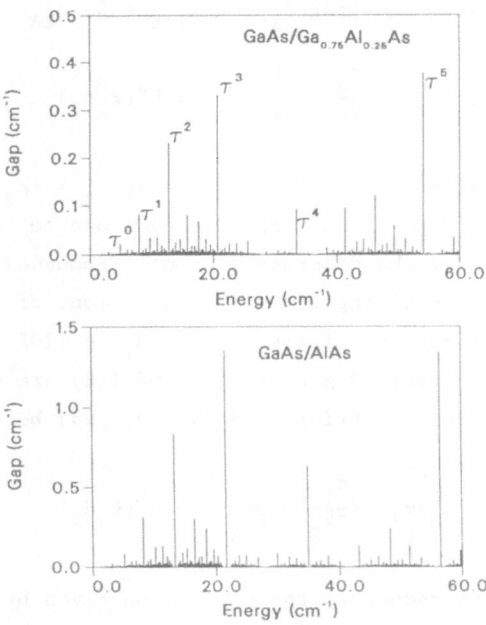

Fig. 9. Phonon frequency gaps versus gap center energies for the Fibonacci superlattice in the text.

In the effective mass approximation[63] the conduction band states in a semiconductor superlattice are determined by solving the Schrödinger equation

$$\frac{-\hbar^2}{2m^*} \frac{d^2\psi}{dx^2} + V(x)\psi = E\psi. \tag{102}$$

Here $V(x)$ is the local conduction-band minimum. Defining a narrow region around each Fibonacci lattice point where V has some reference value leads to the same transfer matrix as for the phonon case[64,65,66] except that the relation between wavevector and energy is

$$E = V_i + \frac{\hbar^2 k_i^2}{2m^*} \tag{103}$$

where V_i is the conduction-band minimum in material "i". Note that k_i is imaginary for classically forbidden energies. For GaAs - $Ga_{1-x}Al_xAs$ systems we take $m^* = 0.068\ m_0$, where m_0 is the electron mass, and $|V_2 - V_1| = 134$ meV, corresponding to $x \simeq 0.2$. A situation similar to that of the discrete model with on-site energies alternating in a Fibonacci sequence (the diagonal model) may be realized by choosing the width of the barriers ($Ga_{1-x}Al_xAs$ regions) to be the same in A and B intervals but allowing the well-widths (GaAs regions) to vary. Letting $d_{1rA} = d_{1rB} = 0$, $d_{2A} = d_{2B} = d_2$ yields

$$I_D = \sin^2[(d_A - d_B)k_1)] \left(\frac{\kappa_2}{2k_1} + \frac{k_1}{2\kappa_2}\right)^2 \sinh^2(\kappa_2 d_2) \tag{104}$$

where we have assumed that $V_1 < E < V_2$ and let $k_2 = i\kappa_2$. (For wide high barriers ($\kappa_2 d_2 \gg 1$, $\kappa_2/k_1 \gg 1$) Eq. (15) can be shown to reduce to the corresponding expression for the discrete diagonal Fibonacci chain.) In Fig. 10 we show the spectrum resulting after 12 generations of the Fibonacci superlattice and I(E) for barrier widths of 20Å, $d_{1A} = 120$Å and $d_{1B} = 80$Å. In Fig. 11 the spectrum after 12 generations and I(E) are shown for $d_2 = 100$Å, $d_{1A} = 24$Å and $d_{1B} = 16$Å. Letting $\kappa_1 = ik_1$ Eq. (99) becomes

$$I_{OD} = \sinh^2((d_A - d_B)\kappa_1) \left(\frac{\kappa_1}{2k_2} + \frac{k_2}{2\kappa_2}\right)^2 \sin^2(k_2 d_2) \tag{105}$$

which can be shown to reduce to the expression given by Kohmoto and Banavar for the off-diagonal discrete Fibonacci chain model in the appropriate limit.

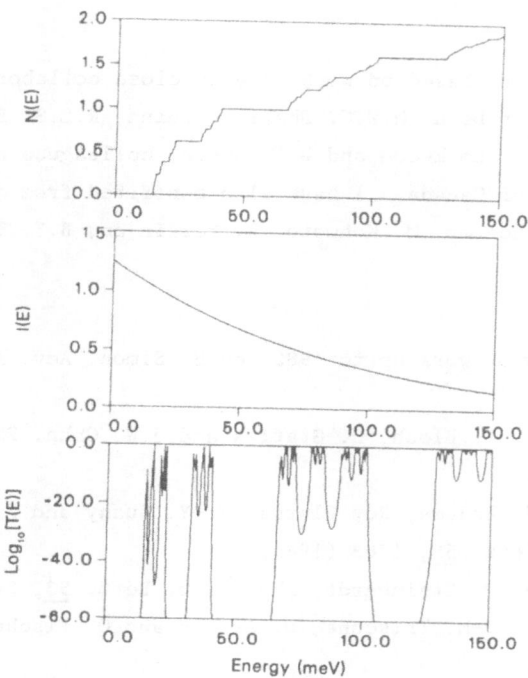

Fig. 10. Integrated density of states for the diagonal Kronig-Penney model.

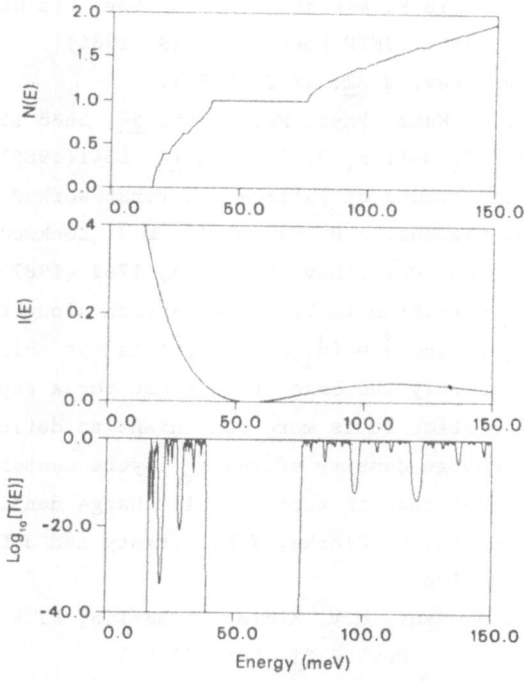

Fig. 11. Integrated density of states for the off-diagonal Kronig-Penney model.

ACKNOWLEDGMENT

These notes are based on work done in close collaboration with
G.C. Aers, J.M. Baribeau, M.W.C. Dharma-wardana, R.L.S. Devine,
D.C. Houghton, D.J. Lockwood and W.T. Moore, colleagues at the National
Research Council of Canada. I have also benifited from discussions with
P. Hawvrylak, M. Holzer, M. Kohmoto, R. Merlin and B.Y. Tong.

REFERENCES

1. For a review of work up to 1982 see B. Simon, Adv. Appl. Math. $\underline{3}$, 463
   (1982).

2. D. Schectman, I. Bloch, D. Gratias and J.W. Cahn, Phys. Rev. Lett. $\underline{53}$,
   1951 (1984).

3. R. Merlin, K. Bajema, Roy Clarke, F.-Y. Juany and P.K. Battacharya,
   Phys. Rev. Lett. $\underline{55}$, 1768 (1985).

4. D. Levine and P. Steinhardt, Phys. Rev. Lett. $\underline{53}$, 2477 (1984).

5. M.G. Karkut, J.-M. Triscone, D. Ariosa and O. Fischer, Phys. Rev. B $\underline{34}$,
   4390 (1986).

6. A. Behrooz, M.J. Burns, H. Deckman, D. Levine, B. Whitehead, and P.M.
   Chaikin, Phys. Rev. Lett. $\underline{57}$, 368 (1986).

7. P. Kramer and R. Neri, Acta Crystallogr. Sect. A $\underline{40}$, 580 (1984).

8. P.A. Kalugin, A. Yu Kitaeu and L.C. Lerutov, Pis'ma Zh. Eksp. Teor.
   Fiz. $\underline{41}$, 119 (1985) [JETP Lett. $\underline{41}$, 119 (1985)].

9. V. Elser, Phys. Rev. B $\underline{32}$, 4892 (1985).

10. M. Duneau and A. Katz, Phys. Rev. Lett. $\underline{54}$, 2688 (1985).

11. R.K. Zia and W.J. Dallas, J. Phys. A $\underline{18}$, L341(1985).

12. The case of the Fibonacci lattice was first worked out in detail in
    M.W.C. Dharma-wardana, A.H. MacDonald, D.J. Lockwood, J.-M. Baribeau
    and D.C. Houghton, Phys. Rev. Lett. $\underline{58}$, 1761 (1987).

13. We take the z direction to be the direction along the growth axis and
    write $\vec{r} = (\vec{r}_\perp, z)$ and $\vec{q} = (\vec{q}_\perp, q_z)$. $F(\vec{r})$ is, at this point $F(\vec{r}_\perp, z+d) = F(\vec{r}_\perp, z)$. Frequently the case of interest for a superlattice is $\vec{q}_\perp = 0$.

14. For this calculation it is more convenient to define the form factor in
    terms of the charge density of the N_x layers centered between 0 and
    $d_x = N_x a_x$, rather than in terms of all charge density between 0 and d_x.

15. J. Todd, R. Merlin, R. Clarke, K.M. Mohanty and J.D. Axe, Phys. Rev.
    Lett. $\underline{57}$, 1158 (1986).

16. C. Colvard, T.A. Gant, M.V. Klein, R. Merlin, R. Fischer, H. Morkoc and
    A.C. Gossard, Phys. Rev. B $\underline{31}$, 2080 (1985).

17. B. Jusserand and D. Paquet, in Semiconductor Heterojunctions and Superlattices, edited by N. Boccara, C. Allan, G. Bastard, M. Lannoo and M. Voss (Springer, Berlin, 1986).

18. The experiments are done in a backscattering configuration which allows scattering only by longitudinal acoustic phonons with wavevectors along the growth direction.

19. R. Merlin, K. Bajema, R. Clarke and J. Todd, in Proceedings of the 18th International Conference on the Physics of Semiconductors, ed. by O. Engström, (World Scientific, Singapore, 1987) p. 675.

20. K. Bajema and R. Merlin, Phys. Rev. B 36, 4555 (1987).

21. R. Merlin, K. Bajema, J. Nagle and K. Ploog, Proceedings of the 3rd International Conference on Modulated Semiconductor Structures, to be published (1988).

22. D.J. Lockwood, A.H. MacDonald, G.C. Aers, M.W.C. Dharma-wardana, R.L.S. Devine and W.T. Moore, Phys. Rev. B 36, 9286 (1987).

23. B. Jusserand, F. Alexander, D. Paquet and G. LeRoux, Appl. Phys. Lett. 47, 301 (1985).

24. V. Matijasevic and M.R. Beasley, Phys. Rev. B 25, 3175 (1987).

25. P.G. Harper, Proc. Roy. Soc. London, Sect. A 68, 874 (1955).

26. D.R. Hofstadter, Phys. Rev. B 14, 2239 (1976).

27. G. Andre and S. Aubrey, Ann. Isr. Phys. Soc. 3, 133 (1980).

28. D.J. Thouless, M. Kohmoto, P. Nightingale and M. den Nijs, Phys. Rev. Lett. 49, 405 (1982).

29. M. Kohmoto, Phys. Rev. Lett. 51, 1198 (1983).

30. C. Tang and M. Kohmoto, Phys. Rev. Lett. 34, 2041 (1986).

31. J.B. Sokoloff, Phys. Rep. 126, 189 (1985).

32. M. Kohmoto, L.P. Kadanoff and C. Tang, Phys. Rev. Lett. 50, 1870 (1983), M. Kohmoto and Y. Oono, Phys. Lett. 102A, 145 (1984), L.P. Kadanoff and C. Tang, Proc. Natl. Acad. Sci. USA 81, 1276 (1984).

33. S. Ostlund, R. Pandit, D. Rand, H.J. Schnellnhuber and E. Siggia, Phys. Rev. Lett. 50, 1973 (1983).

34. M. Kohmoto and J.R. Banavar, Phys. Rev. B 34, 563, (1986).

35. J.P. Lu, T. Odagaki and J.L. Birman, Phys. Rev. B 33, 4809 (1986).

36. F. Nori and J.P. Rodriquez, Phys. Rev. B 34, 2207 (1986).

37. T. Odagaki and L. Freidman, Solid State Commun. 57, 915 (1986).

38. K. Machida and M. Fujita, Phys. Rev. B 34, 7376 (1986).

39. A. Mookerjee and V.A. Singh, Phys. Rev. B 34, 7433 (1986).

40. J.M. Luck and D. Petretis, J. Stat. Phys. 42, 289 (1986).

41. M. Kohmoto, B. Sutherland and C. Tang, Phys. Rev. B 35, 1020 (1987).

42. T. Schneider, A. Politi and D. Wurtz (preprint).

43. I. Procaccia, in Proceedings of the Nobel Symposium on Chaos and Related Problems, Phys. Scr. T 9, 40 (1985).

44. B.B. Mandelbrot, Ann. Isr. Phys. Soc. 225 (1977).

45. T.A. Witten, Jr. and L.M. Sander, Phys. Rev. Lett. 47, 1400 (1981).

46. K.G. Wilson, Sci. Am. 241, 158 (1979).

47. D. Katzen and I. Procaccia, Phys. Rev. Lett. 58, 1169 (1987).

48. Thomas C. Halsey, Morgens H. Jensen, Leo P. Kadanoff, Itamar Procaccia and Boris J. Shraiman., Phys. Rev. A 33 1141 (1986).

49. For further refinements and generalizations see L.P. Kadanoff, "Fractal singularities on a measure and how to measure singularities on a fractal." U. of Chicago preprint, and M. Kohmoto, "Entropy Function for Multifractals.", U. of Utah preprint.

50. The fractal dimension is defined, roughly, by, $M \propto W^{-F}$ where M is the number of subbands and W is the width of each subband. For an elementary introduction to the fractional dimension concept see B.B. Mandelbrot, The Fractal Geometry of Nature (Freeman, New York, 1983).

51. $d\tau^*/dq = \alpha > 0$ so τ^* increases monotonically with q.

52. D. Grecu, Phys. Rev. B 8, 1958 (1973).

53. A.L. Fetter, Ann. Phys. (NY) 88, 1 (1974).

54. S. Das Sarma and J.J. Quinn, Phys. Rev. B 25, 7603 (1982).

55. S. Das Sarma, A. Kobayashi and R.E. Prange, Phys. Rev. Lett. 56 1280 (1986).

56. S. Das Sarma, A. Kobayashi and R.E. Prange, (preprint).

57. P. Hawvrylak and J.J. Quinn, Phys. Rev. Lett. 57, 380 (1986).

58. P. Hawvrylak, Gunnar Eliasson and J.J. Quinn, Phys. Rev. B 36, 6501 (1987).

59. We will assume that the relevant elastic constants may be taken to have the same value in each layer of the Fibonacci semiconductor superlattice so that the change in the local sound velocity is due entirely to the change in mass density.

60. In our previous discussion of plasmons, we took the dielectric constant to the same in each layer of the Fibonacci superlattice. Both effects could be included together.

61. M. Kohmoto, Phys. Rev. B 34, 5043 (1986); Phys. Rev. Lett., 58, 2436 (1987).

62. B.Y. Tong, preprint (1987).

63. G. Bastard and J.A. Brum, IEEE J. Quan. Elec. QE-22, 1625 (1986).

64. A.H. MacDonald and G.C. Aers, Phys. Rev. B 36, 9142 (1987).

65. M.Holzer, preprint (1987).

66. G.C. Aers, unpublished.

APPENDIX A

We want to calculate the maximum and minimum η value for points on the Fibonacci lattice. Denoting η in units $|\Delta\eta_x|$ by $\tilde{\eta}$ we have

$$\tilde{\eta} = k_y\tau - k_x \tag{A1}$$

where k_y and k_x are respectively the number of B intervals and the number of A intervals preceding a given point on the lattice. At the end of i generations of the Fibonacci lattice $\tilde{\eta}$ has the value,

$$\tilde{\eta}_i = F_{i-2}\tau - F_{i-1} = (-)^{i-1}\tau^{2-i} , \tag{A2}$$

where we have used Eq. (4c). The maximum (G_i) and minimum (H_i) values of η after i generations therefore obey

$$G_i = \mathrm{Sup}\left(\tilde{\eta}_{i-1}+G_{i-2}, G_{i-1}\right) \tag{A3a}$$

and

$$H_i = \mathrm{Inf}\left(\tilde{\eta}_{i-1}+H_{i-2}, H_{i-1}\right). \tag{A3b}$$

Eqs. (A3) can be iterated starting from the 4th generation of the Fibonacci sequence ($S_4 = ABA, G_4 = \tau^{-1}, H_4 = -1, \tilde{\eta}_4 = \tau^{-2}$: $S_3 = AB, G_3 = t^{-1}$, $H_3 = -1$, $\tilde{\eta}_3 = \tau^{-1}$). For i odd $\tilde{\eta}_{i-1}<0$ so that $G_{i-2}+\tilde{\eta}_{i-1}<G_{i-2}\leq G_{i-1}$ and Eq. (A3a) implies that $G_i = G_{i-1}$. For i even, on the other hand, $\tilde{\eta}_{i-1}>0$ $\tilde{\eta}_{i-1}+G_{i-2}\geq G_{i-2} = G_{i-1}$ so that

$$G_i = G_{i-2}+\tilde{\eta}_{i-1} \tag{A4a}$$

and

$$\lim_{i\to\infty} G_i = \sum_{k=1}^{\infty} \tilde{\eta}_{2k+1} = \tau \sum_{k=1}^{\infty} \tau^{-2k} = \tau^{-1}(1-\tau^{-2})^{-1} = 1. \tag{A4b}$$

Similarly, for i even $\tilde{\eta}_{i-1}+H_{i-2}>H_{i-2}\geq H_{i-1}$ so that Eq. (A3b) gives $H_i = H_{i-1}$. For i odd, $\tilde{\eta}_{i-1}+H_{i-2}<H_{i-2} = H_{i-1}$,

$$H_i = \tilde{\eta}_{i-1} + H_{i-2} \tag{A5a}$$

and

$$\lim_{i\to 0} H_i = \sum_{k=0}^{\infty} \tilde{\eta}_{2k} = -\sum_{k=1}^{\infty} \tau^{-2k} = -\tau. \tag{A5b}$$

The minimum value of η among the Fibonacci points is thus $-|\Delta\eta_x|\tau = -2w\tau^{-1}$ and the maximum value is $|\Delta\eta_x| = 2w\tau^{-2}$. It follows that all the Fibonacci points will be accepted with the choice $\eta_0 = -2w\tau^{-3}$.

The number of points per unit length projected onto the $\eta = 0$ line from the 2D lattice is

$$2w/a_x a_y = \frac{\tau^2\sin\theta}{a_y} = \frac{\tau^2}{d_A(1+\cot^2\theta)} = (\tau^{-1}d_A+\tau^{-2}d_B)^{-1} \equiv \tau^2/d . \tag{A6}$$

This is the reciprocal of the mean distance between points on the Fibonacci lattice and we may conclude that only the points on the Fibonacci lattice are projected onto the $\eta = 0$ line.

LECTURERS

M. Altarelli

Max Planck Institut fur
  Festkorperforschung
Hochfeld-Magnetlabor, 166X
F-38042 Grenoble Cedex
France

G. Bastard

Groupe de Physiques des Solides de
  l'Ecole Normale Superieure
24 rue Lhomond 75231
Paris Cedex 05
France

G. W. Bryant

McDonnell Douglas Research Laboratories
St. Louis, Missouri 63166
U.S.A.

J. P. Eisenstein

AT&T Bell Labs.
600 Mountain Avenue, 1D-221
Murray Hill, New Jersey 07974
U.S.A.

C. T. Foxon

Philips Research Laboratory
Solid State Electronics Div.
Redhill, Surrey, RH1 5H4
England

S. Girvin

Physics Department
Swain Hall W117
Indiana University
Bloomington, IN 47405
U.S.A.

G. Gumbs

Dept. of Physics
University of Lethbridge
Lethbridge, Alberta T1K 3M4
Canada

C. Kallin

Dept. of Physics
McMaster University
1280 Main Street West
Hamilton, Ontario L8S 4M1
Canada

F. Koch   Physik-Department
                            Technische Universitat Munchen
                            D-8046 Garching
                            West Germany

J. P. Kotthaus   Institut fur Angewandte Physik
                            Universitat Hamburg
                            Jungiusstrasse 11
                            2000 Hamburg 36
                            West Germany

R. B. Laughlin   Lawrence Livermore National Laboratory,
                               L-299
                            P.O. Box 808
                            Livermore, California 94550
                            U.S.A.

A. H. MacDonald   Dept. of Physics
                            Indiana University
                            Swain Hall West
                            Bloomington, Indiana 47405
                            U.S.A.

E. E. Mendez   I.B.M. T.J. Watson Research Center
                            P.O. Box 218
                            Yorktown Heights, New York 10598
                            U.S.A.

R. J. Nicholas   Clarendon Laboratory
                            Parks Road
                            Oxford OX1 3PU
                            England

S. Schmitt-Rink   AT&T Bell Labs.
                            600 Mountain Avenue
                            Murray Hill, New Jersey 07974
                            U.S.A.

F. Stern   I.B.M. T.J. Watson Research Center
                            P.O. Box 218
                            Yorktown Heights, New York 10598
                            U.S.A.

P. J. Stiles   Physics Department
                            Brown University
                            Providence, Rhode Island 02912
                            U.S.A.

H. L. Stormer   AT&T Bell Labs.
                            600 Mountain Avenue ID-458
                            Murray Hill, New Jersey 07974
                            U.S.A.

PARTICIPANTS

C. ARSENAULT
Dept. of Engineering Physics
Ecole Polytechnique
Campus de l'Universite de Montreal
Case postale 6079, succursale A
Montreal, Quebec H3C 3A7
Canada

J. BAARS
Fraunhofer-Institut fur Angewandte
    Festkorperphysik
Eckerstrabe 4
D-7800 Freiburg
West Germany

A. J. BERLINSKY
McMaster University
Institute for Materials Research
1280 Main Street West
Hamilton, Ontario L4S 4M1
Canada

P. J. BRADLEY
University College London
Dept. of Electronic & Elec. Eng.
Torrington Place
London WD1E 7JE
England

M. CAGE
Bldg. 220
National Bureau of Standards
Gaithersburg, MD 20499
U.S.A.

M. CHAMBERLAIN
Dept. of Physics
University of Essex
Wivenhoe Park
Colchester CO4 3SO
England

S. CHARBONNEAU
Dept. of Physics
Simon Fraser University
Burnaby, British Columbia V5A 1S6
Canada

M. P. CHAUBEY
Dawson College
Dept. of Mathematics
Selby Campus
350 Selby Street
Westmount, Quebec H3Z 1W7
Canada

C. Z. CIL
Bogazici University
Department of Physics
80815 Bebek - Istanbul
Turkey

B. CLINTON
Physics Department
Georgetown University
Washington, D.C. 20057
U.S.A.

P. COLERIDGE
National Research Council of Canada
Division of Physics
Building M-23A
Montreal Road
Ottawa, Ontario K1A OR6
Canada

J. COOK
National Research Council of Canada
Division of Physics
Building M-36
Montreal Road
Ottawa, Ontario K1A OR6
Canada

J. M. R. CRUZ
University of Toronto
Dept. of Physics
Toronto, Ontario M5S 1A7
Canada

W. R. DATARS
McMaster University
Dept. of Physics
1280 Main Street
Hamilton, Ontario L8S 4M1
Canada

P. W. EPPERLEIN
I.B.M. Corp.
Forschungslaboratorium Zurich
Zurich Research Laboratory
Saumerstrasse 4
8803 Ruschlikon
Switzerland

E. FENTON
National Research Council of Canada
Division of Physics
Building M-36
Montreal Road
Ottawa, Ontario K1A OR6
Canada

T. GODIN
University of Oregon
Dept. of Physics
College of Arts and Sciences
122 Science I
Eugene, Oregon 97403-1274
U.S.A.

M. GOIRAN
Institut National des Sciences
Appliquees
Departement de Physique
Avenue de Rangueil
31077 Toulouse-Cedex
France

A. GOLD
Physik Department Eng.
Technische Universitat
Munchen, D-8046 Garching
West Germany

F. GREEN
Los Alamos National Laboratory
CMS, MS K765
Los Alamos, New Mexico 87545
U.S.A.

V. GRIDIN
McMaster University
Dept. of Physics
1280 Main Street West
Hamilton, Ontario L8S 4M1
Canada

D. P. HALLIDAY
Dept. of Physics
The University of Nottingham
University Park
Nottingham NG7 2RD
England

J. HAVERKORT
Physics Dept.
University of Eindhoven
Den Dolech-2
Eindhoven
The Netherlands

B. HEINRICH
Simon Fraser University
Dept. of Physics
Burnaby, British Columbia V5A 1S6
Canada

R. HELBING
University of Windsor
401 Sunset Avenue
Windsor, Ontario N9B 3P4
Canada

M. HOLZER
Dept. of Physics
The University of British Columbia
6224 Agriculture Road
Vancouver, British Columbia V6T 2A6
Canada

S. HUANT
C.N.R.S.
25, Avenue des Martyrs
B.P. 166 X
18042 Grenoble Cedex
France

H. E. JACKSON
University of Cincinnati
210 Braunstein (ML 11)
Cincinnati, Ohio 45221-0011
U.S.A.

E. A. JOHNSON
Imperial College of Science and
   Technology
The Blackett Laboratory
Prince Consort Road
London SW7 2BZ
England

J. KEMP
University of Oxford
Inorganic Chemistry Laboratory
South Parks Road
Oxford OX1 3QR
England

G. KIRCZENOW
Simon Fraser University
Dept. of Physics
Burnaby, British Columbia V5A 1S6
Canada

H. KLEIN
Keyano College
University Transfer Department
8115 Franklin Avenue
Fort McMurray, Alberta T9H 2H7
Canada

E. KOTELES
GTE Laboratories Inc.
40 Sylvan Road
Waltham, MA 02254
U.S.A.

R. KUSZELEWICZ
CNET - Laboratoire de Bagneux
196, Av. Henri Ravera
92220 Bagneux
France

D. R. LEADLEY
University of Oxford
Dept. of Physics
Clarendon Laboratory
Parks Road
Oxford OX1 3PU
England

R. LEAVENS
National Research Council of Canada
Division of Physics
Building M-36
Montreal Road
Ottawa, Ontario K1A 0R6
Canada

K. S. LEE
Northeastern University
Dept. of Physics
360 Huntington Avenue
Boston, Massachusetts 02115
U.S.A.

J. LEO
Imperial College of Science and
    Technology
The Blackett Laboratory
Prince Consort Road
London SW7 2BZ
England

J. LEOTIN
Institut National des Sciences
Appliquees
Departement de Genie Physique
Avenue de Rangueil
31077 Toulouse Cedex
France

A. W. F. LO
Solid State Research Group
Dept. of Electrical Engineering
Imperial College
Exhibition Road
London SW7 2BT
England

P. LOLY
University of Manitoba
Dept. of Physics
Winnipeg, Manitoba R3T 2N2
Canada

F. MALCHER
Universitat Regensburg
Inst. fur Physik I - Theoratische
Physik
8400 Regensburg
Universitatsstrabe 31
Postfach 397
West Germany

R. MALLARD
University of Oxford
Department of Metallurgy and
    Science of Materials
Parks Road
Oxford OX1 3PH
England

R. S. MAND
Bell Northern Research Ltd.
Dept. 5C30
P.O. Box 3511, Station C
Ottawa, Ontario K1Y 4H7
Canada

K. MARTIN
School of Electrical Engineering
Georgia Institute of Technology
Atlanta, Georgia 30332
U.S.A.

D. B. MAST
University of Cincinnati
Department of Physics
210 Braunstein (ML 11)
Cincinnati, Ohio 45221-0011
U.S.A.

M. MEUNIER
Ecole Polytechnique
Dept. of Engineering Physics
Campus de l'Universite de Montreal
Case postale 6079, succursale A
Montreal, Quebec H3C 3A7
Canada

D. MORRIS
c/o National Research Council of Canada
Division of Physics
100 Sussex Drive
Ottawa, Ontario K1A 0R6
Canada

R. PANDEY
Physics Department
University of Manitoba
Winnipeg, Manitoba R3T 2N2
Canada

J. A. A. J. PERENBOOM
High Field Magnet Laboratory
Faculty of Science
University of Nijmegen
Toernooiveld, 6525 ED Nijmegen
The Netherlands

L. PFEIFFER
Redgate Road
Morristown, New Jersey 07760
U.S.A.

S. PURCELL
c/o Physics Department
Simon Fraser University
Burnaby, British Columbia V5A 1S6
Canada

S. R. SHARMA
Department of Chemistry
The University of Lethbridge
4401 University Drive
Lethbridge, Alberta T1K 3M4
Canada

M. SINGH
Dept. of Physics
The University of Western Ontario
London, Ontario N6A 3K7
Canada

R. SOLLIE
Institute for Theoretical Physics
University of Trondheim
Norwegian Institute of Technology
N-7034 Trondheim - NTH
Norway

R. STONER
National Research Council of Canada
Division of Physics
Building M-50
Montreal Road
Ottawa, Ontario K1A 0R6
Canada

W. QUE
Dept. of Physics
Simon Fraser University
Burnaby, British Columbia V5A 1S6
Canada

A. SA'AR
Faculty of Exact Sciences
Dept. of Physics and Astronomy
Tel-Aviv University
Ramat-Aviv 69978
Tel-Aviv, Israel

M. SAGLAM
University of California
Department of Physics
Riverside, California 92521-0413
U.S.A.

Y. SHANI
Faculty of Exact Sciences
School of Physics and Astronomy
Tel-Aviv University
Ramat-Aviv 69978
Tel-Aviv, Israel

T. SWAHN
Institute of Theoretical Physics
Chalmers University of Technology
S-412 96 Goteborg
Sweden

J. SZYMANSKI
Telecom Australia
Research Laboratories
762-772 Blackburn Road
Clayton North Vic. 3168
Australia

M. TANAKA
Institute of Industrial Science
University of Tokyo
7-22-1 Roppongi Minato-ku
Tokyo 106
Japan

R. TAYLOR
The University of Nottingham
Dept. of Physics
University Park
Nottingham NG7 2RD
England

R. TAYLOR
National Research Council of Canada
Division of Informatics
Building M-60
Montreal Road
Ottawa, Ontario K1A 0R6
Canada

E. TEKMAN
Dept. of Elect. & Electronics Eng.
Bilkent University
P.O.B. 8, Maltepe
06572 Ankara
Turkey

M. N. TEZEY
Istanbul Technical University
Art & Science Faculty (Fen-ED)
Physics Department
Maslak - Istanbul
Turkey

F. TREMBLAY
University of Cambridge
Department of Physics
Cavendish Laboratory
Madingley Road
Cambridge CB3 OHE
England

N. TRIVEDI
University of Illinois
Dept. of Physics
1110 West Green Street
Urbana, IL 61801
U.S.A.

V. TRZECIAKOWSKI
National Research Council of Canada
Division of Physics
Building M-36
Montreal Road
Ottawa, Ontario K1A OR6
Canada

H. C. TSO
c/o Dept. of Physics
Stevens Institute of Technology
Hoboken, New Jersey 07030
U.S.A.

A. H. UCISIK
Okumus Adam Sokak Oilek Apt.
No. 12 D. 12
Fatih-Istanbul 34260
Turkey

S. E. ULLOA
Dept. of Physics & Astronomy
Clippinger Research Labs.
Ohio University
Athens, Ohio 45701-2979
U.S.A.

K. URQUHART
Dept. of Physics
Simon Fraser University
Burnaby, British Columbia V5A 1S6
Canada

P. VASILOPOULOS
Universite of Montreal
Centre de recherches mathematiques
C.P. 6128, succursale A
Montreal, Quebec H3C 3J7
Canada

M. WHITEHEAD
Dept. of Electronic & Elec. Eng.
University College London
Torrington Place
London WC1E 7JE
England

N. WINGREEN
Cornell University
Lab. of Atomic & Solid State Physics
Clark Hall
Ithaca, New York 14853-2501
U.S.A.

A. ZIEGLER
Universitat Regensburg
Inst. fur Physik I - Theoretische Physik
D-4800 Regensburg
Universitasstrabe 31
Postfach 397
West Germany

Many body effects (continued)
  Green's function, 181-182
  impurity potential, 177-178
  self-consistent exciton
      approximation, 179-181
  quantum Hall effect, integral,
      308-316
  self-energies, 70
  subband structure, 80-82
Many-body energy, optical
    excitation, 85
Mean field theories, 218
Mean free path, thermal conduc-
    tivity at high magnetic
    fields and, 285-288
Measurement methods, 6
Mesoscopic regime, see Quantum
    boxes
Metastable states, QWs in electric
    field, 192-193, 196, 197
MNOSFET, 3
Mobility of 2DEG, variation in, 32
Modulated molecular beam spectro-
    metry (MMBS), 12
Modulated molecular beam studies,
    MBE binary compounds, 13-15
Modulation doped field effect
    transistor (MODFET), 34
Modulation-doped GaAs quantum well,
    Raman excitations, 88
Modulation-doped heterojunction, 89
Molecular beam epitaxy, 1-10, 53
  electronic system, 7-10, 12-35
  fundamental aspects, 12-30
    alloy films, 19-22
    binary compounds, 13-19
    interfaces in AlGaAs-GaAs
        structures, 22-30
  high-purity structures, 30-34
  low dimensional structures, 34-35
  physical systems, 4-7
  types of structures, 3-10
    heterostructures, 3
    MOSFETS, 3
    superlattices, 3-4
Monte Carlo studies, MBE binary
    compounds, 18-19
MOSFETs, 3, 95
  impurity scattering, 177-178
  magnetoplasma modes, 163-173
  physical systems, 4-7
MOS structures
  band diagram, schematic, 96
  IR spectroscopy, 99
  one-dimensional intersubband
      resonance, 121-122
  plasmon dispersion in 2D systems,
      107-109
  transition from two- to one-
      dimensional behavior,
      121-122

MQW, see Multiple quantum wells;
    Superlattices
Multiband KP theories, 258
Multi-barrier structures, resonant
    tunneling, 72-73
Multifractal object, Fibonacci SL,
    360, 361-363
Multiple quantum wells, 24
  electronic excitations in, 153-162
    numerical results, 159-162
    type I heterostructure, 153-158
    type II heterostructure, 158-159
  interface quality, 22-30
  Landau levels, 58-59
  magnetophonon effect, 259, 265,
      266
  optical properties, 25-28
Multi-valley subband structures, 90
Multiwire MOS-devices, channel
    conductance in, 97

Narrow-gap semiconductors, density
    of states, 73-74
Negative resistance diode, 231
Nonlinear optical phenomena, ultra-
    fast, 211-226
  electron-hole distributions,
      214-224
    real, nonthermal, 214-218
    virtual, 218-224
  excitonic, 211-214
  quantum boxes, 143-144
  ultrashort electrical pulse
      generation, 224-225
Non-local behavior, 117
  magnetoplasmon splitting, 116
  and plasmon dispersion, 107-109,
      111
Non-parabolicity
  and cyclotron resonance frequen-
      cies, 111
  cyclotron resonance studies,
      102-103
  GaAs-AlGaAs superlattice, 54
  magnetophonon effect, 258
  in parallel magnetic field, 63
  subband energy determination,
      82, 83
Normal mode, lifetime of, 319
Normal mode spectrum, type II
    heterostructure, 160
N-p dipole, 89
N-type doping, 45, 95

Oblique confinement, 344
Occupancy-onset anomalies, 73-74
Odd-denominator rule, 313-316
One-component plasma, two-dimen-
    sional, 325, 336-338

Single quantum wells (continued)
  envelope function approximation,
      56
  in parallel magnetic field, 63
Singular gauge transformation,
      336-340
Singularities of degeneracy,
      integral quantum Hall
      effect, 301
SL *see* Superlattices
Slab geometry, plasmon modes, 157,
      159, 160, 161, 162
Space charge effects, subband
      structure, 85
Spacer layers, 31
Spatial modulation, grating gate
      structure and, 116
Spectroscopy
  interband, 84-90
  subband, *see* Subband energy
      determination
Split-gate heterostructures, energy
      states in, 137-138
Splitting, 91
  cyclotron resonance, 105, 106
  magnetoplasmon, non-local inter-
      actions and, 116
  plasmon, 112-113
  plasmon excitations, 108, 111
  quantum boxes, 147, 148
Square well, 138
Square well potential, 114, 302
Square well potential hetero-
      structure, 84
Square well states, 69
Stark effect
  optical, 218
  ultrashort electrical pulse
      generation, 224-225
Stark shifts and excitonic effects
  electric field effects, 191-201
    quantum wells, isolated,
        191-199
    seperlattices, 199-201
  excitons
    real ls excitions from virtual
        exciton gas, 221
    states in type I QWs and SLs,
        202-207
    virtual, 221, 223
  isolated QW bound states, 189-
      190
  SL structure and conduction band
      edge profile, 190-191
Sticking coefficients, 13, 14
Stokes shift, 26, 29
Subband energy determination,
      67-93
  current status, 90-91
  methods, transport experiments
      and surface layer
      capacitance, 70-76

Subband energy (continued)
  spectroscopy
    interband, 84-90
    intersubband, 77-84
  structure, theory, 68-70
Subbands
  defined, 67
  depletion charge densities and,
      107, 108
  dispersion relation, 191
  excitonic effects in QWs and SLs,
      203-204
  in laterally confined inversion
      layer systems, 114-115
  mixing, 56, 146
  one-dimensional, wavefunctions,
      135-136
  optical excitation and, 84
  quantum boxes, 146, 149, 150
  spacing in laterally confined
      inversion layer systems,
      114-115
Subband spectroscopy, 77-90
Substrate bias voltage, 1D inter-
      subband resonance in MOS
      devices, 122
Superconductivity, 91, 333
Superfluid helium, 320-322
Superfluidity, off-diagonal long-
      range order in, 334-335
Superlattices, 53
  CHIRP, 234-235
  Fibonacci, *see* Fibonacci super-
      lattices
  interface quality, 22-30
  magnetophonon oscillations,
      265, 266
  miniband formation, 191
  one dimensional, 200
  after optical excitation, 91
  optical properties, 25-28
  in parallel magnetic field, 63
  plasmon splittings, 112-113
  prelayer and, 31
  Stark shifts
    electric field effects, 199-201
    exciton states, 202-207
    structure and conduction band
        edge profile, 190-191
  subband constant energy surfaces,
      61
  transparent, 307-308
  transport lifetime and, 284
  tunneling, 230
Surface layer
  capacitance and conductivity,
      70-76
  intersubband spectroscopy, 77
Surface response function
  type I heterostructure, 155-158

401